Driving the Development, Management, and Sustainability of Cognitive Cities

Kiran Ahuja
DAV Institute of Engineering and Technology, India

Arun Khosla
Dr. B.R. Ambedkar National Institute of Technology, India

A volume in the Practice, Progress, and Proficiency in Sustainability (PPPS) Book Series

IGI Global
DISSEMINATOR OF KNOWLEDGE

Published in the United States of America by
IGI Global
Engineering Science Reference (an imprint of IGI Global)
701 E. Chocolate Avenue
Hershey PA, USA 17033
Tel: 717-533-8845
Fax: 717-533-8661
E-mail: cust@igi-global.com
Web site: http://www.igi-global.com

Library of Congress Cataloging-in-Publication Data

Names: Ahuja, Kiran, 1981- editor. | Khosla, Arun, 1967- editor.
Title: Driving the development, management, and sustainability of cognitive cities / Kiran Ahuja and Arun Khosla, editors.
Description: Hershey, PA : Engineering Science Reference, an imprint of IGI Global, [2019] | Includes bibliographical references.
Identifiers: LCCN 2018044182| ISBN 9781522580850 (hardcover) | ISBN 9781522580867 (ebook)
Subjects: LCSH: Smart cities.
Classification: LCC TD159.4 .D75 2019 | DDC 307.760285--dc23 LC record available at https://lccn.loc.gov/2018044182

This book is published in the IGI Global book series Practice, Progress, and Proficiency in Sustainability (PPPS) (ISSN: 2330-3271; eISSN: 2330-328X)

British Cataloguing in Publication Data
A Cataloguing in Publication record for this book is available from the British Library.

Practice, Progress, and Proficiency in Sustainability (PPPS) Book Series

ISSN:2330-3271
EISSN:2330-328X

Editor-in-Chief: Ayman Batisha, International Sustainability Institute, Egypt

MISSION

In a world where traditional business practices are reconsidered and economic activity is performed in a global context, new areas of economic developments are recognized as the key enablers of wealth and income production. This knowledge of information technologies provides infrastructures, systems, and services towards sustainable development.

The **Practices, Progress, and Proficiency in Sustainability (PPPS) Book Series** focuses on the local and global challenges, business opportunities, and societal needs surrounding international collaboration and sustainable development of technology. This series brings together academics, researchers, entrepreneurs, policy makers and government officers aiming to contribute to the progress and proficiency in sustainability.

COVERAGE

- Eco-Innovation
- Innovation Networks
- Knowledge clusters
- Green Technology
- Strategic Management of IT
- Socio-Economic
- Outsourcing
- Environmental informatics
- Technological learning
- Sustainable Development

The Practice, Progress, and Proficiency in Sustainability (PPPS) Book Series (ISSN 2330-3271) is published by IGI Global, 701 E. Chocolate Avenue, Hershey, PA 17033-1240, USA, www.igi-global.com. This series is composed of titles available for purchase individually; each title is edited to be contextually exclusive from any other title within the series. For pricing and ordering information please visit http://www.igi-global.com/book-series/practice-progress-proficiency-sustainability/73810. Postmaster: Send all address changes to above address.

Titles in this Series

For a list of additional titles in this series, please visit:
https://www.igi-global.com/book-series/practice-progress-proficiency-sustainability/73810

Optimizing the Use of Farm Waste and Non-Farm Waste to Increase Productivity and ...
Leighton Naraine (Clarence Fitzroy Bryant College, St. Kitts-Nevis)
Engineering Science Reference • ©2019 • 223pp • H/C (ISBN: 9781522579342) • US $155.00

Handbook of Research on International Collaboration, Economic Development, and ...
Vasilii Erokhin (Harbin Engineering University, China) Tianming Gao (Harbin Engineering University, China) and Xiuhua Zhang (Harbin Engineering Universit, China)
Business Science Reference • ©2019 • 703pp • H/C (ISBN: 9781522569541) • US $235.00

Handbook of Research on Rural Sociology and Community Mobilization for Sustainable Growth
Qaiser Rafique Yasser (Centre for Rural Economy, Pakistan)
Information Science Reference • ©2019 • 393pp • H/C (ISBN: 9781522571582) • US $245.00

Impacts of Violent Conflicts on Resource Control and Sustainability
Esther Akumbo Nyam (Plateau State University, Nigeria) and Asiru Hameed Tunde (Umaru Musa Yar'adua University, Nigeria)
Engineering Science Reference • ©2019 • 218pp • H/C (ISBN: 9781522559870) • US $185.00

Practice and Progress in Social Design and Sustainability
Kin Wai Michael Siu (The Hong Kong Polytechnic University, Hong Kong) and Yin Lin Wong (The Hong Kong Polytechnic University, Hong Kong)
Information Science Reference • ©2019 • 330pp • H/C (ISBN: 9781522541837) • US $195.00

Handbook of Research on Methods and Tools for Assessing Cultural Landscape Adaptation
Isabel de Sousa Rosa (University of Lisbon, Portugal) Joana Corte Lopes (University of Lisbon, Portugal) Ricardo Ribeiro (University of Lisbon, Portugal) and Ana Mendes (University of Lisbon, Portugal)
Engineering Science Reference • ©2018 • 524pp • H/C (ISBN: 9781522541868) • US $265.00

For an entire list of titles in this series, please visit:
https://www.igi-global.com/book-series/practice-progress-proficiency-sustainability/73810

IGI Global
DISSEMINATOR OF KNOWLEDGE

701 East Chocolate Avenue, Hershey, PA 17033, USA
Tel: 717-533-8845 x100 • Fax: 717-533-8661
E-Mail: cust@igi-global.com • www.igi-global.com

Editorial Advisory Board

Table of Contents

Detailed Table of Contents

Smart cities are composed of interlinked components with constant data transfer and services targeted at increasing the life style of the people. The chapter focuses on diverged smart city components as well as the security models designed to be implemented. The four major paradigms discussed in this chapter are smart grids, building automation system (BAS), unmanned aerial vehicle (UAV), and smart vehicles. Apart from addressing the security concerns of every component, the major highlights of this chapter are architecture, smart environment, industry, lifestyle, services, and digital lifestyle quality. Finally, the chapter focuses on privacy preserving mechanisms, its essence over smart cities, strong architecture related to privacy, preserving mechanism, and various approaches available that can retaliate these issues in a smart city environment.

In this chapter, different approaches are presented for removal of fog from video footage taken in moving cars. The methodology uses different approaches, namely dark channel prior, contrast limited adaptive histogram equalization (CLAHE), the combination of two approaches (dark channel prior and CLAHE), and RETINEX algorithm combined with DWT. The algorithms are implemented in MATLAB R2015a. Moreover, the algorithms are compared based on their computational complexity and a visibility metric which is used for computing the CNR of video frames before and after the application of the algorithm. The chapter discusses which algorithm would provide better performance during night fog and daylight fog. Finally, the safe speed of the driver is calculated based on the time complexity of the algorithm used.

Subhankar Das, Duy Tan University, Vietnam
Anand Nayyar, Duy Tan University, Vietnam

The cognitive city transport plays a vital role in setting the targets of monetary similarity and aggressiveness, social harmony, and reasonable significant development. As per the principle of urban development and versatility, the urban communities and their formative experts, which are dynamic in their considerations, are supporting neighborhood governments in handling portability issues, which shape basic examples over all subjective and dynamic populated urban areas. Interest for traveler and cargo transport is continually expanding in urban communities and towns, adding to movement clog, contamination, and car crashes. Understanding the movement clog is a typical test for governments and transport experts in all these urban formative specialists. As per them, putting the current versatility inclines on a maintainable cutting-edge way requires an aggregate rejig of urban portability.

Gajendra Singh, Dr. B. R. Ambedkar National Institute of Technology, India
Rajeev Kapoor, Delhi Technological University, India
Arun Khosla, Dr. B. R. Ambedkar National Institute of Technology, India

With the growing demands of safety for people and their properties, video surveillance has drawn much attention. These requirements have led to the positioning of cameras almost every corner. Smart video surveillance systems can interpret the situation and automatically recognize abnormal situations, which plays a vital role in intelligence monitoring systems. One vital aspect is to detect and alert generation of suspicious events then to notify operators or users automatically. A long time may pass before an event of interest to take place. In such situations, human attention may get diverted and an event of interest may get missed. In such case, video surveillance systems can effectively improve safety and security for the control and management of public areas or personal life. Independent surveillance systems to replace the traditional (human observer-oriented) systems also can relieve the workload of relative personnel.

Based on the advancements of a smart city, the cognitive city focuses on the communication between the city's stakeholders and cognitive systems to create a human-machine symbiosis in which human and machine can interact directly with each other. To develop the cognitive city means to design this sociotechnical relationship. This requires a variety of approaches, methods, and tools. To this purpose, the authors of this chapter suggest the concept of ontological design. This principle helps to develop a research methodology toolbox that can be applied to create research and development strategies suitable for shaping the relationship between human and environment. This chapter is an essay with the aim to encourage the reader to reflect. Ontological design means that it is necessary to become aware of the influences of today's actions on the future. With the help of an illustrative use case, this chapter wants to demonstrate why and to what extent the concept of ontological design can support urban development.

As the world's population ages, those suffering from diseases will increase. Researchers in electronics, computer, networking, and medical fields need to work more seriously to make the broad vision of smart healthcare/e-health system. To achieve the objectives of e-healthcare in smart cities, there is a need to create new system that allows the acquisition of health information smartly, automatically, and transparently in order to take efficient decisions provided by the supporting system. Such systems may be designed technically by embedded together communication, smart signals, internet of things, network of sensors.

This chapter presents an idea in the contribution to the concept of smart cities is to build a place which provides clean air to breathe. Advancements in the technology is acting as a bane to environmental balance leading to global warming, irregular rain cycles, increasing pollution, disturbed rainfalls. The need of the hour is to find a solution to the issue of increasing pollution that may lead to disastrous consequences. This chapter builds a framework of a cognitive city with attention to the impact and burden over environmental resources. This chapter is a perfect integration of traditional approaches, technological methods, and gamified solution as a remedy to growing pollution that can be and should be deployed in the project of smart city.

Preface

A cognitive city will enable each native and business to get benefits proactively from the city without expecting to sign in and ask for the administration, have access to world-class training sitting at home, have their wellbeing observed consistently, medical issues distinguished before occurrence, permit programmed reporting and remedial of family issues identified with water, power, gas and waste through smart meters. Sensors in cars/autos will identify and report potholes, which will trigger the street bots to advance in and settle the potholes. The conceivable outcomes are boundless. The goal of the book *Driving the Development, Management, and Sustainability of Cognitive Cities* is to provide a platform for domain experts, researchers, practitioners, and students to exchange ideas on technologies and practical experiences related to innovations in cognitive cities development, management and sustainability. It covers the innovative ideas implemented or to be implementable in cognitive cities. The innovative ideas are from various aspects of cognitive cities such as energy, transportation, water, healthcare, pollution, and security with a focus on state-of-the art technologies, methodologies, and best practices.

In this book, we present sustainable solutions for cognitive cities development and management under one umbrella. The collection of innovative ideas in the form of book will help us to identify futuristic issues/solutions for fast development and management.

This book is organized into 10 chapters. A brief description of each chapter is as follows:

To develop a sustainable zero-carbon cognitive city, a setup is required that will lessen outflows of ozone-depleting substances to zero and all practices that discharge ozone-harming substances. To attain this goal, authors in Chapter 1 proposed a framework for zero carbon emission sustainable cognitive cities and ensured that each part of a city is without carbon. They have integrated large and small-scale energy initiatives and solutions along with citywide improvements in energy efficiency, low carbon transportation, and distributed energy resources. The cities will become clusters of smart energy communities, which can exploit the benefits of new energy systems. Rewards and badges are provided to consumers for reducing carbon emissions and saving energy by various means. The proposed framework will help to engage number people in reducing CO2 emissions and achieving the zero carbon cities target by employing gamification.

Smart Cities are composed of interlinked components with constant data transfer and services targeted at improving the life style of the people. Chapter 2 focused on diverged smart city components, as well as the security models designed to be implemented. The four major paradigms are discussed in this chapter i.e. Smart Grids, Building Automation System (BAS), Unmanned Aerial Vehicle (UAV) and Smart Vehicles. Apart from addressing the security concerns of every component, the major highlights of this chapter are architecture, smart environment, industry, lifestyle, services, and digital lifestyle quality. They mainly focused on privacy preserving mechanisms, strong architecture related to privacy and various approaches available which can retaliate these issues in a smart city environment.

In Chapter 3, different approaches are presented for removal of fog from video footages taken in moving cars. The methodology employed different approaches namely dark channel prior, contrast limited adaptive histogram equalization (CLAHE), the combination of two approaches (dark channel prior and CLAHE) and RETINEX algorithm combine with DWT. The algorithms are compared based on their computational complexity and a visibility metric which is used for computing the CNR (Contrast to Noise Ratio). of video frames before and after the application of the algorithm. The chapter presented which specific algorithm would provide better performance during night fog and day light fog. They calculated the safe speed of the driver via time complexity of the algorithm employed.

To overcome the ecological and health issues and to meet the end user's high data rate requirements for the successful deployment of the cognitive cities, the most important technology is Green Communication. Taking into consideration, the importance of green communication technologies, Chapter 4 has presented an intense survey on energy efficient technologies developed for green communication. Some key research challenges pertaining to the use and deployment of these technologies for green communication are discussed. In the end of this chapter, the worldwide standardization efforts for cognitive cities are also highlighted.

Smart city needs self-sustainable energy solution. Chapter 5 reviewed influence of dust on PV (photo-voltaic) modules to give a wider knowledge of this critical issue in the world of solar PV. The inclusive literature review presented in this chapter summarized the noteworthy information regarding dust impact for the last two decades. The impact of dust in various regions of the world, on various types of PV and appraisal on the current research was provided. They find that dust critically influenced the panel performance parameters and it is also responsible for deteriorating panel efficiency. In addition, the study revealed that the amount of dust accumulated on module surface is a function of various parameters such as geographical location, wind direction, exposure time, dust properties and characteristics, etc.

Urban transport plays a basic ingredient in meeting the objectives of economic compatibility and competitiveness, social unison and sustainable substantial growth. In accordance with the principle of urban growth and mobility, the cities & their developmental authorities which are progressive in their thoughts is supporting local governments in tackling mobility issues which form common patterns across all cognitive and progressive populated cities. Localized decision makers need to deliver futuristic and integrated transport policies that optimize the use of all transport modes in the urban network for both passengers and goods. Chapter 6 has provided a snippet of all the new futuristic innovations in making the transport in urban areas a success around the world. It gives certain prototype ideas for the transport problems around all cognitive cities.

Chapter 7 has presented intelligent video surveillance systems to substitute the traditional systems for smart cities that provide effectiveness. With the increasingly growing demands of safety for people and their properties, video surveillance has drawn much attention and this has led to the positioning of cameras at almost every corner. Smart video surveillance systems can interpret the situation and automatically recognize abnormal situations which plays a vital role in intelligent monitoring. Video surveillance system effectively improve safety and security for the control and management of public areas or personal life.

Chapter 8 has suggested the concept of ontological design. This principle helps to develop a research methodology toolbox that can be applied to create research and development strategies suitable for shaping the relationship between human and environment. Ontological design means that it is necessary to become aware of the influences of today's actions on the future. With the help of an illustrative use case, this chapter wants to demonstrate why and to what extent the concept of ontological design can support urban development.

To achieve the objectives of E-health care in smart cities, authors in Chapter 9 has described the need to create new system which allows the acquisition of health information smartly, automatically, and transparently in order to take efficient decisions provided by the supporting system. Such systems may be designed technically by embedding together communication, smart signals, Internet of things, and network of sensors.

Chapter 10 has presented an idea to build smart city where people can breathe in clean air. They proposed a gamified solution for pollution control and air quality monitoring in a smart city. The need of the hour is to find a solution to the issue of increasing pollution that may lead to disastrous consequences. Authors build a framework of a cognitive city with attention to the impact and burden over environmental resources. Proposed framework is a perfect integration of traditional approaches, technological methods and gamified solution as a remedy to growing pollution that can be and should be deployed in the project of smart city.

The proposed book exploits innovative technologies and techniques to enhance commuting, safety, and economic affluence in cognitive city. Students/Researchers/Innovators can create an action plan to lower down the service delivery costs, connection with constituents, and rebooting economy via inspiring case studies from various cities. It will provide implemented solutions for improving emergency services, reducing commute times, and creating public private collaboration. Source schemes will be considered to make economic groups that promote entrepreneurship, emerge new industries, and pull new talent. Cognitive city realization will go forward with more transparency, higher data sharing, and citizen engagement.

Acknowledgment

The editors would like to acknowledge the help of all the people involved in this project and, more specifically, to the authors and reviewers that took part in the review process. Without their support, this book would not have become a reality.

Firstly, the editors would like to thank each one of the authors for their contributions. Our sincere gratitude goes to the chapter's authors who contributed their time and expertise to this book.

Secondly, the editors wish to acknowledge the valuable contributions of the reviewers regarding the improvement of quality, coherence, and content presentation of chapters. Most of the authors also served as reviewers as well as referees; we highly appreciate their multi-tasking.

Kiran Ahuja
DAV Institute of Engineering and Technology, India

Arun Khosla
Dr. B.R. Ambedkar National Institute of Technology, India

Chapter 1
A Framework to Develop a Zero-Carbon Emission Sustainable Cognitive City

Kiran Ahuja
DAV Institute of Engineering and Technology, India

Arun Khosla
Dr. B. R. Amedkar National Institute of Technology, India

ABSTRACT

To develop a sustainable zero-carbon cognitive city, a setup is required that will lessen outflows of ozone-depleting substances to zero and all practices that discharge ozone-harming substances. This progress incorporates de-carbonizing power and zero-emanation transport for environmental change. Zero-carbon urban areas keep up ideal living conditions while wiping out natural effect. Rather than utilizing set up urban communities, numerous engineers are beginning sans preparation with a specific end goal to make a zero-carbon city. While keeping this in view, the authors propose a framework for zero-carbon emission sustainable cognitive cities. They ensure that each part of a city adds to its being without carbon. They integrate large and small-scale energy initiatives and solutions along with citywide improvements in energy efficiency, low carbon transportation, and distributed energy resources. The cities will become clusters of smart energy communities, which can exploit the benefits of new energy systems.

DOI: 10.4018/978-1-5225-8085-0.ch001

1. INTRODUCTION

Urban zones are main source of more than 70% of worldwide energy utilization and CO2 outflows, for the most part from structures. By 2030, a stunning 82 billion square meters (900 billion square feet), a territory generally equivalent to 60% of the aggregate building load of the world, will be fabricated and modified in urban territories around the world. This projection gives an uncommon chance to decrease non-renewable energy source CO2 discharges by setting the whole worldwide Building Sector on a way to crest emanations rapidly, and totally eliminate CO2 outflows by around 2050 (Nunes, Gouveia, Rodrigues & Simao, 2017).

The assessment of pollution generated by an organization's actions makes possible to think about changes in the policies to significantly condense the inclusive carbon emissions effects. A carbon footprint can be estimated by embraced a Greenhouse Gas (GHG) discharges assessment. When the span of a carbon footprint is known, a methodology can be contrived to decrease it, e.g., by innovative advancements, better process and item administration, transformed Green Public or Private Procurement (GPP), carbon catch, utilization systems, and others. The main factors form carbon footprint calculation are shown in Figure 1.

To make a zero-carbon city, a setup is required which diminish emanations of ozone depleting substances to zero and stop all the practices that yield ozone depleting. This required change will contact each part of city administrations and framework including energy age and dispersion, warming and cooling frameworks, building energy productivity, transportation, water and waste administration, and the proficiency of city administrations, for example, road lighting. In the meantime, city activities are being changed by advanced innovations, for example, the Internet of Things (IoT), keen structures, man-made brainpower, mechanical technology, and computerized vehicles.

Mobility portrays the present change in the energy area as the rise of the energy cloud. The energy cloud situation depicts an extreme change of energy advertises as the restricted power lattice offers route to a more unique system of partners, advances, and foundation. It imagines a world in which the power supply is cleaner and more circulated, where computerized change grasps man-made reasoning, the IoT, and blockchain-empowered systems, and where broad jolt of the transportation part implies that power free market activity turn out to be progressively versatile.

This change of the energy part likewise gives the bedrock to the production of the low carbon urban communities without bounds. A standout amongst the most essential improvements for urban communities hoping to change their energy profile is the interlinking of the energy division with structures and transportation.

Figure 1. Represents various fields of urban areas involve in carbon footprint calculation
(Cloud Energy, 2017)

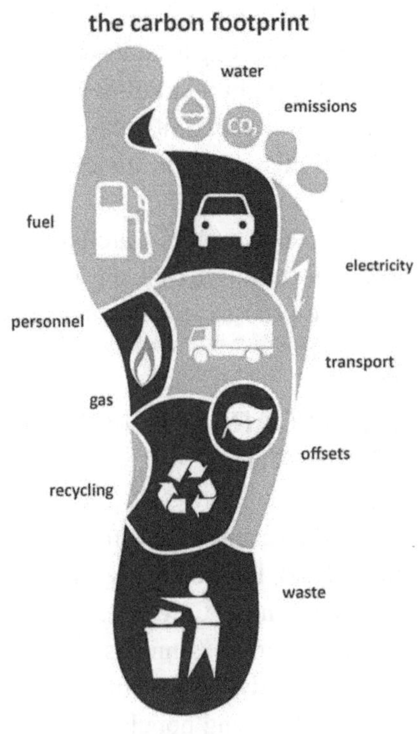

A zero-carbon city should address the part of non-renewable energy sources in space warming and in transportation. Enhancements in energy productivity and the move to inexhaustible assets are fundamental advances however, more significantly, the substantially closer association amongst structures and transportation and the energy lattice will establish the framework for another urban energy cloud (Chen & Ying-Jia, 2011).

The rest of the chapter is organized as follows. In the survey Section 2, the details of the case studies related to the carbon emissions and its effects are presented. In Section 3, challenges in cognitive cities due to emission of CO2 diverse fields are discussed. In section 4, carbon emission mitigation criteria are explained for proposing a framework of zero-carbon cognitive cities. The framework is proposed in Section 5 for making a zero-carbon cognitive city. Barriers of implementation and gamification approach are discussed in Section 6. Finally, the conclusions are drawn in Section 7.

2. SURVEY OF LOW CARBON INITIATIVES AND DIVERSE CO2 CASE STUDIES IN VARIOUS CITIES

To achieve the ultimate goal of making sustainable cognitive cities, number of nations and provinces are before now acting to address the carbon emission problem. There are multiple innovative ideas to build a sustainable low carbon environment. These are also aligned to series of carbon mitigation criteria which are proficient in significant reduction of CO2 emissions. The major case studies along with their respective findings are discussed in Table 1.

The challenges for policy-makers, administrators to decide which criteria are preferable and how to direct effort and investment in diverse areas for achieving zero carbon cognitive cities are discussed in following section.

3. CHALLENGES DUE TO CARBON EMISSIONS FACTORS IN COGNITIVE CITIES

There is a developing focus on urban changes investigate in planning (Newton, 2008; Ernston et al., 2010; Bulkeley, Boto, Hodson & Marvin, 2010; Eames, Dixon, May & Hunt, 2013; Moore, de haan, Horne & Gleeson, 2017) that reflects the extent of difficulties presently confronting urban areas universally. The major problem creating areas are shown in Figure 2. At a worldwide level the primary difficulties are related with a quickly developing and urbanizing populace, assessed to achieve 9 billion by 2050, 66% urban (UN DESA, 2014). Exponential development in various fields of utilization over the previous century (Steffen et al., 2015) are being reflected in proportionate rates of degradation to the world's biosphere. Such is the effect of this human action that it supposedly is debilitating the world's 'planetary limits' that have conveyed generally stable conditions for the advancement of current human settlements up until the present. Expanding centralizations of human-prompted CO2 in the environment prompting a worldwide temperature alteration and environmental change, reflected in anticipated increment in ocean levels, storm floods, flooding and scenes of outrageous warmth (IPCC, 2013).

An increasing carbon emission demonstrates populace demands on the planet more than half comparison to what nature can reestablish. The global footprint network (WWF, 2014) uncovers that 'carbon from consuming petroleum derivatives has been the predominant part of humankind's ecological footprint for the greater part a century and stays on an upward pattern. In 1961, carbon was 36 for each penny of our aggregate footprint; by 2010, it included 53 for every penny. Settled inside this worldwide crystal are sets of national, city and nearby difficulties that have likewise been unmistakably explained somewhere else (Newton, 2008, 2006;

Table 1. Comprehensive review of case studies on carbon emission effects in diverse countries

S.No.	Authors	Case study	Findings
1.	Saige Wang *et al.* 2018	A contextual investigation of *Chongqing city* was led to distinguish the basic parts of three levels that contribute most to add up to carbon discharges, covering 28 monetary divisions amid 2002– 2007. The outcomes demonstrated that discharges of Tier 1 and Tier 2 incorporated into most conventions just involve a little part (27.8%) of the aggregate outflows of every one of the 3 levels, particularly for modern segments.	It is concluded that the current climate control conventions had underestimated the releases in every division and especially, Tier 3 discharges must be joined while defining successful urban administration procedures and atmosphere adjustment arranging. By joining the Three-Tier carbon impression into sectoral input– yield structure, this investigation created carbon division particular classifications to seek after discharges relief pathways inside their own economic action as well as over their supply chain.
2.	Yuan Peng et al. 2018	Authors studied about the low-carbon arrangements rise, advancement, and empowering systems, where *Shanghai* considered for contextual investigation. They built up an expository structure drawing on framework advancement hypothesis and maintainability tests for the same. An aggregate of 186 relevant approaches were chosen and examined, which is supplemented by the meetings with partners in the legislature to pick up a more profound understanding into the strategy settings in Shanghai.	The discoveries give new hypothetical and observational bits of knowledge into the multi-level administration of low-carbon change in urban cities. They found that the city's low carbon activities are installed and coordinated into its existing policy frameworks.
3.	Raffaele Carlia et al. 2018	Authors utilized Analytic Hierarchy Process (AHP) multi-criteria basic decision-making method for application in the smart metropolitan city, with the aim of analyzing the sustainable development of energy, water and natural resources, through an arrangement of target execution markers. In particular, the 35 attributes characterized for the sustainable development of energy, water and environment systems index structure were utilized. The use of the way to deal with the genuine contextual analysis of four metropolitan regions (*Bari, Bitonto, Mola, and Molfetta*) in the city of Bari (*Italy*) demonstrated its implementation for the neighborhood government in benchmarking metropolitan zones and giving choice of the best way to plan the supportable advancement technique of the metropolitan city.	The principle confinement gets from the certifiable vulnerability and vagueness that describe the human decision making. These viewpoints entangle the correct estimation of the pairwise correlations and also the perception of the distinction between close qualities.
4.	Sesil Koutra et al. (2018)	This investigation provided a clear view of the existing apparatuses and strategies contrasting their criteria and key parameters. Authors presented a rearranged methodological evaluation hypothetical instrument (U-ZED) by concentrating on the dedication towards the zero vitality focuses in a future locale.	In a broader point of view, the investigation managed the improvement of a device from building to locale with the main concern to define the context of sustainable and long-haul areas managing the difficulties of 2050 horizon.
5.	YANG Xiu et al 2018	Researchers investigated quantitatively and portrays low-carbon improvement attributes, advancement patterns, and low-carbon advancement pathways of the four kinds of urban areas. As indicated by the present circumstances and goals of emissions, the distinctions being developed stages, difficulties, and open doors for each kind of the city.	This exploration presented suggestions for the low-carbon guide and the medium and long haul (by 2030) discharge drift courses of various kinds of locales in *China*.
6.	Mari Martiskainena et al 2018	Consumers and policy makers encouraging change towards zero carbon private structures by incorporating the ideas of development go-betweens and champions. Authors examined on-screen character setups in three new form housing ventures.	The discoveries demonstrated that the implementation of level wise zero carbon structures at various phases are made possible improvement with changing force, impact and life span. A few delegates take additionally advocating parts, while likewise different on-screen characters champion ventures when intermediation is missing. During a period of constrained strategy bolster for zero carbon housing advancements and intermediation.
7.	Andy Gouldson et al. 2018	Authors concentrating on the low-carbon measures in the structures, transport, and waste divisions. They precisely demonstrated that low-carbon measures can accomplish a scope of advancement needs, for example, work creation, made strides general wellbeing, social incorporation, and enhanced openness.	Investigation recommended that low-carbon interest in urban areas may have a net present esteem of US$16.6 trillion by 2050. This monetary case is an imperative, however not adequate, condition for profound decarbonization.

continued on following page

Table 1. Continued

S.No.	Authors	Case study	Findings
8.	Sieting Tan et al 2017	A system for the assessment of LCC was built up from the points of view of economic, energy, social and living, carbon and environment, urban versatility, strong waste, and water. A far-reaching assessment technique was utilized for LCC positioning by utilizing the entropy weighting factor technique. The benchmark esteems for LCC confirmation were additionally distinguished. The structure was connected to 10 worldwide urban areas to rank their low-carbon levels. The correlation of urban communities at various levels of monetary, social, and ecological advancement improved the all-encompassing of the investigation. The outcomes demonstrated that *Stockholm, Vancouver,* and *Sydney* positioned higher than the benchmark esteem, demonstrating these urban areas accomplished an abnormal state of low-carbon advancement. *São Paulo, London,* and *Mexico City* are still in the moderate change towards LCC. *Beijing* and *New York* lower LCC level than the benchmark esteem because of the poor ecological execution and foundation bolsters caused by concentrated human exercises.	The proposed framework filled as a rule for the institutionalization of LCC and further recognized the key parts of low-carbon administration for various urban areas.
9.	Luka Lugaric et al. (2016)	A structure introduced which incorporates energy, economics and natural city frameworks in the basic leadership process, looking at related hypothetical difficulties and application restrictions. The strategy is connected on the city of *Sisak in Croatia* which has created plans to execute a few activities outfitted towards making a keen energy city.	The model empowered recreation and appraisal of effects of individual undertakings focusing on the advancement of a energy through execution, and utilized an apparatus for assessing nearby improvement options inside the limit of neighborhood assets.
10.	Zhanglan Wu et al 2016	Here, investigation breaks down the effect of government natural control on the low-carbon city change process by embracing *Shenzhen* as the exploration protest. One of the world's most youthful super urban areas, Shenzhen additionally happens to be the city with the most minimal carbon discharge power in *China.* The staggered administration structure is a helpful component with which to measure divisions of obligation and assets. Employments staggered natural administration to clarify the approach for managing the atmosphere in the city of Shenzhen. Endeavoring to investigate green low-carbon advancement way for the entire nation, Shenzhen gives useful experience to nations to adapt to worldwide environmental change.	Low-carbon city progress needs economic help, and for *China's* waterfront urban communities like Shenzhen, they don't worry about it the endowments for the new energy vehicles, or the foundation of the carbon outflow exchanging framework, there is most likely that the Shenzhen government has spent extensive labor and money related assets.
11.	Naoum Tsolakisa et al 2015	Authors demonstrated that the eco-city urban rises drastically and showed a perfect worldview for practical urban improvement. Subsequent to contrasting elective strategies for eco-city, they endeavored to survey the maintainability of the eco-city with the System Dynamics (SD) recreation-based procedure. SD methodological system is proposed, as a way to help chiefs, nearby governments and directors' planning and receiving viable strategies for checking and evaluating the economical execution of eco-urban areas.	They framed a structure and tried with two contextual investigations i.e. individually and socially discoveries about the proficiency of the eco-city.

Newton & Doherty, 2014; Australia, 2010) and in mix with worldwide exogenous pressures and the truth of a limited planet (Krugman, 2010; Daly, 1996; Bartlett & Satterthwaite, 2016) are what build up the logic and drivers for urban change.

The assets that are streaming into the environments of developing and developed nations and additionally the utilization straightforwardly from the occupants of these urban communities demonstrates that a noteworthy ecological cost is as of now being paid to convey high amounts of urban livability (Newton, 2012).

Figure 2. Challenges faced by cognitive cities by carbon emissions from various fields

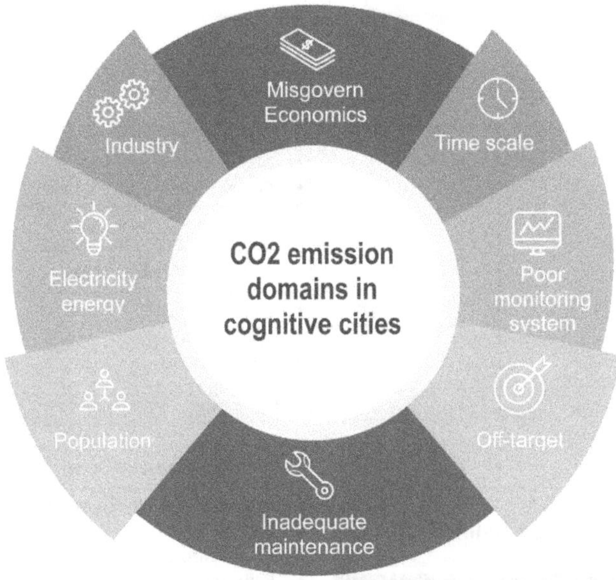

The type of urban areas and their environments hold a vital component to conveying large amounts of livability however at extraordinarily unique dimensions of asset utilization and GHG outflow. For instance, the group of low-thickness vehicle situated urban communities in North America and Australasia remain in checked complexity to Europe's progressively minimized travel arranged urban communities. From both an ecological and worldwide social equity viewpoint (Newton & Meyer, 2016), the test for Australia is to drastically recoil the outflow factor of their urban communities by essentially re-molding and re-planning their constructed surroundings and related social practices, while looking after livability. Urban areas in creating nations, for example, China are developing quickly, as is populace riches and emanation factor. A key inquiry is which city arranging technique will urban areas, for example, China pursue? Some recommend they have been following the North American model (Seto, n.d.), while all the more as of late, a few pundits are proposing a move to increasingly minimized, blended use, and open transport-arranged advancement (Normile, 2016; Newton, 2017).

Figure 3. Major factors participate in carbon emission in diverse fields of cognitive cities

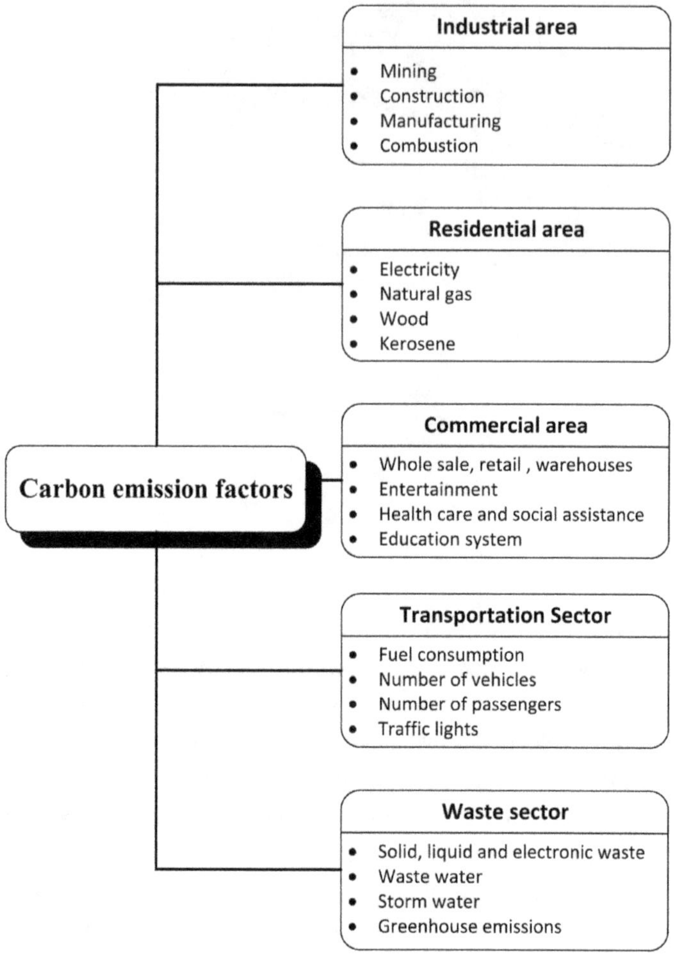

The major factors involve in carbon emission are urban residential areas, commercial areas, industry, transportation, and urban waste sector as shown in Figure 3. Recently, systems are coordinated toward decreasing Greenhouse Gas (GHG) emissions depended upon fuel switching, control gadgets employed, or elective clean advancements. However, emissions can be affected through modification of area/sector utilize designs. Adjustment of area/sector usage as it influences the living, working, and transportation designs must be considered also. The major factors considered for reduction of carbon emissions are described in following section.

4. REDUCTION CRITERIA

Low carbon city is a city which considers the low carbon economy as the headway. People living there consider the low carbon life as the beginning and the direct trademark. Low carbon society is also considered as the city association's advancement standard and the plan. The improvement of the low carbon city incorporates into various fields, for example, economy, population, society, resource and environmental conditions. It is a complicated system framework. Some specific fields which chiefly participate in industry transport development and energy are important for understanding the low carbon city. In the same time, regardless of whether it is a low carbon city is specifically identified with urban morphology, space appropriation, method for utilizing area, and advancement model of city.

The scale of reducing carbon emission challenge is presently with the end goal that change must be transformative, not incremental (Girardet, 2014). For urban communities, a regenerative change is required in diverse fields as shown in Figure 4. Retrofitting of urban communities and their basic foundations turns into a center piece of this procedure (Newton & Tucker, 2011; Glackin & Newton, 2016). Urban areas have the capacity to draw from a pipeline of creative advances, items and procedures that can be substituted while existing applications hint at inability to perform at required levels. The volume of inventive advancements accessible for urban framework changes is substantial (Newton & Bai, 2008; Newton, 2007, 2017; Ryan et al., 2016). The major mitigation criteria which help in reducing carbon emission in diverse sectors of cognitive cities are presented in Fig. 4 and alteration in different sectors of cognitive cities to achieve this target are discussed as follows:

4.1. Transportation System

Numerous advances will be acknowledged, including supplanting of oil based cars with half and halves and sun based electric or sun based hydrogen vehicles; shrewd transport frameworks and independent vehicles to advance neighborhood and between modular transport and limit urban movement congestion; auto sharing administrations; and coordinated urban land-utilize transport planning centered on dynamic transport (such as walking, public transport, and cycling) ; and rapid rail for between local travel and in addition joining common place urban communities into extensive metropolitan agglomeration economies (LVM, 2017).

- **Transportation Operations:** A standout amongst the most unmistakable contender for a 'snappy win' is driver training. At an insignificant cost, the potential decrease can be as high as 7%. Better road management system and support (think for instance of enhancing the moving opposition of streets or improving activity streams) can lessen CO2 emanations by 4%. Keen and associated transport systems utilizing smart transport frameworks hold extraordinary potential, for example truck platooning. Reconsidering most extreme weights and measurements to consider the cross-outskirt, expansive utilization of high-limit vehicles, for example, the European Modular System (EMS).

- **Fleet Renewal:** The innovation of new vehicles is simply part of a greater picture. New vehicles require a little vechicle fleet as comapred to older one. It is vital to take a whole vehicle fleet instead of simply new vehicles.
- **Vehicle Related Measures:** There are larger number of elements than simply the vehicle alone that decide CO2 emissions –, for example, vehicle length and weight, trailer outlines, elective powers, driver conduct, transport tasks, foundation, or a superior usage of the vehicle etc. An integrated approach would draw on this full range of solutions for diminishing CO2 outflows all the more successfully.
- **Alternative Fuels:** Acceptance of alternative fuel vehicles is ascending in market. Advancing more prominent market take-up of option powertrains is in this way key to guaranteeing further decreases in normal CO2 emissions. Governments should expand their help if there is a critical increment in deals, both as far as building the important charging framework and impacting customer decisions. Meanwhile, enhancing conventional innovations, for example, the internal combustion engine, will remain an essential supporters of CO2 decrease. It is likewise critical to energize the advancement of low carbon-concentrated powers by means of the Renewable Energy Directive.

4.2. Smart Buildings and Construction Transition

A number of execution criteria based essentially on wellbeing and least first cost arrangements, to eco-effective buildings whose life cycle execution is surveyed on various levels for all intents and purposes at configuration organize, is fabricated off-site and collected nearby and whose continuous task and support and dis-get together is advanced by means of building data models. The main factors help to mitigate the carbon emissions in buildings and construction areas are as follows:

- **Residential Area:** Turn off the lights when natural light is adequate and when the room is empty. Keep the temperature on moderate setting while present in the room. Utilize windows carefully, if temperature control is on, close them. If there is a need of minimal outside air, off the heater or AC. Limit the plug load by cut down the quantity of apparatuses are running and will save energy. For instance, share your minifridge with roommates and limit the quantity of printers in office. Phantom power control the gadgets continuously utilizing energy notwithstanding when shut down, this is valid for any charger, TV, printer, and so on. Utilize an electrical extension to effortlessly unplug these gadgets when not being used. Give it a rest to PC by shut down when not utilizing it. A PC turned off utilizes something like 65% less energy than a PC left on or sit on a screen saver. Accept the stairs as frequently as conceivable on the grounds that lifts devour power. Attempt to clean up full heaps of clothing and utilize the brilliant hues cycle at whatever point conceivable. The less hot water utilize as the less energy is expected to warm the water. Switch to Compact fluorescent light bulbs (CFLs) as these uses 75% less energy than glowing and last up to 10 times longer (Georgetown University, n.d.).

- **Office Area:** Conserve paper by print and duplicate on opposite sides, spare single-sided pages for notes, and print just what requires. Make a couple of moves to a reusing focus to store aluminum jars, plastics, glass, office paper, daily paper, cardboard. Pickup of office gadgets, mass metal and surplus furniture can be organized by reaching offices work administration. Advance reuse and give utilized PDAs and chargers, furniture apparel and cleaning and school supplies. Get a little exercise by walking or riding a bicycle to grounds if the distance is sensible. Walk from transport or metro stop to office or lab. Workforce and staff carpoolers pay decreased or no expense to stop on grounds, contingent upon size of carpool (Georgetown University, n.d.).
- **Government Area:** The solar powered networks program is a $5 million program to help local reactions to environmental change and help convey bring down power costs for network associations. Government in selected areas will have the capacity to introduce housetop solar panels, solar heated water, and solar oriented associated battery frameworks for government-possessed structures. The national atmosphere strength and adjustment technique requires for dealing with the dangers of a variable and changing atmosphere to support the network, economy and condition. It incorporates an arrangement of standards to on how individuals can adjust to environmental change successfully and to fabricate flexibility and in addition laying out the Government's vision for what's to come.

4.3. Smart Water System

A centralized hybrid, integrated urban water system is required which incorporates recycled waste water collection and decentralized stormwater as additional sources of supply.

- **Waste Water System:** Water-dissolvable poisons and dangerous substances from waste, landfills, and other waste offices, if not controlled and treated properly, can enter the surface or groundwater, and therefore be devoured by people. In uncontrolled landfills or open dumpsites where no physical obstruction (e.g. liner) exists between the waste and the ground, leachate from deteriorating waste is a noteworthy wellspring of groundwater and soil contamination.
- **Water Consumption:** There has been advance in CO2 outflow decrease through the water-sparing execution of restroom apparatuses, for example, toilets and showers. CO2 discharges related with water utilization which can be lessened by 25% by 2020 through the appropriation of the utilization of water-sparing installations. To advance the water-sparing, bowls and thermally protected baths has been added to the Government's energy saving arrangements.

There has been advancement in CO2 emission decrease through research into the water-sparing execution of bathroom apparatuses, for example, toilets and showers. CO2 discharges related with water utilization can be lessened by 25% by 2020 through the reception of the utilization of water-sparing installations. To advance the replacement of current fixtures and flow installations with water-sparing tiolet bowls and thermally protected baths has been added to the Government of energy saving arrangement.

- **System Leakage:** Leakage happens in all circulation and supply system, and is represented by the state of the pipes, the operating pressure and how well leakage issues are observed and tended to. From a financial point of view, the cost of any measures to diminish leakage must be not as much as the cost to deliver the water from another source. Target levels for diminishing leakage are set that adjust the necessities of water clients and the environment.
- **Water Treatment Policies:** In building landfills with landfill gas gathering frameworks, a liner is introduced to go about as a boundary between the waste and the encompassing ground, and in addition a top, to keep landfill gas and leachate from getting away past the landfill border. In this way, introducing a landfill gas accumulation system in new landfill stages has the additional advantage of limiting water contamination. The main purpose of a landfill gas utilisation system is to collect landfill gas and generate energy, it offers the added benefit of minimising water pollution and thus prevents health risks linked to waterborne diseases and conditions. These co-benefits can be substantial considering the severity of the health risks particularly in low- and middle-income countries.

4.4. Waste System

The waste management sector represents an expected 3–5% of worldwide anthropogenic emissions. In spite of the fact that this probably won't be considered as critical as the commitments from different areas, for example, transport and structures, the waste segment is in an extraordinary position to be at the front line of worldwide outflows decrease. Although low levels of emissions are discharged through waste treatment and disposal, waste prevention and recovery can possibly maintain a strategic distance from outflows in every other division of the economy, for example, material extraction, mining, forestry, transport, agriculture, assembling, and energy generation. The environmental change moderation capability of the waste material in developing nations is three times higher than the one in developed countries. The main features which help in reducing CO2 emission are as follows:

- **Waste Reduction Policies:** There is a developing mindfulness among researchers and policy creators of the interface between waste administration arrangements and strategy to handle environmental change. It is additionally perceived that confining wastage to increasingly practical dimensions and applying asset efficiency can viably decrease GHG emanations connected to environmental change, and in addition offer different benefits of a financial and social nature.
- **Waste Recycling:** The climate benefits from more sustainable waste management practices are significant, such as reduction in landfill gas emissions, reduction in natural materials extraction and manufacturing, replacement of virgin materials with recycled or recovered wastes, replacement of fossil fuel energy with energy from waste and carbon bound in soils through compost application. A number of low-carbon measures can deliver these climate benefits, such as waste prevention, waste recovery, recycling, composting of organic waste, energy from waste, anaerobic digestion of waste with electricity and/or heat recovery, mechanical biological treatment (MBT), landfill gas utilisation or flaring.
- **Municipal Waste Production:** Net emissions of ozone harming substances from the management of municipal waste are anticipated to decay from a pinnacle of around 55 million tons CO2-counterparts every year in the late 1980s to 10 million tons CO2 equivalents by 2020. This is because of two separate advancements. From one perspective, waste enter via. administration offices are anticipated to keep on growing as waste per capita increments and waste gathering is additionally moved forward. This drive coordinate outflows of ozone harming substances from the waste administration division

up. Landfilling represents 60% of the aggregate in 2020, and reusing and burning around 20% each.

- **Green Land Use Policy:** As far as land utilize, the effects of atmosphere strategy are less pronounced, yet at the same time perceptible. As a matter of first importance, the expansion in the utilization of present-day bio-vitality prompts an expansion of land for bio-vitality. Besides, the expected endeavors on Afforestation, Reforestation and avoided Deforestation (ARD) prompt a little increment in timberland territory.

4.5. Industry Sector

The ways that the industry sector can reduce CO2 emissions include (Evans, 2018):

- **Measuring Carbon Footprint:** Coal, tar sands and fossil fuels coal are the main petroleum products which sufficiently contribute to the measure of CO2 important to trigger irreversible environmental change. Organizations that endeavor to change from coal to more economical energy sources, for example, wind or sun powered power, can diminish CO2 outflows enormously.
- **Carbon Capping:** To keep away from carbon emissions, carbon capping is a methodology that would essentially put an expense on carbon dioxide discharges by pitching licenses to produce the gas. Each huge scale maker, or association, will have a limit on the proportion of ozone draining substance that it can exude. The firm ought to have an "emissions permit" for every ton of carbon dioxide it releases into the earth. These licenses set an enforceable most distant point, or best, on the proportion of ozone draining substance contamination that the firm is allowed to emanate.
- **Rewarding Green Commutes:** Urging agents to change to open transportation, carpooling, biking, telecommuting and other inventive ways to deal with extra energy and decline ozone depleting substance outpourings while in travel to and from work can incorporate and have enormous effects. Managers can offer suburbanite benefits that address compelled or exorbitant ceasing, reduce movement blockage, improve workers selection and limit the regular impacts related with drive-alone driving.
- **Investing in Renewables:** There are choices in case attempt new energy successful building activities aren't doable, or an organization fundamentally can't stand to put sunlight based fueled sheets on structures. The balance of carbon impressions through the headway of elective endeavors, for instance, sun powered or wind energy or reforestation, speaks to a strategy for lessening a carbon impression and is routinely known as carbon offsetting.

Figure 4. Represents the diverse areas and factors contribute to mitigate the CO2 emissions in cognitive cities

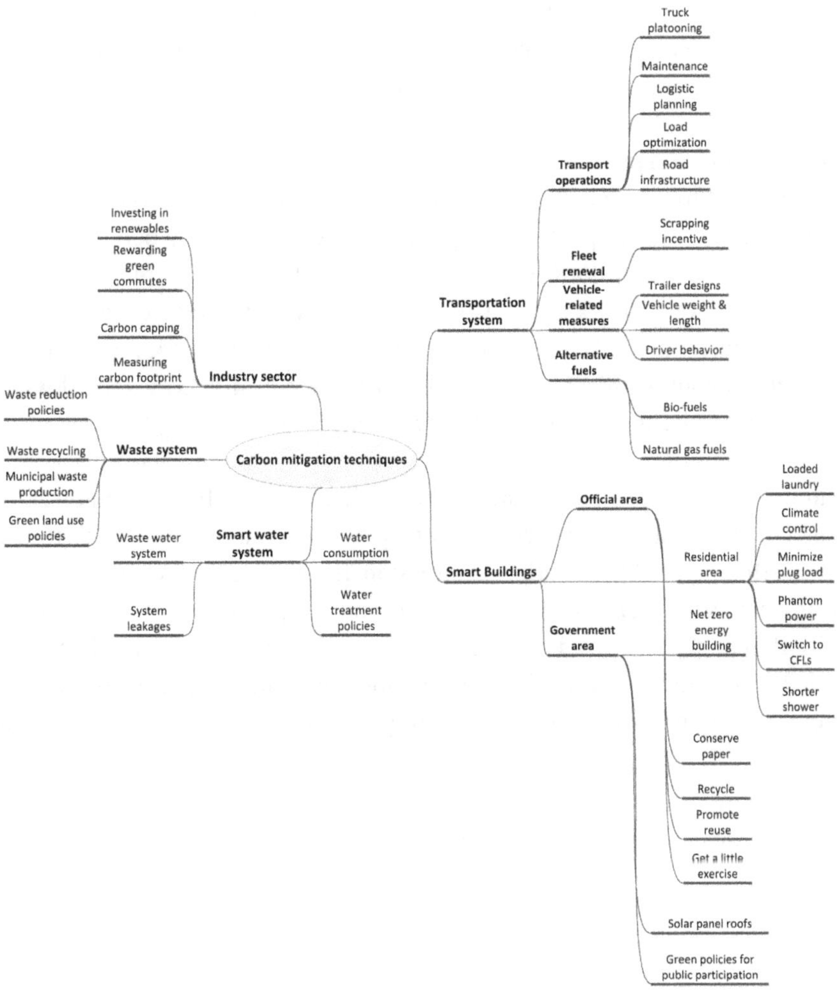

5. PROPOSED FRAMEWORK FOR ZERO-CARBON COGNITIVE CITIES

There are many synergies between aspirations for urban development and the imperative for climate action are conclusively demonstrated in the above sections. In this section, we propose a low-carbon congnitive city framework, which includes buildings, transportation, food, industry and electricity sectors as shown in Figure 5.

A gamification approach is employed over these sectors to achieve the the target of low carbon cognitive city. This systematic approach specially works on the behavioral activities of the persons to achieve an exclusive reduction of CO2 emissions in the diverse fields. We mainly focused on housing, transportation and food domain for behavioural interventions for achieving zero carbon cities in future.

- **Gamification Approach:** When it comes to employ gamification to convince youngsters to assume more prominent liability for their activities, couple of thoughts has had the planet-saving aspirations with exciting benefits that urges clients to deal with their own carbon footprints. For example, maintainable exercises like 1,300 kg (2,860 lbs) of waste was redirected which adds up to 191 packs of junk. Individuals spared 163,400 liters (43,166 gallons) of water, enough to fill 864 baths and 106,148 kWh of energy spared which is sufficient to light a football arena for eight games.

Gamification approach joins internet based life, and portable encounters. Clients are urged to "buzz" an alarm each time they play out a maintainable activity which will lessen the extent of their carbon impression, from utilizing a reusable coffee cup to reusing a sheet of paper and give rewards and badges in lieu of their respective behavioral change for reducing CO2 emissions. Different behavioral factors are identified from literature especially in the domain of housing, transportation and food for applying gamification approach to get zero carbon future cognitive cities.

Figure 5. Proposed framework for making zero carbon cognitive cities

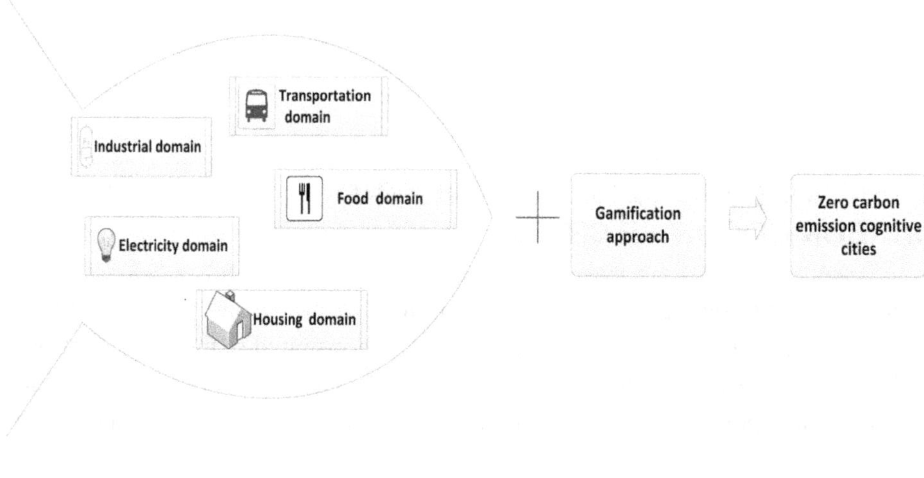

Housing Domain

A couple of conduct choices are recognized from the literature for applying gamification approach in housing area. Predominantly prefer to diminish the request of energy for warming by either decreasing room temperatures, lessening the measure of warmed space, decreasing the utilization of room warming by any means. Other options are either upgrading the settings of water radiators as well as the space warming offices or identify with ventilation conduct. However, the main options exhibited in the Fig. 6 has longer impacts of various social choices regularly considered together and data is compiled to check consolidated impacts. From a theoretical perspective, behavior measures in connection to energy usage in domestic areas are classifications into efficiency as well as curtailment behaviors.

The first incorporates one-shot practices like the choice on and interest in hardware utilized, i.e. the energy source and the machine for producing energy. Seeing that these practices don't require a proceeding with change in conduct and are as of now more frequently secured by models, they are not considered in this investigation (Rohde, Dutschke Gigli & Bles, 2012; Feng & Sauer, 2004). The second classification focuses on repetition, once adapted, typically ongoing endeavors to spare energy by changes in ordinary conduct, i.e. the task of machines, favored room temperatures, use designs with respect to opening windows, and so on. A portion of these conduct measures suggest a difference in schedules without evolving way of life (e.g. streamlined task of warming establishments without diminishing the room temperature), others suggest more noteworthy changes (e.g. decreased room temperature).

Transportation Domain

General classes of behavioral mitigation options in transport can be distinguished for implementation of gamification approach are as follows: Using more eco-friendly autos; individuals could lessen the atmosphere effect of their versatility conduct by utilizing more eco-friendly autos. E.g. they could decide for a littler auto or an electric auto. Note that this examination has avoided changes in buy conduct which don't essentially influence the manner in which the item could be utilized. In this manner, behavioral changes like purchasing an eco-friendlier car (e.g. because of a more proficient engine) from a similar size or purchasing autos running on elective powers (e.g. biofuels, flammable gas) are not considered (Barth & Boriboonsomsin, 2008; Skinner et al., 2010). Making utilization of the auto in a more productive manner; by utilizing traveler autos in a more effective manner GHG outflow decreases of street transport could be figured it out. Productivity estimates that could be connected are: applying an eco-friendlier driving style, auto pooling, sharing autos, and so on. Using more practical modes; a move to movement modes with generally low GHG emanations per traveler kilometers (e.g. strolling, cycling, open transport) could add to de-carbonization of transport (Schroten, 2012; ES Energy Information Administration, 2015; Drexel, 2016).

Other social measures are taking an interest in auto sharing activities or make utilization of aggregate transport programs composed by businesses for driving treks. Reducing travel remove; individuals could lessen the quantity of kilometers they travel from various perspectives: working at home, living close to the activity, less occasions (to far-away nations), consolidating different excursions, and so on.

Food Domain

Behavioural change options that reduce carbon emissions are as follows:

Change a diet as carbon discharges related with meat are significantly higher than outflows related with plant protein sources. Consequently, a change to a veggie lover eating routine would lessen emissions. Dairy and eggs have emanations like meat. Henceforth, a decrease of animal protein would diminish carbon outflows. Changing to a solid eating routine would therefore lessen the general nourishment

Figure 6. Represents points to be considered for implementing gamification approach on diverse domains

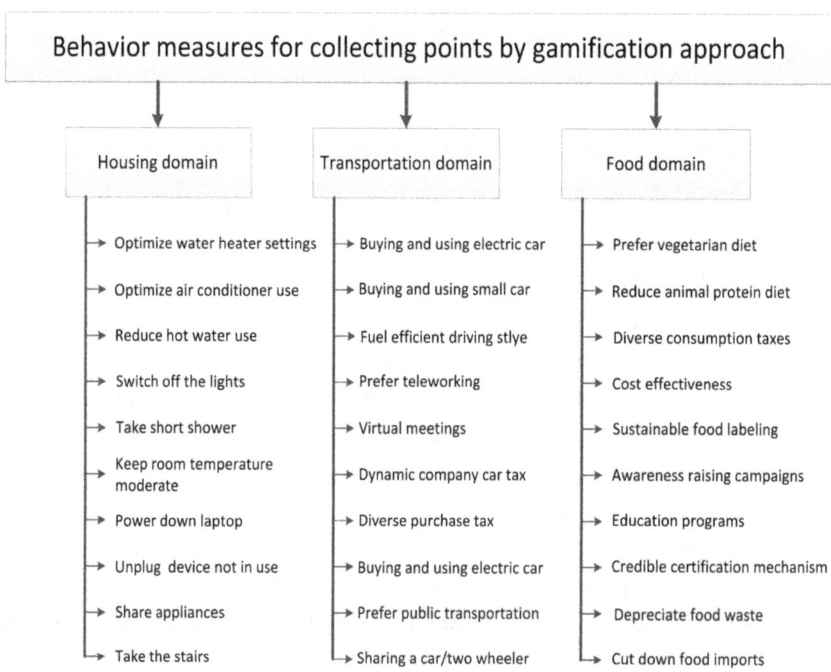

utilization and could likewise diminish the utilization of meat items, in this way bringing down carbon emanations. Nourishment wastage can be separated into the class of unavoidable waste (inedible remains) and waste which could be abstained from (discarding lapsed sustenance, leaving palatable nourishment on the plate) [54]. By lessening waste, the aggregate nourishment utilization is decreased and furthermore are carbon discharges. A few vegetables developed in nurseries and items which are transported over long separations require more energy contribution to their life cycle than privately created as well as regular nourishment. Lessening energy and fuel utilize by another arrangement of alternatives identified with the nourishment part would diminish energy and fuel utilize. Fuel utilize can be decreased by more concentrated utilization of the home conveyance of staple goods benefit. It is obvious that items including more transport, and cooling require more energy input and, in this way, produce more carbon emissions.

6. BARRIERS TO ACHIEVE ZERO CARBON ENERGY SAVING BY BEHAVIOR INTERVENTIONS

In this section, we considered a coordinated approach for the investigation and discussion of boundaries and strategy instruments. The behavioral choices within reach are reliant in the sense, that these help each other so as to diminish carbon emissions in diverse areas. Approach instruments are in this way not recognized per conduct relief alternative, but rather for the mix of practices going for decreasing warm vitality utilization at home. Policy instruments are not only identified as behavioral mitigation options, but for the combination of behaviours aiming at reducing carbon emissions in cognitive cities.

- **Hosing Domain:** While considering energy saving behavior in housing domain, a refinement of decrease and productivity practices must be made, the last tending to interests in generally surprising expense effectiveness advancements in buildings. The focus lies on short duration behavior, which is driven by day by day habits and schedules which shows itself as a major aspect of individuals' ways of life: diminishing space heating temperature, enhancing indoor regulator settings, improving ventilation conduct. A categorization of obstructions as per a given structure was useful for distinguishing basic examples and attributes for the different social relief alternatives. To the most essential hindrances towards private energy sparing have a place restricted insight, as absence of information and mindfulness about one's own particular energy utilization. Besides, blocking variables can be perspectives that tend to block genius natural dispositions, correlations with key other individuals (that normally go about as a driver) or the attribution of obligation to other people, sunk energy costs, connected to conduct schedules and the absence of direct energy utilization feedback.

- **Transport Domain:** The principle barriers are at both the person as societal level. At the individual level buyers have negative state of mind towards electric and plug hybrid autos. Electric autos are seen as less alluring than traditional autos on numerous measurements: exhibitions (e.g. run), unwavering quality, costs, picture, and so on. Also, electric and hybrid vehicles may challenge portability related propensities for individuals, driving individuals to change their ways of life. For instance, because of the long energizing time the utilization of electric autos is seen to be less adaptable than the utilization of ordinary autos. At the societal level basic boundaries (the poor accessibility of charging framework and the predetermined number of electric and cross breed vehicle models) and financial hindrances (high buy costs) are the principle obstructions for a take-up of electric and hybrid vehicles.

A few statistic variables may influence the resistance of individuals to purchase a small car. Household measure; a couple of individual families will probably purchase a small car than families with youngsters (Choo & Mokhtarian, 2004; Kitamura et al., 2000), since for the last the interior space of the car will be even more a key factor in the car buy choice. Ladies are relied upon to probably purchase a small car than men (Choo & Mokhtarian, 2004). Ladies are less worried to the status of car and, what's more, will be more worried on the natural execution of the car (Johansson-Stenman & Martinsson, 2006). It might be normal that elders taught individuals will probably persuade smaller cars (Choo & Mokhtarian, 2004). As indicated by Johansson-Stenman and Martinsson (2006) these individuals are less worried on the status of their car. Also, ecological concern is regularly anticipated that would be higher for elder taught individuals than for low-educated individuals. Individuals living in the city are more eager to purchase a smaller car than individuals living on the wide open. Since the last gathering will utilize the car all the more frequently, they incline toward a more agreeable car than subjects. Moreover, small cars are more down to earth in the urban rush hours. The social/mental boundaries allude to individuals' impression of the disadvantages of teleworking: social disengagement, inclination for workaholic behavior, unfriendly effects on vocation, stirring up private and expert life, and so forth.

- **Food Domain:** For vegetarian and diminished animal protein diets, learning, propensities and social hindrances are the most essential obstructions. It is likely that, once information levels, habits and culinary societies have changed, items for meat and animal protein items will wind up accessible in the sustenance benefit part and in dinners and items that are instant and simple to deliver. Situational and infrastructural boundaries are less imperative than learning, habits and social obstructions. One critical inquiry is whether the obstructions are similarly essential for a veggie lover eat less carbs and a diminished animal protein count calorie. Since changing to a veggie lover eating routine establishes a major change when contrasted with most shoppers' present eating methodologies, though lessening animal protein admission to six days/seven days comprises a more restricted change and basically leaves the eating regimen flawless on six out of seven days, we accept that habits and social barriers are somewhat more imperative for a vegetarian eat less than for diminishing animal protein consumption.

Learning based barriers can be tended to through communication, e.g. mass media campaigns and food labelling. However, there is as yet the way that food decisions are in expansive part constant. A first significant strategy that tends to this comprises of school-based intercession programs. Habits grow from the get-go throughout everyday life, and it is along these lines essential to enable youngsters to create solid and maintainable habits at a youthful age. Another manner by which habits can be focused on is by utilizing 'upstream' interventions, for example, charging meat/animal protein utilization with consumption charges (Faber et al., 2012).

7. CONCLUSION AND FUTURE WORK

Buildings and transportation contribute major part of carbon emissions in current situations. Substantial emissions are derived from other domains includes industrial processes, solid waste transport and disposal, fleeting emissions from landfills, water and waste water transport and treatment, and street lighting. Electricity generation intersects substantially, but not completely with other domains as electricity is used largely in buildings, transportation, waste and water management as well as in industrial processes and public works. Further, a new framework is required to reduce carbon emission to achieve the target of zero carbon cities. The most unprecedented changes in the particular buildings of the city will be the expansion of building systems from free applications focused on the movement, broader arrangement of energy and environmentally friendly structures. Additional options include for reducing carbon emissions are during shopping, preparation and storage of food, these options also overlap with transport options and with more general energy saving options. In this chapter, the focal point of investigations is coordinated diverse areas/fields for accomplishing utilization designs that are manageable with respect to environmental change. By applying proposed framework, in which carbon emission mitigation techniques along with gamification approach are integrated to achieve zero cognitive cities. Rewards and badges are provided to consumers for reducing carbon emissions and saving energy by various means. This framework will help to engage more number people in reducing CO2 emissions and achieving the zero carbon cities target.

REFERENCES

An, F., & Sauer, A. (2004). *Comparison of passenger vehicle fuel economy and greenhouse gas emission standards around the world.* Pew Center on Global Climate Change.

Australia, I. (2010). *State of Australian cities 2010.* Canberra: Infrastructure Australia.

Barth, M., & Boriboonsomsin, K. (2008). Real-world carbon dioxide impacts of traffic congestion. *Transportation Research Record: Journal of the Transportation Research Board, 2058*(1), 163–171. doi:10.3141/2058-20

Bartlett, S., & Satterthwaite, D. (2016). *Cities on a finite planet: Towards transformative responses to climate change*. Routledge. doi:10.4324/9781315645421

Birkeland, J. (2014). Resilient and sustainable buildings. Resilient Sustainable Cities, 146(159).

Bulkeley, H., Broto, V., Hodson, M., & Marvin, S. (Eds.). (2010). Cities and Low carbon transitions. Routledge.

Carli, R., Dotolia, M., & Pellegrino, R. (2018, November). Multi-criteria decision-making for sustainable metropolitan cities. *Journal of Environmental Management, 226*(15), 46–61. doi:10.1016/j.jenvman.2018.07.075 PMID:30107310

Choo, S., & Mokhtarian, P. L. (2004). What type of vehicle do people drive? The role of attitude and lifestyle in influencing vehicle type choice. *Transportation Research Part A, Policy and Practice, 38*(3), 201–222. doi:10.1016/j.tra.2003.10.005

Daly, H. (1996). *Beyond growth.* Boston: Beacon Press.

Drexel, A.J. (2016). *Options for Achieving Deep Reductions in Carbon Emissions in Philadelphia by 2050.* Prepared by Drexel University for The Philadelphia Mayor's Office of Sustainability.

Eames, M., Dixon, T., May, T., & Hunt, M. (2013, October). City futures: Exploring urban retrofit and sustainable transitions. *Building Research and Information, 41*(5), 504–516. doi:10.1080/09613218.2013.805063

Energy, C. (2017) What is a carbon footprint? *Cloud Energy*. Retrieved from: https://cloudenergy.com.ng/blog/what-carbon-footprint

Ernston, H., van der Leeuw, S. E., Redman, C. L., Meffert, D. J., Davis, G., Alfsen, C., & Elmqvist, T. (2010). Urban transitions: On urban resilience and human-dominated ecosystems. *Ambio, 39*(8), 531–545. doi:10.100713280-010-0081-9 PMID:21141773

Evans, M. (2018). 7 ways to reduce CO2 emissions pollution in industry. *The Balance Small Business*. Retrieved from: https://www.thebalancesmb.com/reducing-co2-emissions-3157794

Faber, J., Schroten, A., Bles, M., Sevenster, M., Markowska, A., & Smit, M. (2012). *Behavioural climate change mitigation options and their appropriate inclusion in quantitative longer-term policy scenarios*. Delft: CE Delft.

Faber, J., Sevenster, M., Markowska, A., Smit, M., Zimmermann, K., Soboh, R., & van 't Riet, J. (2012). *Behavioural Climate Change Mitigation Options: Domain Report Food*. Report Delft.

Georgetown University. (n.d.). Simple tupes to reduce your carbon footpting. *Georgetown University*. Retrieved from: https://sustainability.georgetown.edu/getinvolved/thingsyoucando

Girardet, H. (2014). *Creating regenerative cities*. Routledge. doi:10.4324/9781315764375

Glackin, S., & Newton, P. (2016). *Engaging the Greyfields: community engagement and co-design in residential redevelopment of public housing. Instruments of Planning*. Taylor & Francis.

Gouldson, A., Sudmant, A., Khreis, H., & Papargyropoulou, E. (2018). *The economic and social benefits of low-carbon cities: a systematic review of the evidence*. Coalition for Urban Transitions.

IPCC Fifth Assessment Report. (2013). *IPCC*. Retrieved from: http://ipcc-wg2.gov/oeAR5/images/uploads/WGIIAR5-Glossary_FGD.pdf

Jin-Yi, C., & Ying-Jia, H. (2011) Low-carbon urban planning of the resource-based cities. In *Multimedia Technology (ICMT), 2011 International Conference on* (pp. 1102-1105). IEEE.

Johansson-Stenman, O., & Martinsson, P. (2006). Honestly, why are you driving a BMW? *Journal of Economic Behavior & Organization*, *60*(2), 129–146. doi:10.1016/j.jebo.2004.08.006

Kaźmierczak, J., Hilarowicz, A., & Wolny, M. (2018). *Introduction to the research project "smart city: a holistic approach."* Retrieved from: https://www.researchgate.net/publication/326147170_Smart_City_Holistic_Approach

Kitamura, R., Golob, T. F., Yamamoto, T., & Wu, G. (1999). *Accessibility and Auto Use in a Motorized Metropolis Center for Activity Systems Analysis*. Institute of Transportation Studies, UC Irvine.

Koutra, S., Ioakimidis, C.S., Gallas, M-A & Becue, V. (2018, March). Towards the Development of a Net-Zero Energy District Evaluation Approach: A Review of Sustainable Approaches and Assessment Tools. *Sustainable Cities and Society*.

Krugman, P. (2010) The finite world. *The New York Times*, 26.

Lugaric, L., & Krajcar, S. (2016). Transforming cities towards sustainable low-carbon energy systems using emergy synthesis for support in decision making. *Energy Policy*, *98*, 471–482. doi:10.1016/j.enpol.2016.09.028

LVM. (2017). Vehicle fleet renewal supports the reduction of emissions. *LVM*. Retrieved from: https://www.lvm.fi/en/-/vehicle-fleet-renewal-supports-the-reduction-of-emissions-955841

Martiskainen, M., & Kivimaa, P. (2018). Creating innovative zero carbon homes in the United Kingdom—Intermediaries and champions in building projects. *Environmental Innovation and Societal Transitions*, *26*, 15–31. doi:10.1016/j.eist.2017.08.002

Moore, T., de haan, F.J., Horne, R., & Gleeson, B J. (Eds.). (2017). Urban sustainability transitions: Australian cases – International perspectives. Springer.

Newton, P. (2017). Framing new retrofit models for regenerating Australia's fast-growing cities. In Retrofitting Cities for Tomorrow's World. Academic Press.

Newton, P. W. (2006). *Australian State of the Environment Report: Human Settlements Theme Commentary*. Department of Environment and Heritage, Canberra. Retrieved from www.environment.gov.au/soe

Newton, P. W. (2007). Horizon 3 Planning: Meshing Liveability with Sustainability, Environment and Planning B. *Environmental Development*, *34*, 571–575.

Newton, P. W. (2008). *Transitions: Pathways Towards Sustainable Urban Development in Australia*. Dordrecht: Springer.

Newton, P. W. (2012). Liveable and Sustainable? Socio-Technical Challenges for 21st Century Cities. *Journal of Urban Technology*, *19*(1), 81–102. doi:10.1080/10630732.2012.626703

Newton, P. W. (2017). Innovation for a Sustainable Low Carbon Built Environment. *Procedia Engineering*, *18*, 16–32. doi:10.1016/j.proeng.2017.04.161

Newton, P. W., & Bai, X. (2008). Transitioning to Sustainable Urban Development. In P. W. Newton (Ed.), *Transitions: Pathways to Sustainable Urban Development in Australia*. Dordrecht: Springer.

Newton, P. W., & Doherty, P. (2014). The challenges to urban sustainability and resilience. Resilient Sustainable Cities, 7(18), 7-18.

Newton, P.W. & Meyer, D. (2016). 11 Built Environment and Human Behaviour Challenges to Sustainable Consumption. *Ethics and morality in consumption: Interdisciplinary perspectives, 8*, 189.

Newton, P. W., & Tucker, S. N. (2011). Pathways to Decarbonising the Housing Sector: A Scenario Approach. *Building Research and Information*, *39*(1), 34–50. doi:10.1080/09613218.2010.531085

Normile, D. (2016). China rethinks cities. *Science*, *352*(6288), 916–918. doi:10.1126cience.352.6288.916 PMID:27199411

Nunes, V., Gouveia, J.P., Rodrigues, A.M., & Simão, T. (2017). INSMART–towards the new distribution systems operators potential roles in low carbon future and integrated frameworks for smart cities. *CIRED-Open Access Proceedings Journal, 2017*(1), 2797-2799.

Peng, Y. & Bai, X. (2017). Experimenting towards a low-carbon city: Policy evolution and nested structure of innovation. *Journal of Cleaner Production, 174*, 201-212.

Rohde, C., Dütschke, E., Gigli, M. & Bles, M. (2012). *Behavioural Climate Change Mitigation Options Domain Report Housing*. Report Delft.

Ryan, C. (2016). *Visions and Pathways 2040*. Sydney: CRC for Low Carbon Living.

Schroten, A. (2012). *Behavioural Climate Change Mitigation Options: Domain Report Transport*. Report Delft.

Seto, K., Sanchez-Rodriguez, R., & Fragkias, M. (2010). The new geography of contemporary urbanisation and the environment. *Annual Review of Environment and Resources*, *35*(1), 167–194. doi:10.1146/annurev-environ-100809-125336

Skinner, Van Essen, Smokers, & Hill. (2010). *EU Transport GHG: routes to 2050.* Towards the decarbonisation of the EU's transport sector by 2050.

Steffen, W., Broadgate, W., Deutch, L., Gaffney, O., & Ludwig, C. (2015). The trajectory of the Anthropocene: The great acceleration. *The Anthropocene Review, 2*(1), 81–98. doi:10.1177/2053019614564785

Tan, S., Yang, J., Yan, J., Lee, C., Hashim, H., & Chen, B. (2017). A holistic low carbon city indicator framework for sustainable development. *Applied Energy, 185*(2), 1919–1930. doi:10.1016/j.apenergy.2016.03.041

Tsolakis, N., & Anthopoulos, L. (2015). Eco-cities: An integrated system dynamics framework and a concise research taxonomy. *Sustainable Cities and Society, 17*, 1–14. doi:10.1016/j.scs.2015.03.002

UN DESA. (2014). *2014 Revision of the world urbanization prospects*. New York: United Nations.

U.S. Energy Information Administration. (2015). *Natural Gas*. Energy Information Agency. Retrieved August 26, 2015, from: http://www.eia.gov/dnav/ng/ng_pri_sum_a_EPG0_PRS_DMcf_a.htm

Wang, S., & Chen, B. (2018). Three-Tier carbon accounting model for cities. *Applied Energy, 229*, 163–175. doi:10.1016/j.apenergy.2018.07.109

Wu, Z., Tang, J., & Wang, D. (2016). Low carbon urban transitioning in Shenzhen: A multi-level environmental governance perspective. *Sustainability, 8*(8), 720. doi:10.3390u8080720

WWF Living Planet Report 2014. (n.d.). Available: http://wwf.panda.org/about_our_earth/all_publications/living_planet_report/

Yang, X., Wang, X.-C., & Zhou, Z.-Y. (2018). Development path of Chinese low-carbon cities based on index evaluation. *Advances in Climate Change Research, 9*(2), 144–153. doi:10.1016/j.accre.2018.05.004

Chapter 2
Cyber Security Challenges for Smart Cities

Anand Nayyar
Duy Tan University, Vietnam

Rachna Jain
Bharati Vidyapeeth's College of Engineering, India

Bandana Mahapatra
https://orcid.org/0000-0002-1388-5413
SOA University, India

Anubhav Singh
Bharati Vidyapeeth's College of Engineering, India

ABSTRACT

Smart cities are composed of interlinked components with constant data transfer and services targeted at increasing the life style of the people. The chapter focuses on diverged smart city components as well as the security models designed to be implemented. The four major paradigms discussed in this chapter are smart grids, building automation system (BAS), unmanned aerial vehicle (UAV), and smart vehicles. Apart from addressing the security concerns of every component, the major highlights of this chapter are architecture, smart environment, industry, lifestyle, services, and digital lifestyle quality. Finally, the chapter focuses on privacy preserving mechanisms, its essence over smart cities, strong architecture related to privacy, preserving mechanism, and various approaches available that can retaliate these issues in a smart city environment.

DOI: 10.4018/978-1-5225-8085-0.ch002

1. INTRODUCTION

A Smart City is a term used to denote the urban culture and modern facilities available to the people living in towns. These cities designated as urban areas use various kinds of electronic data collection sensors that relays information used to manage the assets and resources efficiently. It includes the data collected from electronic gadgets, the customers, assets that is being processed and analysed in order to manage and track various systems, e.g. the traffic, transportation, power plants, water supply network, waste management, law enforcement, information systems, schools, schools, libraries, community hospitals etc (Bartoli et al., 2011).

The concept of smart cities is a formal hybridization of ICT (Information and Communication Technology along with physical devices attached to the network in order to optimize the efficiency of city operations along with services connected to the residents. The smart city technology allows the city workers and officers for direct interaction with the community and city infrastructure in order to monitor the day to day happenings of the city. With the increase in the standard of lifestyle maintained by people with its overly increasing demands to ease the daily affairs, the technical the technical world has always encountered new challenges and benchmarks to overcome (Khurana, Hadley, Lu & Frincke, 2010; Baig et al., 2017). In order to meet the social objectives and physical realm with enhancement of the class of life leads to evolvement of what is called 'SMART CITIES'. An overtly increase in the population of cities poses another major challenge giving rise to several complex problems like city administration or the governance. The continuous rise in demand and comparatively shortage of resources mandates the need of smart and more 'astute' arrangements to address the rising prerequisites in these urban situations.

The world faces several complicated problems and challenges for its survival and attaining a basic sustainability in day to day life. To fulfil the increasing demand of people and meet the objectives of the social and the physical realm in order to enhance of the class of life leads to evolvement of "smart cities". This concept gained widespread framework since it offers solutions to all kinds of challenges. In recent years, it has been noted, global urbanization has undergone drastic expansion, which does not show any sign of undergoing reduction in the coming years. In these circumstances there are huge related endeavours to encourage a better and feasible standard of living. This can be best dealt with propelling and innovative technologies as a response and solutions to these crucial challenges. This encourages the worldwide notch companies for drifting their technologies towards smarter and astute innovation. Cisco, Google, HP, Dell, IBM and more global companies got triggered to launch their own smart and connected devices, smart projects "smart cities" as the response to these challenges (Yan, Qian, Sharif & Tipper, 2012). The market of a smart city is estimated to have worth of several billions of dollars by 2020. Therefore, important city models encompass various diverse aspects of smartness and completely focused on the importance of incorporation and collaboration across several domains. In an overall it can be said that "A city, in the end, is an arrangement of frameworks", and "any models that endeavour to characterize its dynamic nature must likewise have the capacity to speak to the assorted variety of those components (Wicaksono, Rogalski & Kusnady, 2010; Antonini, Barenghi, Pelosi & Zounouz, 2014).

Built over the spines of propelling data, communication and control technologies, smart cities are equipped with intelligent transportation, smart energy, intelligent healthcare and smart buildings, models and related devices. These are programmed to automatically identify the demands of people and improves the quality of life (Mohammed et al., 2014).

Though the Smart city benefits people and the city in various ways, such as environment, industry, living services, and energy, but are equally susceptible to threats related to security and privacy issues. The privacy concerns here regarding the data stored may be related to resident's personal information or any other private data which may be extremely crucial to be disclosed. The chapter is regarding the various cyber security issues and challenges encountered while meeting the milestones of a smart city requirement.

2. ENTITIES OF SMART CITIES

Smart cities are assorted and interconnected segments that constantly exchange information and encourage enhanced lifestyle for the residents. The main goal is to make the cities safe, secure, green, clean, more efficient, future-compatible making the use of advanced, networks, sensors and integrated materials (Bartoli et al., 2011; Vattapparamban et al., 2016). The key parts of smart cities are Smart grids, Automation Systems, Unmanned Aerial Networks (UAVs) and Smart Vehicles. This has also led to an increase in the dangers and criminal abuses in the smart city system. But being prepared for the upcoming potential threats to the architecture and infrastructure appropriate measures have been taken by identifying and analyzing the past incidents and accordingly prepared for any future incidents.

With increasing population in urban cities due to migration from small cities, towns and village in search of a better lifestyle, education and work has led to strategic management to remove any sort of burden on urban cities. This also involves the better use of energy to reduce the pollution, increase the efficiency bringing an energy revolution. Smart cities also have the capability of using communication and sensor to enhance electrical, transport, and other strategic tasks supporting daily lifestyles, thus enhancing security/reliability concerns.

What's more important is that since the presumptions and necessities for the smart city infrastructure are quite different, it must be designed in a different way which also introduces an integration problem. Figure 1 shows the various components of smart infrastructures.

2.1. Smart Grids

Smart grids use the intelligent transmission and conveyance systems to transmit power. This approach makes the electrical system more reliable, efficient and secured enhancing the transmission framework's productivity, the electric-system operations capability, etc. A smart grid converts a centralized network into a decentralized producer-controlled network which is consumer-interactive (Khurana et al., 2010). Imparting data regarding energy consumption to their efficacy provider through communication channels with the help of smart meters makes the system vulnerable to network-based threats. This information is thereafter used for examining the energy consumption patterns and supply/demand administration. The conventional data management applications are not designed to handle huge amount of scaled information produced by the framework. Hence, cloud computing can be used for storage and processing of these data (Baig et al., 2017; Pauli, 2016).

The new transmission system transforms the electric grid to enable public policy objectives, thus increasing transferring capability which results in a huge market efficiency for the benefit of the customer. In the meanwhile, electricity needs are changing and growing fast. With data centers and storage units alone need about 2,500 megawatt hours (MWh) of annual global demand. According to researchers the world's electric supply will need to triple by 2050 pointing several weaknesses in the current grid system. Thus, to modernize the current grid system to smart

Figure 1. Smart city entities

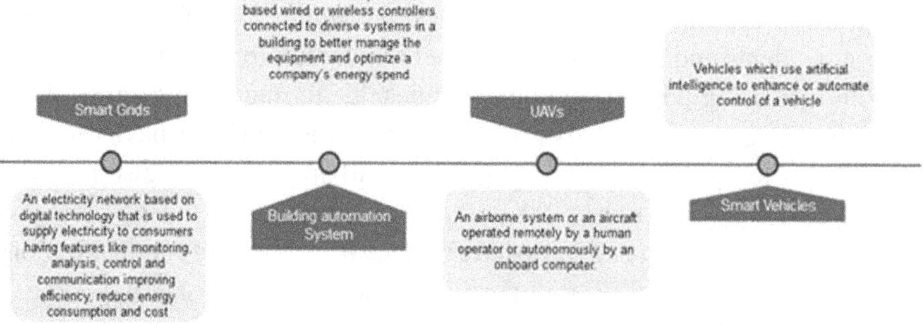

grid, we need to rebuild the entire electric power system following the aspects of generations, delivery and consumption (Chapman, 2016).

The main function of smart cities in smart grids are:

1. It educates and teaches customers about their vitality use, expenses and elective choices empower them to make independent choices about how and when to utilize the power and fills.
2. It modernizes control frameworks through self-mending plans, mechanization, remote control & checking and foundation of microgrids.
3. It provides safe, secure and dependable integration of dispersed and sustainable energy sources (Novotne, Kuchta & Kadlec, 2014).

Figure 2 shows the architecture of a standard smart grid in a smart city.

2.2. Building Automation System

Building automation system is a network of microprocessor based wired and wireless controllers, connected to diverse systems in a building enhancing the company's energy band. Current building automation resolutions for private or business area provides a huge potential for the complete change of vitality. Constructing an automation

Figure 2. A standard smart grid architecture
Source: https://www.sciencedirect.com/science/article/pii/S1742287617300579#bbib28

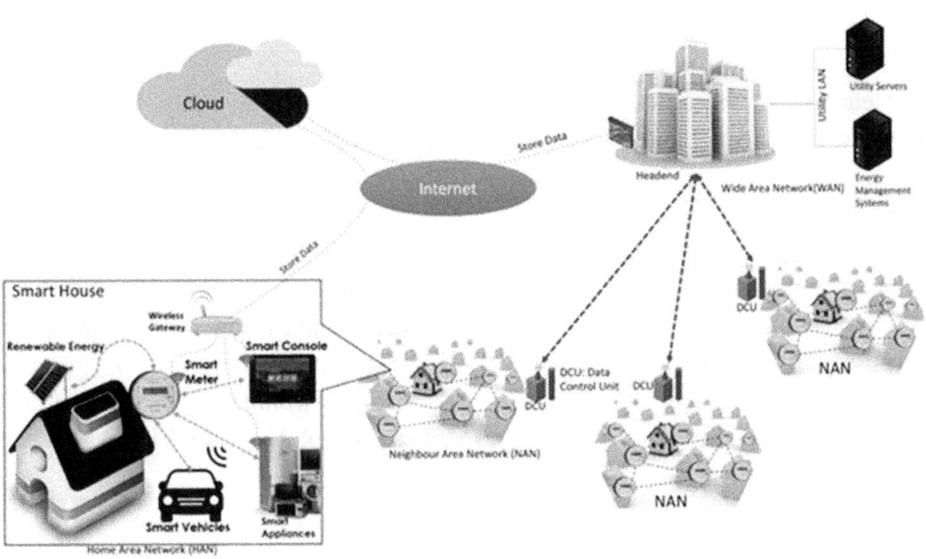

framework, ease the day to day life greatly. They provide better protection, e.g. well, like automatic alarming in case of fire in the building (GitHub, n.d.; Petit et al., 2015).

The concept of smart building was introduced two decades ago to equip the building with energy efficient networking devices like sensors, actuators, controllers, central unit and their interfaces, which previously was limited to concepts like renewable energy resources using wireless communication technology.

Building automation allows the user to monitor the power consumption, individual power loads, providing customers various information and allowing them to participate in the process of decision making. It can be used as a communication interface between devices at home and commercial premises (Yan, Xu & Liu, 2016).

Buildings furnished with a BAS can be commercial, institutional, residential, or industrial, where workplaces and markets come under commercial buildings, institutional building incorporates libraries, schools and medical centers, the residential building incorporates apartments and industrial building include plants or machine shops. Private houses of today are infrequently equipped with BAS despite the fact, that solar system on the roof requires a control framework provided by the manufacturer. The main challenge in BAS is to monitor and maintain building framework with a specific end goal to keep the building functional and accessible for building users. Numerous buildings with existing frameworks are observed through the manual information gathering or with an intricate arrangement of the shield system.

BAS is targeted to improve customer experiences and process, expanding a building's operational viability, decreasing vitality utilization, and eventually improving the budgetary execution of the building with an increase in its connectivity, BAS remains unprotected from the threat that is met by the IT-based protocols and networks. This involves Denial of Service (DoS) attack where the attacker may take control over the whole building (Antonini et al., 2014).

Hence, BAS protocols may be regarded unsafe since it highly trusts on the installed sensors and controllers and lacks source authentication is non-existent.

2.3. UAVs

UAVs are unmanned aerial vehicles also called as drones without any pilot controlled wirelessly by the controller. It was originally designed to be used for military purposes but later on, it got popular in civil applications also where it was used in traffic and mob management, surveillance systems, environmental and agricultural management, disaster management etc. (Mohammed et al., 2014). Figure 3 shows UAV or Drone.

Figure 3. UAV (or drone)
Source: https://www.itweb.co.za/content/2j5alrMQageMpYQk

Smart Cities aim at creating a design for the sustainable environment so that people can live, grow, work and play. These simple factors constitute a smart city.

In the recent times use of UAVs by government institutions, companies and individuals have grown enormously which has brought serious security and technical concerns to be addressed. Drones having several benefits can also be used for malicious activities, e.g. spy on someone, stealing trade secrets, doing cyber-attacks, etc (Vattapparamban et al., 2016; Granzer, Praus & Kastner, 2010).

Today, governments are highly diverted towards developing smart cities seeing the huge increase in population. Taking into account the future, which generates new tasks and expectations from smart cities. This has increased the need to emphasis on ICT services and smart solutions which can be dealt with by UAVs.

The components influencing smart cities are

1. **Technology**: Smart city relies mainly on smart technology, which is evolving daily according to the needs which includes software, hardware, and network technologies which ease the lifestyle of people.
2. **Economy**: Smart cities need and create a powerful and reliable economy in the city whose outcomes are job creations, economic development, proper education, high manufacturing capabilities and huge market growth.

3. **Environment**: The key goal of smart cities is to provide a healthy and sustainable nature of the present and preserve it for the future also. It includes effective preservation policies, including its growth.
4. **ICT Infrastructure**: Enactment of ICT infrastructure in smart city's growth is contingent upon availability and performance. Advancements in ICT include increased efficiency, making the life cheaper and better quality.

2.4. Smart Vehicles

With the increasing number of road accidents, a large number of people are killed every hour in some place or the other. This requires an increasing need for smart vehicles on the road. To counter these accidents governments and manufacturers have introduced new technologies like airbags, anti-blocking brake systems, etc., which has reduced rates of accidents though the numbers are still high.

Latest cars are now equipped with fast processor, smart wireless communication network, USB, Bluetooth, and IEEE 802.11 interfaces.

Figure 4 shows us the onboard instrumentation of a smart vehicle. It has an Event data recorder which is similar to the black box of airplanes recording all major data of the vehicle undergoing a crash. Front-end radar can detect any obstacle before 200 meters and can be used for adaptive cruise control. Smart Vehicle's GPS coordinates and direction can control the next traffic light, whereas the dashboard shows a countdown clock (Baig et al., 2017).

For running smart vehicles on the road, it is important to improve our traffic monitoring system. By traffic monitoring it means pre-detection of accidents, enhancing traffic flow, synchronizing traffic lights using efficient traffic techniques, etc. Traffic management not only include avoiding any accidents but also having

Figure 4. Smart vehicle's on-board instrumentation
Source: https://csdl-images.computer.org/mags/sp/2004/03/figures/j30491.gif

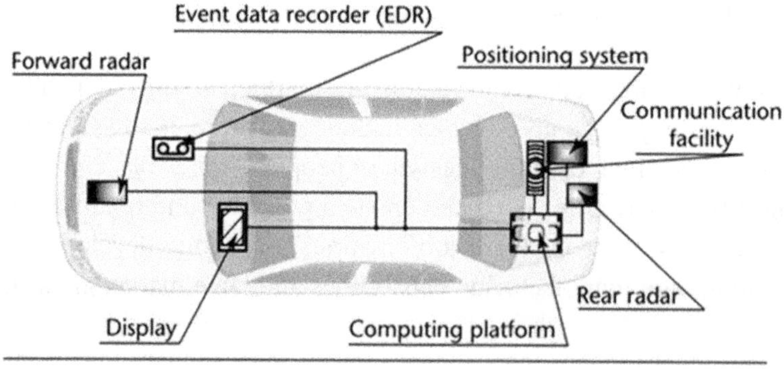

proper response system post accidents. Smart vehicles would also ease the mechanism of the payment process at tolls, fuel stations and parking.

3. SMART CITY ARCHITECTURE

A city is said to be a smart when its major services like healthcare, parking, education and electricity grids are controlled through ICT infrastructure to ease the operations. By smart city architecture, we mean an efficient way to deal with various issues like air pollution, scarcity of resources, waste management, traffic management and human health concerns. All these issues require enhanced capabilities to handle components of this technology.

Foundation of ICT infrastructure makes the base of a smart city. ICT infrastructure comprises and represents the novel e-infrastructure that will allow access to multi-disciplinary data, products, and services to different stakeholders (Petit et el., 2015; Botta et al., 2016).

For designing a smart city architecture, one has to understand the environment of its existence and their elements of interaction. Figure 5 shows us the components of smart city architecture comprising six layers- “Goals”, “People”, “Ecosystem”, “Soft Infrastructure”, “City Systems” and “Hard Infrastructures”.

1. **Goals**: Goals are the target upon which the architecture of the smart city is decided e.g. creation of jobs, promote sustainability, etc. The goal defined directly or indirectly affects the population whether they are residents, workers or visitors. Architects and designers of the Smart Cities should keep in mind to create infrastructures and services that would become part of the fabric, ecosystem of communities and people.
2. **Soft Infrastructure**: It is necessary that there is an infrastructure that will establish an interactive tunnel between the experience of the smart city and the population like governance system which measures overall progress and the performance of individual city systems against the objectives of Smart City.
3. **City System**: City systems give a life support for cities which include transport, educate and provide healthcare facilities to the people. They are the real aspects of a smart city. A designer has to take these into account as the constraint of the city.
4. **Hard Infrastructure**: Transport, information & communication, technology & utilities, network & spaces, and buildings are included in hard infrastructure.

Figure 5. Smart city architecture

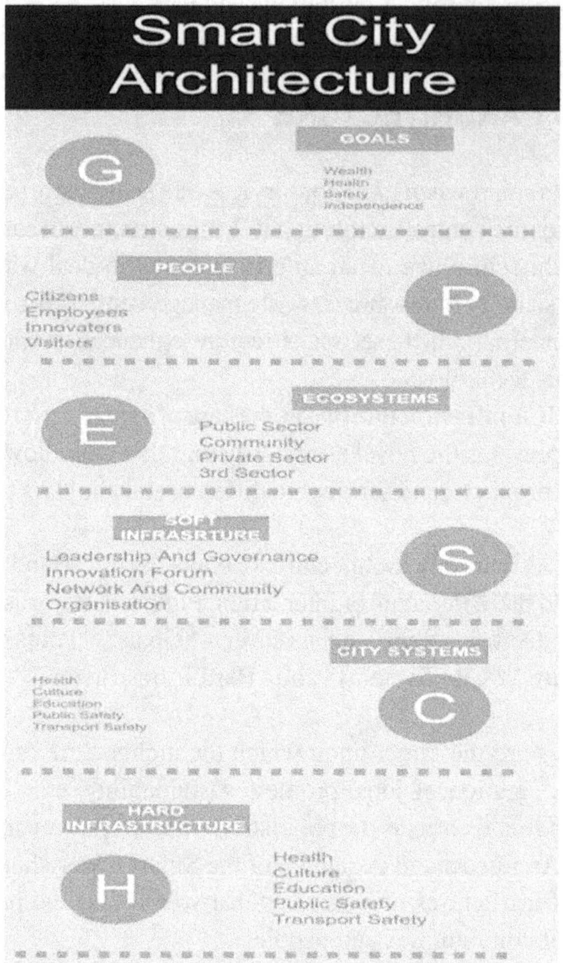

Generally, departments in the city function independently and share data with each other whenever needed. ICT technology and infrastructure, mainly focus on its activity, creating a chaos in the execution of services and the overall response time of department get reduced.

On the other hand, modern smart city model shares data via open data model based on real-time data which improves the overall efficiency of city services. This technique makes the smart city more responsive towards a problem which ultimately enhances resource planning. Events and trends can be predicted via city-wide data which affects the overall data collected (Bawny & Shamsi, 2015; Zhou et al., 2013).

5. SECURITY IN SMART GRID

The smart grid requires creating as well as installing of broad computer and communication foundation that supports increased situational awareness allowing fine-grained command and control.

One of the greatest challenges encountered in building the smart city is Cyber security of the system.

Like other technologies Smart grid also has several vulnerabilities and challenges. Vulnerabilities may enable attackers to penetrate the network gaining access to control software and modifying the framework.

To secure a smart city factors to be considered are (Kosmatos, Tselikas & Boucouvalas, 2011):

1. **Privacy**: Smart-grid communications have unintended outcomes for the client usage and practices. Smart-grid shares client's electricity usage information through the smart meter which acts as an information-rich side channel, exposing client's usage behaviour. This client's data can be employed by the attackers. Google introduced Google Power Meter service which receives a customized web page that visualizes local usage of the customers. Proper analysis of the data can lead to disclosing of a client's personal information. For example, analysis of electricity pattern can lead to a disclosure of energy usage and also when they're at home or work. Also, an increase in power usage can indicate the change in business operations. An attack of this kind of personal information could lead its exposure to, hacker's hand.
2. **Availability**: Availability means system cannot deny access or use to approve clients. For smart grid systems, this indicates all the IT elements like safety system, surveillance system, and operator workstations are free from denial-of-service (DoS) attacks, which endeavours to delayed or corrupt data broadcast in order to make information unattainable to a smart grid.

The network traffic in a smart grid communication network is time-critical. The communication channel has to be connected by an attacker in order to launch a DoS attack. Hence, it is a basic necessity to consider the DoS attack on the smart grid and thus design powerful countermeasures to such attacks (Yan et al., 2012; Gershenfeld, Krikorian & Cohen, 2004).

6. THREATS AND ISSUES IN AUTOMATION SYSTEM

In building automation based on the parameter, the activities and processes are executed automatically offering an easier path to control and monitor the building conditions (Wicaksono, Rogalski & Kusnady, 2010; Chao, 2011). There are many types of security implications in Building automation systems which have brought BAS to a variety of threats like denial of services (DoS) attack, physical damage, etc. Due to increased availability of BAS to shared systems, BAS is presented with the same threats encountered by customary IT-based systems and conventions. Denial of services can occur on a large scale due to the supply constraint of BAS devices.

Building automation system is mainly dependent upon Information Technology (IT) infrastructure where most of the facility control standards have been introduced in the last century. So, there is an urgent need for an IT-friendly solution which is consistent with standard IT infrastructure (IT Business Edge, n.d.; Madakam, Ramaswamy & Tripathi, 2015).

Both the Field of building automation and building automation systems (BAS) has been undergoing stable evolvement in the past years due to the introduction of networked control systems. Modern BAS is developing new building services, enhancing control methods and communication systems, managing intricate arrangements thus improving security aspects.

Following are some of the features in the building automation system (IT Business Edge, 2018):

1. **Climate Control**: Works on the parameter like room temperature, humidity and overall air quality. These parameters are needed to be synchronized in order to maintain services. Latest BAS use, air quality sensors to enhance the volume and speed of fresh air.
2. **Visual Comfort**: Has to deal with lighting conditions in the building. This includes automatic control of light like dimming or increasing the brightness when required. Constant light control (CLC) involves controlling of light-level in a room temperature.
3. **Alarm Management**: Used to monitor the building parameter and generate the alarm in case of an emergency. Necessary actions are then taken accordingly. BAS includes safety equipments and systems like gas and smoke detectors. These include presence detectors, breaking glass detectors, installing CCTV and audio surveillance.

Maintenance of BAS is a necessary issue which includes routine equipment checks and reporting of problems. A building can be optimized with regards to cost, comfort and energy efficiency. As the level of integration has increased, each data point is prone to IP-based network attack which includes attackers from the public Internet, attacks within building highlighting issues like integrity, authenticity and privacy (Zhang et al., 2012).

According to Benjamin Freas, Principal Research Analyst at Navigant Research in building automation system, there has to be better sensing and automation making the building automatic to take decisions. This can be solved by integrating BAS with the smartphone enhancing the user's comfort and experience.

In November 2011 the industrial control system was hacked which was used for wastewater management and was found that a three-character password was used. These threats can pose serious issues, e.g. interrupting facility operations, causing tenant discomfort, equipment or facility damage, etc. So, everyone influenced by the building automation system should consider these systems seriously (IT Business Edge, n.d.; Sun et al., 2010).

So, in an increasing facility between automation and information technology one has to protect these shared IT systems from vulnerabilities and threats. Building Automation Internetworks should also provide protection schemes commensurate with, and adhering to, the security controls already established in the IT domain etc. (Granzer, Praus & Kastner, 2010; Yun & Yuxin, 2010).

If prevention is not possible with reasonable effort, mechanisms trying to detect security attacks, reporting them and then trying to minimize them should be adopted, similar to intrusion detection systems (IDS) that shows any abnormal system response and tries to prevent them via isolating the attacking source (Miorandi et al., 2012).

7. THREATS AND ISSUES IN UAVs

Recent popularity of drones has been eye catching for many attackers. With increase in the number of drones, it is becoming difficult to differentiate between safe and threat causing drones (Vattapparamban et al., 2016; HP, n.d.).

There are numerous threats to drones which can be related to cybersecurity, privacy and forensics. To deal with these threats US has established the Federal Aviation Administration (FAA) regulations. The FAA carries rules like UAS weighing above 0.55 has to be registered with the government system. They also have a smartphone application B4UFLY which delivers drone awareness to operate the drones (Vattapparamban et al., 2016).

UAV GPS is highly vulnerable to attacks. These UAVs are attacked using the information from GPS functionality which makes it susceptible to attacks known as "GPS spoofing". The spoofed GPS-signal originates from a local transmitter with a strong signal which leads to a re-creation of GPS-signals. Figure 6 shows the GPS spoofing through signal spoofer.

Figure 6. GPS spoofing attack
Source: https://securepositioning.com/wp-content/uploads/2017/04/spoofing_attack.jpg

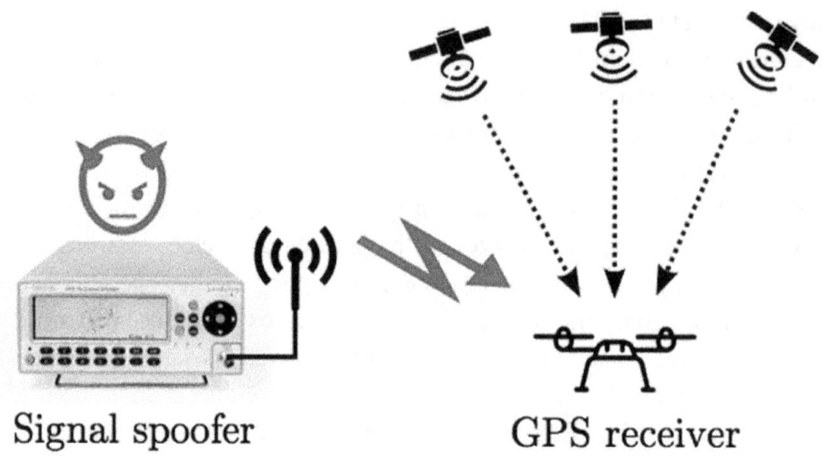

Apart from spoofing there are many kinds of threats. Following are some threats connected to drones.

1. These drones are sometimes used for surveillance, trying to break the privacy of the people. Many countries have strict policies regarding this like Sweden has banned the use of camera drones without permission. Therefore, privacy concerns around unmanned aircraft will improve by adopting necessary policies by law enforcement. (Morandi, D. et. al, 2012).
2. Research has shown the major security flaws in some popular drone models where they proved that anybody can link and see the drone video feed, as the data was unencrypted. By using root access and modifying SSID of the wireless network of the drone they can gain control of the drone. (GSMA)
3. IoT devices lack authentication, several IoT devices have weak or no user authentication, which makes drone prone to be hacked by the attacker within the Wi-Fi signal range of the device. This type of attack is commonly known as a Fly-away attack.
4. Drones like AR. Drones 2.0 and Parrot Bebop drone is highly vulnerable to Takedown Flying Drone Attack which has the TELNET port open by default. ((GSMA), Zeinab and Elmustafa, 2017).

All the communication between the drones and the controller is via Wi-Fi network. According to an IBM researcher drone with un-encrypted on-board chips can be attacked from a distance of about 2 kilometers ("The Register", 2018). AR. Drone 2.0 is a popular drone model which creates an access point (AP) and can be connected to the smartphone in the name of ardron2_ followed by a random number. This AP has open access to be connected and requires no authentication from the user side. The default gateway address is 192.168.1.1 which can easily be available to connect since it is left open on this system. All this action can be performed easily via Raspberry Pi using Bash scripts, making it easily hackable, demanding the need to make it secure ((Chapman, 2016), Zeinab and Elmustafa, 2017).

In the de-authentication attack pattern, it is very easy to launch an attack on the 802.11 Wi-Fi networks. The difference between wired and wireless connectivity is that the attacker needs the accessibility of the signal in order to launch the attack. As discussed previously these types of attack disconnects the owner's access point (AP) while being launched. Generally, these attacks are made by the company's security tester as routine penetration testing, but it can be used by the attackers to hack other drones. AP connections are made through the exchange of various frames, which involve an exchange of authentication and association request frames as shown in Figure 7.

Privacy is considered to be a major issue in drones since they can easily be used to capture information and photos through cameras on them which makes the privacy of people at risk. Therefore, governments have issued several rules and regulations

Figure 7. De-authentication attack
Source: https://opensourceforu.com/2015/04/beware-it%C2%92s-easy-to-launch-a-wireless-deauthentication-attack/

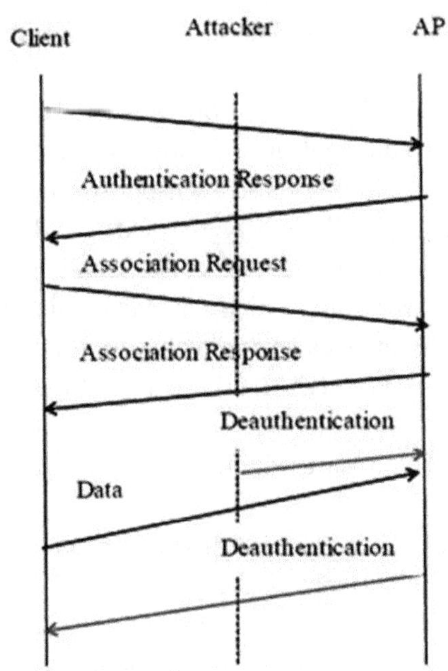

on privacy for UAVs. Using UAVs, one can inspect on the other's territory illegally, which can be used for personal benefits. In many cases it was found continuous flights of UAV was used to reveal and get the trade secrets of competitors that could destroy their business.

Drones used in smart cities, extensively present a serious security threat to lines of communication and lots of hard work is required to harden their security (Jun, 2014).

8. THREATS AND ISSUES IN SMART VEHICLES

The inclusion of new features and use of new technology in smart vehicles has resulted in security concerns over its use. These threats can be an interception, physical, abuse and loss of information threats. Loss of information can be due to the leakage of data to the connected cloud in case the vehicle is resold. Attackers can also inject malicious code into the operating system of the cloud which could cause leakage of data.

Smart Vehicles are vulnerable to vehicle sensors attack which can be global positioning (GPS) spoofing/jamming, light detection and Ranging (LiDAR) sensor relay/spoofing attacks ((Jun, 2014), Sarita and Manik, 2011).

1. GPS jamming/spoofing attacks: GPS spoofing can be easily done through cheap software like software defined radio (SDR). This calls for a need to engross into more research for developing advanced systems against GPS spoofing as there are no such commercial systems available ((Github, 2017), Sarita and Manik, 2011).
2. Camera Sensor Attacks:Petit et al. by blinding the camera with laser and LED matrix befuddled the auto controls in MobilEye C2-270 camera sensor. There is no commercial anti-blinding camera sensor in the market (Petit et. Al, 2015).
3. MMW Radar Attacking:Millimeter wave (MMW) radar has the 76.5GHz band automotive radar worldwide. There are also jamming and spoofing attacks in contradiction of MMW radars, which can lead to blinding and malfunction of the smart vehicle. This attack also doesn't have any anti-jamming or a spoofing system in the market ((Petit et. Al, 2015), Yan, Xu, and Liu, 2016).

Considering the threats to smart vehicles, attacks are listed as:

1. Loss of data and sensitive information via deletion or alteration by attackers in the cloud, harms the privacy of the user.

2. Malfunctions/Failures of the power supply in the parts of devices due to software can create a great loss in companies or institutes where data is required every time and its gap can hamper the process.
3. Physical threats glitching/ side channel which includes fault injection inside a channel breaking central locking systems, side channel attacks to leak information and gaining unauthorized access to debug interfaces (Miorandi et. Al, 2012).
4. In Denial of Service attacks attackers systematically or selectively jams the signal of the vehicle. Here the central issue is lack of an effective solution to this problem with the aim that one would not be able to see car overriding its driver ((Miorandi et al., 2012), Furness, 2008).

9. IOT TECHNOLOGY- BASIC CONCEPTS AND ISSUES

The term Internet of Things was first introduced in Chicago in September 30 when Auto-ID Centre launched EPC network. For the first time Auto-ID Centre mentioned IoT in its book about Electronic Product Code by David Brock in 2001. IoT makes real world data more accessible through network and internet making it an efficient business tool enabling a convenience in daily life (Medaglia and Serbanati, 2010).

The IoT covers tremendous aspects according to the demands and requirements of the people, by allowing a good communication between the objects, machines and everything that can lead to the delay of the internet and the web in the physical realm. It is quite different what actually internet provide us today, by means of several spatially distributed and innovative technologies that have been combined into IoT, that is referred to as the Future Internet of things (IoT).The various widespread technologies with embedded identification and capabilities that have been linked with physical entities, by appropriate knowledge and communication technologies so that from these existing technologies a new class of invention has been evolved and having much widespread applications and services. The IoT sensors use several types of connections like Wi-Fi, ZigBee and Bluetooth with the inclusion of wide area connectivity using various technologies such as GSM, GPRS, 3G, and LTE. By the invasion of IoT, the world will become smarter day by day in every aspect, since the IoT would provide smart cities, smart homes, smart healthcare, smart buildings, that leads to great applications like smart grid, smart energy, transportation, management of waste and various social issues too ((“IT Business Edge”, 2018) “Network World”, 2018). The smart devices signify the utilization of each bit of data that we share or use in the everyday routine without accessing much storage and space. These connected devices utilise the given data to cooperate with us and the complete the desired tasks. Smart devices fill the gap between the physical social realms and digital world so that productivity, quality of life will enhance (Madakam, 2015).

A smart city is another very crucial application of IoT that increases peculiarity among the population. Smart waste management and monitoring, energy management, water drainage system distribution of water among the remote areas using the innovative technologies for the betterment of the society, security issues regarding the urban areas are few the vital examples of the internet of things ((“Barcelona”, 2013) GSMA, 2018).

Pollution, shortage of resources like energy, water, traffic control are some major problems in human life that lead to the deployment of the society and obstruct human growth and welfare. These pertaining problems has been analysed and solved by IoT. ((Sun et. Al, 2010), (Yun and Yuxin, 2010), (Miorandi et. Al, 2012). For example, the technology like cellular communication to send alerts via messages to various authorities or public services when a bin needs to be emptied. Various sensors that are installed using web applications helps in detection that there are free parking slots or not across the city ((Zhang et. Al, 2012), “IT Business Edge”, 2018). Figure 8 shows the Three-layer IoT architecture.

Critical Issues encountered by Internet of Things are as follows:

1. **Communication**: The major concern in IoT is regarding the communication aspect, how the smart or networked devices communicate with each other (Jun, 2014).

Figure 8. Three-layer IoT architecture
Source: https://ai2-s2-public.s3.amazonaws.com/figures/2017-08-08/add4396a95b96927dbef194b2657800312553e21/2-Figure1-1.png

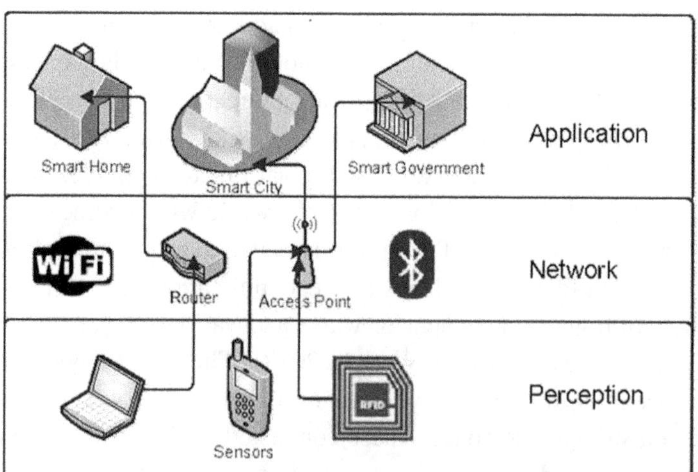

2. **Tracking**: Device tracking and monitoring in IoT has a widespread approach, it's not only limited to the connected devices at rest, but, also to the devices that are not static like cars, flying devices in gadgets like drones (Sarita and Manik, 2011). Nowadays, every device or electronic gadget in use be it market or sector can be connected to the internet (Miorandi et. Al, 2012). The major issue is this how the authors monitor these devices, what could be the best possible solution and evolution for hijacking, tracking or outages these devices.
3. **Data Rate**: Another issue regarding the amount of data transmitted to cloud per hour is not only dependent on the number of connected devices that are growing, but also depend on the amount of data that has been sent. Four out of five devices having an internet connection and continuously sending 25 GB of data to the cloud per hour will increase stress on the network. Currently the major focus point is, how the companies and its clients will get assurance that the performance they are getting is the best performance available and the gadget makers ensure that devices are perfectly optimized.
4. **Security and Privacy Issues**: The issue of security attracts great attention millions of users and billions of connected devices. The vital thing is that at least the authors get to know when the incident is occurring (Medaglia, 2010).
5. **Alarm generation:** Huge information apparatuses can help here to generate alarm when assaults conceivably happen or are in advance. The researchers likewise require observing frameworks to guarantee dangers are tended to continuously. ("Network World", 2018).

10. CLOUD COMPUTING- CONCEPTS AND ISSUES

Today, cloud computing is emerging as one of the largest emerging technology in the computer world which has an exceptional change in IT industry by efficiently managing the resources and database of an institution and individual through an online server. Cloud Computing has turned IT industry into a productive and responsive business by introducing new ways and database platform.

Considering cloud computing and Internet of Things (IoT), they are very varied technologies and their integration is known as the Cloud IoT model (Botta et. Al, 2016). IoT is largely described by genuine little things, widely distributed, with constrained capacity and handling limit involving issues like reliability, performance and security. It also has a large number of welfares and capabilities pertaining to capacity and processing power ((Zhou et. Al, 2013), Chao, 2011).

Internet of Things (IoT) consists of physical objects having electronics, software, and sensors which does sensing and control of remotely existing network infrastructure. IoT is a stage, interconnecting everyday objects over the internet, which are an inexhaustible source of data. IoT makes the technology available and accessible in the hands of a client through smart devices whenever and wherever demanded. It also takes the responsibility of utilization of a tremendous measure of information collected by means of IoT Senor and network into a valuable data to empower real-time decision-making system (Soucek and Zucker, 2012).

Cloud computing is a model for on-request access to a shared pool of configurable assets that can be effortlessly provisioned as and when needed. It empowers the framework software to be arranged as required for singular application necessities providing access to reliable, high-performance clusters and storage without buying and keeping up with modern hardware. It's an evolving sub-domain of computer security, information security and network security consisting of a wide set of policies, technologies, and controls organized to secure the data, applications and interlinked infrastructures. (Samy, 2013).

Cloud computation is defined as the use of cloud computing logistical capitals at software level using services transported over the Internet. Cloud Computing is a technology which can be set as a base technology in the use of IoT. Whereas, mobile computing is known to integrate cloud computing to mobile devices, increasing memory, storage, energy and computational power.

With the rapid and independent growth of IoT and cloud computing, there has been an immense increase in the capabilities and resources of Cloud computing in order to compensate for its technological constraints. IoT has also increased the efficiency to deal real-world things in a more distributed and dynamic manner. ((Valente and Cardenas, 2017), Geisler, 2013).

Due to the integration of IoT and cloud computing gaps like limited application and storage of IoT has been filled. This integration has also increased the same concern in IoT as they were in cloud computing, such as trust level between the service provider and the customer, maintaining service level of agreement (SLAs) and third-party control issues. Cloud can provide all information and the application wirelessly to the user required for IoT connectivity (Novotny, 2018).

The integration of cloud and IoT can provide few more benefits like:

1. Can be used in HTTP protocol to transfer data between IoT things and cloud computing application.
2. Useful in connecting sensors and sharing sensor readings thus lowering the security risks.
3. Data can be transferred between M2M by the support of IoT technologies.
4. Helps in processing and computation of above collected and transmitted data.

IoT comprises not only complex devices such as mobile phones, but daily objects such as food, clothing, paper, furniture, etc. (Kosmatos, Tselikas, and Boucouvalas, 2011). Dealing with cloud technology brings us to deal with the cloud security concerns like the protection and privacy of the data.

Big firms like (Amazon, Google, Facebook, etc.) implemented this model for distributing facilities over the Internet, gaining both economic and technical benefits. It also provides an energy efficient way of managing the increasing demand of energy consumption by supplying more services from the same input or may be less producing a more output ((McKinsey Global Institute, 2018), Zhang et. al., 2017).

Cloud computing can provide an effective way to implement the IOT service management and composition as well as application exploiting the output data. Here cloud acts as an intermediate layer between the applications and the things hiding all the complexity and functionalities. Also, it is characterized with devices, technologies and protocols solving various security problems adding some more benefits ease-of-access and use (Uckelmann, Harrison, Michahelles, 2011).

11. APPROACHES TO ENFORCE SECURITY AND PRIVACY ISSUES IN SMART CITIES

After the introduction of the concept of IoT for the smart cities, there had been a tremendous research in the field which has brought in lot of advancement regarding ideas to develop smart cities. Building smart cities would ease the daily life of people as well as it will provide a sustainable way of development and growth. In other words, smart cities provide a better way to manage and counter attack risks faced in day to day life. Apart from enormous benefits, these advancements have also brought new ways to hack and disrupt smart city. These systems, like doors, grills and fences also have some vulnerability, which has to be dealt with ((Stergiou et. Al, 2018), Amin, 2015).

The existence of Internet of Things (IoT) is intertwined with the objective to provide conventional and innovative services in an improvised such that it can find beneficial applications in the domains we dig into most. As we know the word 'Smart' indicates the top-tier amenities, these dominions will serve the society with improved lifestyle and sustainable development. Thus, requires a top-notch security which can be achieved by a robust and levelled architecture, taking civil privacy as the primary topic of concern. Privacy and reliability are the major focus here since they are vulnerable to a number of attacks.

The architecture basically includes data confidentiality and authentication, access check over IoT, mutual trust among end-users and things, and implementation of legal policies regarding security and privacy. Thus, narrowing usage of any traditional approach which is not compatible with IoT standards. A mechanism capable of auditing trails and executing quick responses in order to prevent malicious exploitation of weak node through the end to end encryption system, as well ensuring regular software patching, is what can be called reliable and beneficial (Morvaj, Lugaric, and Krajcar, 2011).

The Internet of Things has become a term to describe devices having senses and communicating capabilities through local network or internet. The IoT finds its way almost into everything from smartwatches to devices in factories. As IoT is an emerging technology and in the near future, it can be implemented in lots of devices building up a "smart city". Now, this is a security concern also as these devices are continuously working on data (might include private data). So, security needs to be a top priority for these devices.

Hacking a device and extracting data or controlling the device is a danger to society and there must be a preventive approach to combat security and privacy issues in the world of IoT. A security solution must check whether the firmware of the device is not tampered of changed, must receive and send data securely and must protect the device from cyber-attacks ((Uckelmann, Harrison, and Michahelles, 2011), Morvaj, Lugaric, and Krajcar, 2011).

Cities nowadays encounter the problem of standardization that would decrease the cost of maintenance, costs of exchange, e-services and distribution of capitals among people. Smart city models are defined based on their applications, provided services, financial engineering and public-private partnerships. For achieving these goals city authorities have to embrace activities and methodologies that make smart cities smarter.

Few security goals to be attained while designing a smart city are (Stergiou et. al, 2018):

1. **Secure Communication:** As communication is one of the essential parts of IOT and safe communication is a prime goal. So, every part of data transfer must be secured using encrypted communication (SSH, SSL etc.). And the use of proper modern algorithms for encryption must be used.
2. **Secure Booting:** Detecting that the firmware has been tampered or not, can be achieved using cryptographically signed code from the manufacturer. Firmware checking should be done during boot process.
3. **Code Updates:** Proper methodology should be adapted for upgrading a system's core security-related code and continuous manufacturer support for any issues.

4. **Preventing Access:** By encrypted data storage, and encrypted communication. One can prevent any type of successful cyber-attacks.
5. **Authentication:** The device will be in touch with several other devices and there must be proper authentication procedure like a strong password (at least) or any authentication protocol.
6. **Protection against Cyber-Attacks:** In every stage of modern communication firewall is used so here as well an embedded firewall can play a crucial role in protecting against any attacks because firewall can limit communication to only known ones.
7. **Intrusion Detection:** There are several techniques by which a hacker can try millions of operations or passcode (login attempts) without the attack being reported. Devices must be able to detect several invalid login attempts and respond to such potentially harmful activities to the owner or manufacturer.

Devices in IoT are in wide range with simple to fairly complex. These can be originating from consumer gadgets to IOT based industrial automation. The devices are different from a standard PC or mobile phones. Some use a specialized operating system like slimmed down version of Linux for a particular task. Some are just coded with instructions where most cases, these devices are designed for minimum process cycle or memory usage providing no support for traditional security measures. So, these devices have some specialized task-code or OS on it, which make them vulnerable to attacks as these embedded systems can't be upgraded to newer versions via the internet or something until the technician directly upgrade them into the system core ((Furness, 2008), Novotny, 2018). The model of Data Forensics and Intrusion investigation can be utilized to monitor the network. IDS work on primitive or more precisely, in simple terms. It checks upon malicious or anomalous element and debars access via confiscating IP address and reporting to the IoT network administrators.

While both reliability and Security, are aspects aimed at while constructing the highest level of a security system; Cognitive IoT can be looked forward to. Thus, the future directions and conclusions point towards improving the security of alert traffic, alert correlation and autonomic management systems. These protective measures will certainly act as the vault for a safe digital and cyber domain of the celebrated ambitious project of smart cities.

Public security is also a necessary concern in smart cities which is firmly identified with human trace information since public security is about individual action and social events in major cases. Human traces could be used to infer lots of social practice, get-togethers, bad-conduct of people and development of lawbreakers in a fiasco. With a view of human and group conduct data mined from the following information, public security could be improved in numerous perspectives, for example: distinguishing trouble making activity of people. Most individuals have routine courses each day, for instance, a tedious example of movement between home and work every day.

The best way to secure smart city structures has been to use a secure and reliable defense system and to try and prevent access to enable renewal if any compromise occurs. This can be done by using access controls like (user/password, two-stage authentication, face-detection), having end-to-end strong encryption and actions to ensure routine software patching and the ability to respond with urgent updates to close exploits as they occur and auditing trails can be a lot beneficial. In complex distributed systems with many mechanisms, these explanations need to work similarly across the entire foundation just as the weakest connection (Ehlers, Howerton, and Speegle, 1996).

12. CONCLUSION

This chapter has given a broad-brush explanation of smart cities as an efficient way to tackle and manage uncertainty and urban challenges through the real and efficient delivery of services to the population fighting against the vulnerabilities and threats, including making the city most advanced, reliable and secure. A smart city is majorly composed of several entities such as smart grids, building automation systems, unmanned aerial networks (UAVs) and smart vehicles. The goal of these entities is to make smart cities safer, green, and efficient. The chapter further explains the possible threats and issues associated with these entities. Based on these entities smart city architecture is designed to enhance the capabilities to handle smart cities enabling advanced ICT infrastructure. This infrastructure is key determinants of the safety of cities. The author presents a six-layer architecture - “Goals”, “People”, “Ecosystem”, “Soft infrastructure”, “City System”, and “Hard infrastructure”.

Continuous attacks are faced by the smart city architecture every second so there is a need for developing a more sophisticated method to fight against them. This chapter brings critical issues of IoT Integrating cloud computing and Internet of Things (IoT), introducing Cloud IoT model interconnecting data with the internet.

The chapter sheds more lights on smart cities as these are the critical element of society requiring top-level of security, considering the privacy-preserving mechanisms as an essential matter of concern focusing on privacy and security preventing illegal access to information and all types of attacks which will make architecture vulnerable. In addition to this, chapter includes open questions regarding privacy and security of smart cities and hoping that it will help in such conversations, where more effective research efforts would be observed.

REFERENCES

Agrawal & Das. (2011). *Internet of Things – A Paradigm Shift of Future Internet Applications*. Academic Press.

Amin, M. (2015). *Smart Grid. Public Utilities Fortnightly*.

Antonini, A., Barenghi, A., Pelosi, G., & Zonouz, S. (2014). Security challenges in building automation and SCADA. *2014 International Carnahan Conference on Security Technology (ICCST)*, 1-6. 10.1109/CCST.2014.6986996

Baig, Z. A., Szewczyk, P., Valli, C., Rabadia, P., Hannay, P., Chernyshev, M., & Syed, N. (2017). Future challenges for smart cities: Cyber-security and digital forensics. *Digital Investigation*, *22*, 3–13. doi:10.1016/j.diin.2017.06.015

Bartoli, A., Hernández-Serrano, J., Soriano, M., Dohler, M., Kountouris, A., & Barthel, D. (2011). Security and privacy in your smart city. *Proceedings of the Barcelona smart cities congress*, 292.

Bawany, N. Z., & Shamsi, J. A. (2015). Smart city architecture: Vision and challenges. *International Journal of Advanced Computer Science and Applications*, *6*(11), 246–255.

Botta, A., De Donato, W., Persico, V., & Pescapé, A. (2016). Integration of cloud computing and internet of things: A survey. *Future Generation Computer Systems*, *56*, 684–700. doi:10.1016/j.future.2015.09.021

Chao, H.-C. (2011). Internet of Things and Cloud computing for future Internet. In Ubiquitous Intelligence and Computing. Springer. doi:10.1007/978-3-642-23641-9_1

Chuah, J. W. (2014). The Internet of Things: An Overview and New Perspectives in Systems Design. *International Symposium on Integrated Circuits*. 10.1109/ISICIR.2014.7029576

Ehlers, G. A., Howerton, R. D., & Speegle, G. E. (1996). *U.S. Patent No. 5,572,438*. Washington, DC: U.S. Patent and Trademark Office.

Furness, A. (2008). A Framework Model for The Internet of Things. *GRIFS/CASAGRAS Workshop*.

Furness, A. (2008). A Framework Model for The Internet of Things. *GRIFS/CASAGRAS Workshop*.

Geisler, K. (2013). *The relationship between smart grids and smart cities*. IEEE Smart Grid Newsletter.

Gershenfeld, N., Krikorian, R., & Cohen, D. (2004). The Internet of things. *Scientific American, 291*(4), 76–81. doi:10.1038cientificamerican1004-76 PMID:15487673

Granzer, W., Praus, F., & Kastner, W. (2010). Security in building automation systems. *IEEE Transactions on Industrial Electronics, 57*(11), 3622–3630. doi:10.1109/TIE.2009.2036033

Khurana, H., Hadley, M., Lu, N., & Frincke, D. A. (2010). Smart-grid security issues. *IEEE Security and Privacy, 8*(1), 81–85. doi:10.1109/MSP.2010.49

Kosmatos, E. A., Tselikas, N. D., & Boucouvalas, A. C. (2011). Integrating RFIDs and smart objects into a Unified Internet of Things architecture. *Advances in Internet of Things, 1*(01), 5–12. doi:10.4236/ait.2011.11002

Madakam, S., Ramaswamy, R., & Tripathi, S. (2015). Internet of Things (IoT): A literature review. *Journal of Computer and Communications, 3*(05), 164–173. doi:10.4236/jcc.2015.35021

Medaglia, C. M., & Serbanati, A. (2010). An Overview of Privacy and Security Issues in the Internet of Things. *The Internet of Things*, 389-395.

Miorandi, D., Sicari, S., De Pellegrini, F., & Chlamtac, I. (2012). Internet of things: Vision, applications and research challenges. *Ad Hoc Networks, 10*(7), 1497–1516. doi:10.1016/j.adhoc.2012.02.016

Mohammed, F., Idries, A., Mohamed, N., Al-Jaroodi, J., & Jawhar, I. (2014). UAVs for smart cities: Opportunities and challenges. *International Conference on Unmanned Aircraft Systems (ICUAS)*, 267-273. 10.1109/ICUAS.2014.6842265

Morvaj, B., Lugaric, L., & Krajcar, S. (2011). Demonstrating smart buildings and smart grid features in a smart energy city. *Proceedings of the 3rd International Youth Conference on Energetics (IYCE)*, 1-8.

Petit, J., Stottelaar, B., Feiri, M., & Kargl, F. (2015). *Remote attacks on automated vehicles sensors: Experiments on camera and lidar*. Black Hat Europe.

SamyKamkar. (2013). *SkyJack*. Retrieved from https://github.com/samyk/skyjack

Software-Defined GPS Signal Simulator. (n.d.). Retrieved from https://github.com/osqzss/gps-sdr-sim

Soucek, S., & Zucker, G. (2012). Current developments and challenges in building automation. *e & iElektrotechnik and Informations technik, 129*(4), 278-285.

Stergiou, C., Psannis, K. E., Kim, B. G., & Gupta, B. (2018). Secure integration of IoT and cloud computing. *Future Generation Computer Systems*, *78*, 964–975. doi:10.1016/j.future.2016.11.031

Sun, Q. B., Liu, J., Li, S., Fan, C. X., & Sun, J. J. (2010). Internet of Things: Summarize on Concepts, Architecture and Key Technology Problem. *Journal of Beijing University of Posts and Telecommunications*, *3*(3), 1–9.

Uckelmann, D., Harrison, M., & Michahelles, F. (2011). An architectural approach towards the future internet of things. *Architecting the Internet Of things*, 1-24.

Valente, J., & Cardenas, A. A. (2017). Understanding security threats in consumer drones through the lens of the discovery quadcopter family. *Proceedings of the 2017 Workshop on Internet of Things Security and Privacy*, 31-36. 10.1145/3139937.3139943

Vattapparamban, E., Güvenç, İ., Yurekli, A. İ., Akkaya, K., & Uluağaç, S. (2016). Drones for smart cities: Issues in cybersecurity, privacy, and public safety. *International Conference Wireless Communications and Mobile computing (IWCMC)*, 216-221. 10.1109/IWCMC.2016.7577060

Wicaksono, H., Rogalski, S., & Kusnady, E. (2010). Knowledge-based intelligent energy management using building automation system. *IPEC, 2010 Conference Proceeding*, 1140-1145. 10.1109/IPECON.2010.5696994

Yan, C., Xu, W., & Liu, J. (2016). *Can you trust autonomous vehicles: Contactless attacks against sensors of self-driving vehicle?* DEF CON.

Yan, Y., Qian, Y., Sharif, H., & Tipper, D. (2012). A survey on cyber security for smart grid communications. *IEEE Communications Surveys and Tutorials*, *14*(4), 998–1010. doi:10.1109/SURV.2012.010912.00035

Yun, M., & Yuxin, B. (2010). Research on the architecture and key technology of Internet of Things (IoT) applied on smart grid. *International Conference on in Advances in Energy Engineering (ICAEE)*, 69-72.

Zeinab, K. A. M., & Elmustafa, S. A. A. (2017). Internet of Things applications, challenges and related future technologies. *World Scientific News*, *2*(67), 126–148.

Zhang, K., Ni, J., Yang, K., Liang, X., Ren, J., & Shen, X. S. (2017). Security and privacy in smart city applications: Challenges and solutions. *IEEE Communications Magazine*, *55*(1), 122–129. doi:10.1109/MCOM.2017.1600267CM

Zhang, M., Zhao, H., Zheng, R., Wu, Q., & Wei, W. (2012). Cognitive internet of things: Concepts and application example. *International Journal of Computer Science Issues*, *9*(6), 151.

Zhou, J., Leppanen, T., Harjula, E., Ylianttila, M., Ojala, T., Yu, C., & Yang, L. T. (2013). Cloud things: A common architecture for integrating the internet of things with cloud computing. *IEEE 17th International Conference on Computer Supported Cooperative Work in Design (CSCWD),* 651-657.

Chapter 3
Fog Removal Algorithms for Real-Time Video Footage in Smart Cities for Safe Driving

Neetu Sood
Dr. B. R. Ambedkar National Institute of Technology, India

Milin Kaur Saini
Dr. B. R. Ambedkar National Institute of Technology, India

Indu Saini
Dr. B. R. Ambedkar National Institute of Technology, India

Parul Bhoriwal
Dr. B. R. Ambedkar National Institute of Technology, India

Tarannum Awasthi
Dr. B. R. Ambedkar National Institute of Technology, India

Tanveer Kaur
Dr. B. R. Ambedkar National Institute of Technology, India

ABSTRACT

In this chapter, different approaches are presented for removal of fog from video footage taken in moving cars. The methodology uses different approaches, namely dark channel prior, contrast limited adaptive histogram equalization (CLAHE), the combination of two approaches (dark channel prior and CLAHE), and RETINEX algorithm combined with DWT. The algorithms are implemented in MATLAB R2015a. Moreover, the algorithms are compared based on their computational complexity and a visibility metric which is used for computing the CNR of video frames before and after the application of the algorithm. The chapter discusses which algorithm would provide better performance during night fog and daylight fog. Finally, the safe speed of the driver is calculated based on the time complexity of the algorithm used.

DOI: 10.4018/978-1-5225-8085-0.ch003

INTRODUCTION

Humanity has witnessed a tremendous increase in population along with large-scale urbanization in the past decade. As the economic status of people continues to improve considerable investment is done in the field of technology applications for enhancing the lives of people. The smart cities are developed with the intent to provide an improved lifestyle to its citizens using data and technology-driven solutions to tackle real-life problems. Various data collection and processing techniques are employed in the public systems to extract useful information and aid in the management of various systems like traffic, law enforcement, schools. The word 'smart' in smart city refers to the application of intelligence in the above systems.

Security and safety of people is an important indicator of the development of a city and cannot be neglected. It is not uncommon to witness cases of trains getting derailed, flights getting delayed and collision of vehicles on road due to poor visibility. The bad climate conditions considerably decrease the visibility of the drivers which makes it almost impossible to drive. Owing to the steady rise in population in the cities there has been heavy traffic due to which effective management of Traffic and Transportation Systems becomes important. Fog, mist, rain etc are examples of some natural phenomenon which are a major cause of road casualties. Similarly, in the industrial areas smog (a mixture of smoke and fog) is a commonplace phenomenon which considerably degrades human vision.

The main aim of this chapter is to devise such a system which can provide clear footage of the road to the driver and thus assist in driving. The system is simulated using the MATLAB R2015a graphical user interface.

This chapter starts off with introductory definitions of image and video processing along with the fields associated with them. This is followed by the need for fog removal and literature review of algorithms which have been implemented in the past for fog removal in images. The chapter further implements these algorithms and their combinations on video feeds which have been captured from moving cars

Two major constraints need to be considered for real-time traffic monitoring:

- The quality of picture after fog removal is good enough for further processing such as object identification and tracking.
- The algorithm must not be too complex for real-time processing.

This chapter focuses on three major algorithms namely Dark Channel Prior, CLAHE and Retinex algorithm in the process of fog removal which has been used previously mainly on images A comparative analysis of the techniques provided with respect to time delay and visibility. Visibility is defined in terms of CNR (Contrast to Noise Ratio). The chapter develops and analyses the above-mentioned techniques for fog removal in video feeds. This will give the driver a clear view of the road while driving thereby helping in the prevention of accidents

The chapter begins with the dark channel prior technique. In this technique, the least value for each pixel in the image is calculated using minimum filtering. A model of the foggy image is considered which consists of atmospheric light A and transmission map t. These are estimated for the reconstruction of a fog-free image. In the end, guided filtering is applied to remove residual fog from the video frame.

The Contrast Limited Adaptive Histogram Equalisation (CLAHE) algorithm establishes a maximum value to clip the histogram and redistributes the clipped pixels equally to each gray level. It limits the noise while enhancing the contrast. Firstly, the background image is extracted from the video sequence followed by which the moving pixels are estimated and bounded into foreground images. Secondly, the foreground and background images are defogged respectively by CLAHE. Lastly, the foreground and background images are fused into the new frames.

The retinex algorithm deals with compensation for illumination effects in images. The aim is to decompose a given image S into two different images, the reflectance image denoted by R, and the illumination image denoted by L, such that at each point (x,y) in the image domain S(x,y)=R(x,y).L(x,y). The method implemented here makes use of the Retinex algorithm to enhance the image, then 'haar' wavelet transform is used to enhance the details of the image. Finally, a clear image which has the fog removed can be obtained after reducing the non-important coefficients.

Nowadays almost every car is installed with a camcorder or dashcam to record the road scene while driving. Facing the front direction, these camcorders are set up in a car behind the windscreen. Apart from the fact that fog degrades visibility sometimes the windscreens are also covered with a thin layer of mist during winters. This thin layer of fog contaminates the videos captured with the camcorders and resulting in a low quality recorded video. Some objects inside the car will be reflected off the windscreen and form a light layer of reflected images in the captured videos. The fog and reflections degrade the quality of the video and make the design of portable video-based driving assistant system a challenging task.

The algorithms with their MATLAB implementations are provided. A fully functional user-friendly Graphical User Interface has been built to illustrate the functionality of the methods and assist in their comparison.

The algorithms are compared based on their performance and complexity. The time delay introduced due to the complexity of the algorithm applied will aid in the calculation of the safe speed of the driver. This safe speed is the speed at which a driver should drive to receive the clear video of the road in real time.

BACKGROUND

What Is an Image and What Is Video?

In a layman's language, an image is simply a visual picturization of a scene, objects etc. However, for the task of processing and extracting useful information, images are modelled using 2-D matrices on which various operations are carried out. A pixel is the smallest unit in an image. Each pixel is unique in terms of location and value. A video is formed when several images are combined in a sequence to from a moving picture. A composite video is formed with the combination of luminance, brightness, and chrominance of an image in a single channel.

- **Image Processing:** Image processing encompasses techniques used to sector the object of interest from an image. This is done by enhancing the object in question while suppressing the background (for instance the edges in an image) and extract useful information from them.
- **Video Processing**: Video processing involves all the methods of image processing. Along with the two dimensions of an image the third dimension of time is also considered. Special video filters are used wherein the input and output signals are video files or video streams. For instance, the quality of video processing is different in TV sets of different manufacturers which determines their price
- **Machine Vision**: Machine vision is the technology that enables devices to automatically capture and identify stationary or moving images. It is used in industry for auto-inspection of products and in satellites for auto-imaging. Machine understanding comprises computer vision and related fields.
- **Computer Vision**: It is a sub-part of Artificial Intelligence and Computer Science wherein computers are given a visual understanding of the world. It means making the computer perform advanced functions like those a human can do, e.g., face and object identification tasks. These tasks can be performed using one or multiple cameras.

Need for Fog Removal

Before we proceed towards discussing and analysing the algorithms for removing fog from videos it becomes imperative to realize its importance. Bad weather conditions like fog and mist lead to reduced visibility which is a major cause in road accidents. According to a report on "Road traffic injury prevention" published by WHO, "to see and be seen" is a fundamental prerequisite for the safety of all the road users.

The report analysed the crashes in the state of Victoria, Australia and found out that insufficient visibility caused 65% of them and was the only factor in 21% of them.

Poor visibility plays a key role in three major types of crashes. Fog is listed among the top twenty-five causes of car accidents. As image processing advances clarity in images can be easily obtained using certain algorithms. Moreover, high-tech cars are equipped with cameras on dashboards which can be utilized for embedding these enhancement algorithms.

The above facts necessitate the need for processing real-time video footage for the drivers to provide them with a clear view of the road ahead. As the field of image processing advances what better use can it be put to than assisting the drivers during driving n bad climate conditions?

Fog Removal Algorithms: A Historical Review

Videos captured during extreme winter conditions are often degraded due to the presence of fog or mist. These types of videos can be generally termed as foggy videos. In addition to that, there are so many other reasons for haziness of videos captured in moving cars. It may be a sandstorm, rain, fog, haze etc. Video restoration methods can be classified as additional information approach, multiple image approach and video sequences approach.

First is the additional information approach that requires some knowledge about the depth of foggy video which needs to be explicitly provided by the user or some additional operator. The second one is the multiple image approach which requires multiple images for the same scene. Its disadvantage is that it needs expensive hardware and lots of image data for processing. Lately, researchers focus on the single image defogging methods and lots of work is being done to remove haze using a single image as input. This method has an advantage over the other two methods due to its simplicity of data collection. The poor visibility during fog or mist is due to the presence of tiny atmospheric particles that absorb and scatter light.

A video is made up of many frames. It can be treated as a collection of many images continually moving at a speed. Continuous work is being done in image enhancement and deblurring the videos. Videos captured in moving cars during foggy weather are not very useful as the objects are blurred and concealed in the fog. Fog affects driver visibility which may lead to mishaps. Different types of techniques are used for video frame restoration from foggy videos discussed here. This chapter implements the technique to extract fog free videos from foggy videos which are captured in moving cars using video processing. Removing fog and restoring high quality defogged video is a challenging task.

RT Tan (2008) came up with a method for haze removal from an input image considering two ideas. First, a clear image has better contrast than a foggy image. Second, airlight variation depends on the object's distance from the observer. Jisha John and M. Wilscy (2008) separately processed the background and foreground pixels by wavelet fusion. Experimental results provided better visibility and colour fidelity. Another work proposed by Fattal (2014) restored the visibility from foggy images by finding the transmission map of the scene. He assumed that the transmission medium and the surface shading are uncorrelated. However, this method fails in case of heavy fog conditions. JP Tarel and N. Hautière (2009) proposed a method for restoration of visibility from a single colour or grey level image. The main advantage when compared with other approaches was that its speed and complexity were a linear function of the number of pixels n in the image.

He et al (2011) observed that except the sky regions, at least one spectrum in the outdoor RGB image exhibits very low intensity within patches of an image. This observation led to the Dark Channel Prior (DCP) concept by which we can effectively estimate the fog density and further restore the scene radiance. Eventually, improved methods based on DCP in which visibility restoration can be captured in hazy conditions were proposed. Single Image Fog Removal Algorithm based on an Improved Dark Channel Prior Method (2013) improved upon Dark Channel Prior given by Kaiming He, et al (2011). The work mainly focused on removing fog in foggy videos obtained by surveillance cameras. It enabled the use of surveillance cameras in all weather conditions. It improved upon the computation time by 28.5% as well as contrast ratio by 41.8%. It also removed the fog efficiently without the influence in the night.

The Histogram equalization technique finds tremendous use in the enhancement of images degraded by fog or mist. Z. Xu, X. Liu, X. Chen (2009) compared the Adaptive histogram equalization (AHE) and CLAHE algorithms and came to the conclusion that CLAHE produces better quality result since it uses 'Rayleigh Distribution' parameter to create a bell-shaped histogram.

Zhang et al. (1999) adopted an improved RETINEX-based method to remove fog in a traffic video. Experimental results showed that the proposed method can not only remove the fog but also enhance the clarity of the traffic video images. The retinex method used for enhancement of hazy images can be divided into two categories: single-scale RETINEX (SSR) and multi-scale RETINEX (MSR). An SSR algorithm has been proposed by Jobson et al. based on the center/surrounding RETINEX method. The essence of this algorithm is to obtain the reflection image by estimating the ambient brightness. To keep a good balance between the dynamic range compression and the colour constancy, Rahman et al. extended the SSR algorithm to multiple scales and proposed an MSR algorithm. Since the reflection image has little dependence on the intensity of the illumination, the RETINEX algorithm can easily realize image dehazing.

Fu et al (2013). proposed a variation framework for RETINEX to process the reflection and the illumination from a single underwater image by decomposing, enhancing and combining after colour correction

Xie et al. (2010) presented an approach for Improved single image dehazing using dark channel prior and multi-scale *retinex* where the main improvement lied in the automatic and fast acquisition of transmission map of the scene. Multi-scale Retinex algorithm was implemented on the luminance component in YCbCr space to obtain a pseudo transmission map. Combined with the haze image model and the dark channel prior, a high-quality haze-free image can be recovered. Compared with the original method, the algorithm had two main advantages:

- No user interaction is needed
- Restoring the image much faster while maintaining comparable dehazing performance.

P. Sankaran et al. (2012) in their work mention that though DCP approach gives very good results, it is computationally complex and hence proposed a RETINEX based approach that apart from giving good results and is also computationally simpler.

Xie et al. (2010), combined the characteristics of both DCP and Multi-Scale RETINEX (MSR) technique to restore the visibility. However, this method failed to restore the colour and edges during extreme foggy conditions. This is because of the hindrance of the restoration ability by colour cast problems and insufficient estimation of fog density by transmission map.

Xu (April 2015), presented, single image fog removal based on local extrema. His work is aimed at increasing the visibility of images degraded due to foggy conditions. It utilizes the atmospheric scattering model to analyse the presence of fog. Here, multiscale tone manipulation algorithm to understand the contrast details and estimation of an atmospheric veil. The results were compared with the traditional methods and found that the proposed method restored back the edges in clear videos more than the real colours

W Wang (October 2015). gave a multiscale fusion method to remove the haze from a single image. The approach used two atmospheric veils with different scales which are derived from the hazy image. Then, an adaptive local similarity-based wavelet fusion method was applied for preserving the significant scene depth property and avoiding blocky artefacts. G Chen (2007), proposed a fog removal algorithm to assist artificial intelligence in an object using a physics-based image restoration approach. Then a novel background maintenance algorithm based on the Unscented Kalman Filter (UKF) to subtract the background from the defogged video was introduced. Finally, moving objects are segmented by background differencing. C Chen (2016), addressed the problems of most image contrast boosting algorithms produce by proposing an approach for handling image artefacts such as noise, colour aliasing and blocking.

Khosla (2017), presented an algorithm based on Dark Channel Prior (DCP) concept followed by a post-processing technique based on visibility enhancement scheme for foggy images. The algorithm preserved sharp details and maintained the colour quality of the defogged image. PB Karthik and R Sandhu (July 2017) proposed the colour attenuation prior technique to remove haze from images. A simple method to remove haze from video by making use of MATLAB software.

The above algorithmic approaches provide an essence of how image enhancement methodologies came up, developed and were improved upon by the researchers. Image processing is a burning field of research and a lot of research is being done on improving the visibility of images using latest computer vision techniques using minimal human interference.

MEASURING THE PERFORMANCE OF ALGORITHMS

The algorithms implemented in this chapter are implemented on videos recorded in the month of January to capture dense morning fog. The videos are captured in moving cars to simulate the cameras in smart cars which will be the device for providing fog- free video footage of the road to the driver. The delay calculation procedure will aid in displaying safe speed to the driver

The programs used for fog removal must provide a certain level of performance according to the application. For surveillance purposes, the primary focus would be on improving the clarity obtained in the footage. For use in ADAS (Advanced Driver Assistance Systems), it is required that the haze removal is done in real time. Thus, the algorithms are studied for two parameters- delay and visibility parameter. The delay parameter is crucial since the algorithms used to solve the fog-removal problems are computationally expensive. In case of fog removal, we have defined two parameters:

Visibility Metric

The visibility metric is based on the calculation of contrast to noise ratio, which can be used to judge the quality of an image. The contrast to noise ratio is used to compare the quality of hazy and the dehazed image. This metric is computed with the help of two Gaussian kernels.

Visibility Metric Function Pseudocode

epsilon=1e-1

Algorithm 1. Pseudocode to calculate CNR in Matlab

```
procedure Compute VM (IM,sigma1, sigma2)
halfsize1 = ceil(-norminv(epsilon/2, 0, sigma1)) size1 = 2 *
halfsize1 + 1
halfsize2 = ceil(-norminv(epsilon/2, 0, sigma2))
size2 = 2 * halfsize2 + 1
Application of gaussian filtering
gaussian1 = fspecial('gaussian', size1, sigma1)
gaussian2 = fspecial('gaussian', size2, sigma2)
IM = abs(IM - imfilter(IM, gaussian1, 'replicate'))
10:         noise = IM - imfilter(IM, gaussian1, 'replicate')
lsd = sqrt(imfilter(noise.^2,gaussian2,'replicate'))
lsd = lsd./max(lsd(:)).  255
CNR = sum(lsd(:))/(size(lsd,1)  size(lsd,2))
end procedure
```

Time Complexity

The time interval between the display of two output frames after the application of the desired algorithm will give the time complexity of the algorithm.

Pseudocode

Here, the command tic starts a stopwatch timer. The code records the internal time at the execution of the tic command. The elapsed time is displayed with the toc command.

Algorithm 2. Pseudocode to measure the time complexity of an algorithm

```
  REPS = 1000   minTime = Inf nsum = 10 tic
for each i from 1 to REPS
do tStart = tic
```

```
//insert the required algorithm code whose complexity is to be
measured
tElapsed = toc(tStart)
minTime = min (tElapsed, minTime)
end for
averageTime = toc/REPS
```

GUI Design

A GUI can be developed for use by the driver of the car. For testing the algorithms and provide a simulation of how the fog removal algorithms would work in real time, a suitable GUI was designed and implemented by the authors. The GUI would display the video feed with fog removed as well as the parameters used to measure the performance. To make it user-friendly, clear and unambiguous labelling of control buttons was done so that there would be no confusion in understanding their function.

The GUI design consists of two video frame canvas – one shows the unprocessed live video feed and the other shows the output obtained after fog removal. Options to rotate and adjust the display have been provided. A recorded video can be played, or a camera feed can be given as input.

The algorithm can be selected by choosing the appropriate option and the defogged video along with delay and visibility, which were defined earlier, are also displayed. This allows easy comparison of the algorithms as well as convenience in determining their performance. A MATLAB GUI was created to implement the design described above and thus the real-time system was simulated.

ALGORITHMS FOR FOG REMOVAL

Fog removal algorithms are broadly classified into two types:

- Multiple image dehazing
- Single image dehazing

Multiple image dehazing algorithms use multiple images in the fog scene to estimate transmission map and then restore the image scene by optical module. The restoration process enhances both contrast and visibility. The demerit of these methods is they usually need to wait for a long time until weather conditions change. In addition to this, Complex and lengthy calculations need to be done to estimate the depth information.

The defog algorithms based single image method will only need one image to estimate the transmission map in the fog scene. These are preferred as that they do not need to wait for weather conditions to change and need less computation time.

Dark Channel Prior

Consider the following model to describe a foggy image:

$$I(x) = J(x)t(x) + A\,(1 - t(x)) \tag{1}$$

where

$I(x)$ is the observed intensity,
$J(x)$ is the scene radiance,
A is the global atmospheric light,
$t(x)$ is the medium transmission describing the portion of the light that is not scattered and reaches the camera.

In our algorithm, we obtain the radiance $J(x)$, atmospheric light A and transmission $t(x)$. With a specific end goal to re-establish the picture degraded by fog, He et al. provided the Dark Channel Prior (DCP) idea. This idea relies upon the suspicion fog free open-air pictures displays less than one shading channel with a low power an incentive as to the patches of the picture, which don't contain sky.

For an arbitrary image J, its dark channel J^{dark} is given by

$$J^{dark}\left(x\right) = \min_{y \in \varphi\left(x\right)} \left(\min_{\propto \left\{r,g,b\right\}} J^{c}\left(y\right) \right) \tag{2}$$

where $J^{dark}\left(x\right)$ is the dark channel $J^{c}\left(y\right)$ is the arbitrary image

Atmospheric Light Estimation

The atmospheric light A is calculated from the dark channel. The environmental light is the measure of light that is reflected into the camera from the scene. This can be considered as the highest intensity pixels lying in the dark channel of the foggy video

A local patch is considered in which the component t is estimated. This estimation is done by maximizing the visibility of the patch under a constraint such that the intensity of J(x) is less than the intensity of A.

The global atmospheric light A is found by tracking back from 0.1% maxima of the dark channel to the maximum of the corresponding pixels in the input image. If a beam of light is emitted into an atmospheric medium, when the incident light goes through a unit area, the vitality of the light will then be attenuated. It can be expressed by Airlight estimation model.

Transmission Map Estimation

The transmission map gives an idea about the medium between the scene and the digital camera. From the transmission map, we came to know about the density of the fog in the image. The method to estimate the normalized transmission map $\bar{t}(x)$ is shown below

$$I^c(x) / A^c = \bar{t}(x)\{J^c\left(x\right) / A^c\} + 1 - \bar{t}(x) \tag{3}$$

The operation of dark channel is applied on both sides of equation (3)

$$min\ (minI^c(y) / A^c) = \bar{t}(x)min\ (min\{J^c\left(y\right) / Ac)\} + 1 - \bar{t}(x) \tag{4}$$

For an outdoor clear video of the dark channel prior will be close to zero. This can be illustrated as follows

$$min\ (minI^c\left(y\right) / A^c = 0 \tag{5}$$

The estimate of transmission map can be derived from the above (4) & (5) as follows as follows

$$\bar{t}(x) = 1 - min\ (minI^c\left(y\right) / A^c) \tag{6}$$

The results may be seen unnatural if all the fog were removed, so He et al. added constant ω $(0 < \omega < 1)$ in (6) as

$$\bar{t}(x) = 1 - \omega min\ (min\{I^c\left(y\right) / A^c)\} \tag{7}$$

Transmission Map Refinement by Guided Filtering

The estimated transmission map is refined using guided filtering. The guided filter is a replacement to soft matting because soft matting has high computational complexity. It is an edge-preserving smoothing filter. Edge-saving smoothing is a video/image processing technique that smoothens away noise or textures while keeping sharp edges. Median, bilateral, guided, and anisotropic diffusion filters are different examples of this filter. These filters can be utilized as a part of numerous applications like medical or Satellite imaging. The edges are imp features that it preserved as sharp and not degraded in denoising. Edge-preserving filters are designed to automatically limit the smoothing at "edges" in images/videos measured, e.g., by high gradient magnitudes. Guided filter enhances the sharpness of the mist video and removes the blurring noises from videos.

Estimating the Scene Radiance

After all the above estimations the scene radiance is calculated using equation (10). The recovered scene radiance is displayed as dehazed image.

Workflow for DCP

See Figure 1.

Figure 1. Workflow for dark channel prior

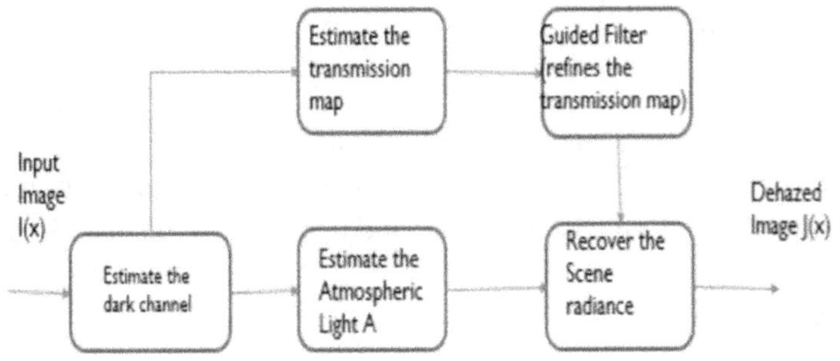

MATLAB Simulation Program

Contents

The Dark Channel Prior algorithm is implemented through the following sequence of steps in MATLAB

1. The original video is loaded into the MATLAB
2. The video is processed frame by frame and on each frame dark channel prior is applied separately
3. The algorithm is implemented using a separate function which calculates the dark channel of each frame
4. The function for estimating transmission map is called. The transmission map t(x) is given as:

$$\tilde{t}\left(x\right) = 1 - \omega * J^{dark}\left(x\right) \tag{8}$$

5. The calculation of Atmospheric Light A is done by tracking back from 0.1% maxima of the dark channel to the maximum of the corresponding pixels in the input video frame
6. Then guided filter is employed which refines the transmission map
7. Finally, the scene radiance J(x) which is the defogged image is recovered according to the following equation

$$\hat{J}\left(x\right) = \frac{I\left(x\right) - A}{\max\left[t\left(x\right), t_o\left(x\right)\right]} + A \tag{10}$$

8. The output video is then displayed to the driver
9. As each frame is processed the delay between consecutive frames and the visibility of each frame is calculated and displayed to the user in the GUI

MATLAB Code

Algorithm 3. Pseudocode for implementation of DCP in MATLAB

```
while hasFrame(obj) do
vidFrame = readFrame(obj)
result = dehaze(double(vidFrame), 0.75, 15)
```

```
im3 = rgb2gray(double(result))
CNR3 = ComputeVM(im3, 5, 5)
return CNR3
Display image(result/255)
diff = toc(tStart) - tElapsed
tElapsed = toc(tStart)
set(handles.edit4, 'String',diff)
end while
```

Algorithm 4. Pseudocode for calculation of Dark Channel

```
Procedure get_dark_channel(image, win size)
[m, n, ] = size(image)
padsize = floor(win size/2)
paddedimage = padarray(image, [padsize padsize], Inf)
dark channel = zeros(m, n)
for each j from 1 to m do
for each i from 1 to n do
patch = padded image(j:j+(win size-1),i:i+(win size-1),:)
dark channel(j,i) = min(patch(:))
end for
end for
return dark channel
end procedure
```

Algorithm 5. Pseudocode for calculation of Atmospheric Value

```
Procedure get_atmosphere(image, darkchannel)
[m, n,-] = size(image) npixels = m * n
nsearchpixels = floor(npixels * 0.01)
darkvec = reshape (dark channel, npixels, 1)
imagevec = reshape (image, npixels, 3)
[ -, indices] = sort(darkvec,'descend') accumulator = zeros(1,
3)
for each k from 1 to n search pixels do
accumulator = accumulator + imagevec(indices(k),:)
end for
atmosphere = accumulator / n search pixels
```

```
return atmosphere
end procedure
```

Algorithm 6. Pseudocode for Transmission Estimation

```
Procedure get transmission _estimate (image, atmosphere, omega,
win size)
[m, n,] = size(image)
rep_atmosphere = repmat(reshape(atmosphere, [1, 1, 3]), m, n)
trans est = 1 - omega * get darkchannel (image./ rep
atmosphere, win size)
end procedure
```

Results of Dark Channel Prior

In the case of outdoor foggy videos captured from the car, air light will be present as a result the foggy video frame is brighter than the original foggy view. Thus, the dark channel of foggy videos has higher intensity at denser fog and lower intensity at thinner fog locations. So, the dim channel of the foggy video will provide an approximate idea about the thickness of fog in the videos.

Figure 2. Video implementation of dark channel prior
CNR for Original Video = 5.9277 Calculations for DCP Delay: 2.65635s Visibility (CNR): 12.9581

Contrast Limited Adaptive Histogram Equalisation

A histogram is a picturization tool which shows the repetitiveness of a variable x with the aid of rectangles with different areas. Similarly, the histogram of an image I depicts the varying intensity values of various pixels in the image. Histogram equalization is a technique that spreads the most continuous power pixel values. This is an enhancement method which is mostly useful for images with both visible or not visible background and foreground. Its main application lies in photography and the field of medicine where it used to get a great perspective in X-ray. Its primary advantage is its invertible nature. So, if HE function is identified, then the original histogram can be retrieved. A challenge for this equalization is that it can increase the noise contrast of the background. There are various types of HE like AHE and CLAHE.

The CLAHE method is one of the approaches we have used to solve the problem of fog removal. The CLAHE algorithm can be used to apply adaptive histogram equalization. For this, the histogram is clipped. The pixels which have been clipped are redistributed to each level. The main difference from the AHE algorithm is that CLAHE can limit the enhancement of noise. Contrast Limited Algorithm provides better contrast as compared to dark channel prior approach. CLAHE is also computationally less expensive and thus has lower delay values. In the CLAHE approach, each frame is converted to the foreground and background frame. Both frames are defogged separately using the algorithm. The frames are then finally fused again to form an output frame.

CLAHE can be used to remove haze from foggy images. After dividing the frame into foreground and background, the image is converted from RGB to LAB. The intensity component is defogged but the hue and saturation components are not changed. And then the new LAB image is obtained. Finally, the new LAB image is converted back to the RGB image. Then the processed frames are combined to form the output video.

MATLAB Simulation Program

Contents

The algorithm for CLAHE is implemented by the following sequence of steps:

1. The input video is loaded into Matlab GUI and is processed frame by frame
2. Each frame is converted to LAB format
3. Adaptive histogram equalization is then implemented using the adapthist command in MATLAB

Figure 3 .Workflow for CLAHE

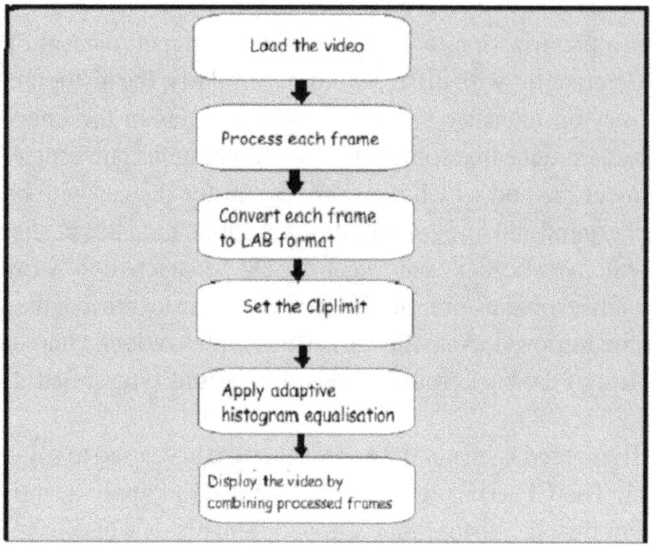

Figure 4. Comparison of original image and CLAHE processed image

4. The clipping limit is set for contrast limited adaptive histogram equalization.
5. The frame is then converted to double format for display
6. The output video is then displayed to the user
7. As each frame is processed the delay between the consecutive frames and the visibility of each frame is calculated.

Figure 5. Histogram of original image

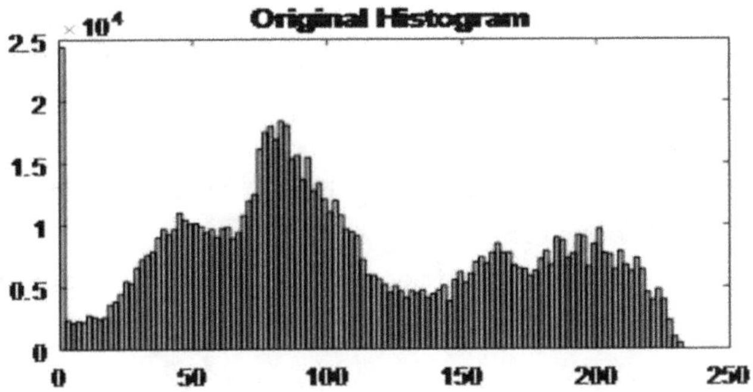

Figure 6. Histogram of CLAHE enhanced image

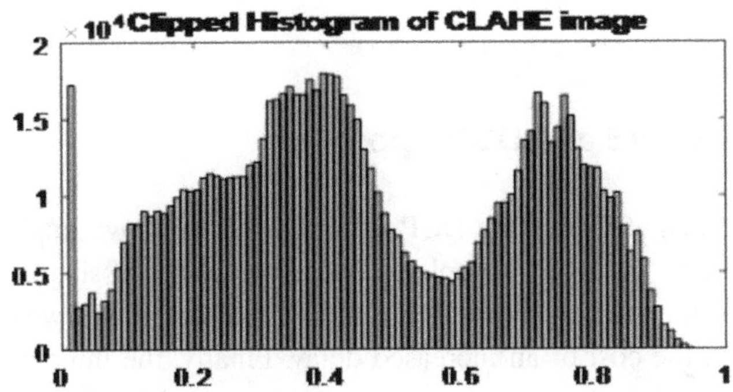

MATLAB Code

Algorithm 7. Pseudocode for Implementation of CLAHE in MATLAB

```
while hasFrame(obj) do
vidFrame = readFrame(obj) Read each video frame till the end of
video
LAB = rgb2lab(vidFrame) Convert the image from RGB to LAB
format
L = LAB(:,:,1)/100;
Application of adaptive Histogram equalisation with ClipLimit
L = adapthisteq(L,'NumTiles',[8 8],'ClipLimit',0.005);
LAB(:,:,1) = L*100;
J = lab2rgb(LAB) Reconversion of image from RGB to LAB
im2 = rgb2gray(double(J));
```

```
CNR2 = Compute VM(im2,5,5) // for calculating the visibility of
each frame
set(handles.edit3, 'String', CNR2) // Display the value of CNR
for the purpose of simulation
Display the Enhanced video frame
diff = toc(tStart) - tElapsed //  The differnce between started
and end time will give the
                              //time to execute the algorithm
tElapsed = toc(tStart)
end while
```

Results of CLAHE

See Figure7.

Combined CLAHE and DCP Approach

To add up the advantages of the DCP and CLAHE, we have implemented their combination. First, the DCP is implemented as a pre-processing algorithm. The result of DCP is then fed into the CLAHE algorithm. The visibility is considerably increased but at the cost of an increased delay. Finally, the three approaches are compared based on delay between the successive frames and the contrast to noise ratio.

Figure 7. Video implementation of CLAHE
CNR for Original Video = 5.9277 Calculations for CLAHE Delay: 0.609761s Visibility (CNR): 43.4573

MATLAB Simulation Program

Contents

1. The original video is loaded in MATLAB
2. DCP is used for pre-processing in the frames
3. CLAHE is then applied to the pre-processed frames
4. Finally, the resulting video is displayed

MATLAB Code

Algorithm 8. Pseudocode for combined implementation of DCP and CLAHE in MATLAB

```
while the video still has frames do
vidFrame = readFrame(obj);
DCP implementation
result = dehaze(double(vidFrame), 0.75, 15)
CLAHE implementation
LAB = rgb2lab(result/255)
L = LAB(:,:,1)/100;
L = adapthisteq(L,'NumTiles',[8 8],'ClipLimit',0.005)
LAB(:,:,1) = L*100
J1 = lab2rgb(LAB)
im4 = rgb2gray(double(J1))
CNR4 = Compute VM(im4, 5, 5)
set(handles.edit3, 'String', CNR4) Display the value of CNR
image(J1, 'Parent', ax) Display the video frame by frame
diff = toc(tStart) - tElapsed
tElapsed = toc(tStart)
set(handles.edit4, 'String',diff)
end while
```

Results of the Combined DCP and CLAHE

See Figure 8.

Figure 8. Video implementation of combined DCP and CAHE
CNR for Original Video = 5.9277 Calculations for the combination of DCP and CLAHE Delay: 3.14s Visibility (CNR): 50.1563

RETINEX Approach Combine With DWT

Land and McCann came up with the idea of retinex approach in image processing in the year 1971. The term ‘retinex’ was coined by Edwin Land to represent spatial image processing for ‘colour constancy’. This means that similar colours will look the same under different lighting conditions. The approach is based on the thought that the original image comprises of light image L (x, y) and object reflection properties R (x, y).

RETINEX algorithm shows an improvement in fog expulsion from video by reducing the effects of incident light on the video. RETINEX hypothesis presents a simple model about how the human visual framework sees particles brightness and colours. It is a methodology which overcomes any issues between videos obtained and the viewer observation of scenes. For enhancing the traffic visibility in a foggy climate, RETINEX algorithm depends on the changing scale for mist expulsion with a depth outline. There is a requirement of haze picture dark channel prior treatment to acquire the evaluated depth map. In the end, every part of the videos is transformed using single scale retinex.

By comparing customary picture upgrade algorithms, the retinex can accomplish an adjustment in the dynamic Gray range pressure, edge improvement and colour constancy.

where

F (x, y) is the captured image
S (x, y) is the image in the log domain
I (x, y) is the enlightenment part, which depends on the ambient light and connected with the dynamic scope of the picture.

Figure 9. Workflow of RETINEX method

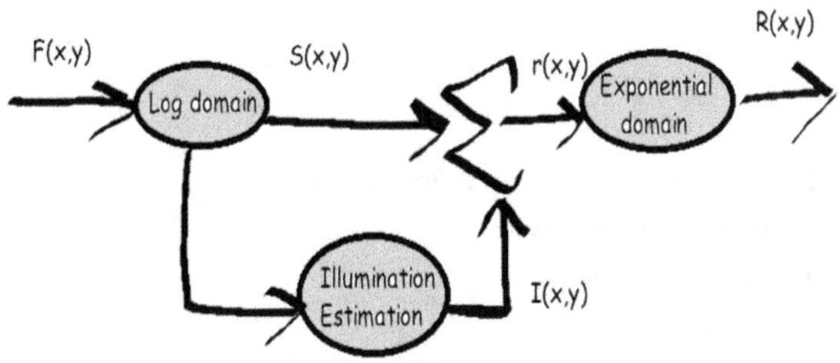

$r(x, y)$ is the combination of the image in the log domain and the illumination image
$R(x, y)$ is the reflection component, which depicts the reflection of the surface of an object and is identified with the characteristic idea of the picture

There are two divisions of the RETINEX method used for enhancement of foggy videos called

1. Single-Scale RETINEX (SSR)
2. Multi-Scale RETINEX (MSR)

Jobson et al. in his work propose an SSR algorithm which depends on the centre/ surrounding RETINEX method. Using the algorithm author estimated the ambient brightness using which reflection image is obtained. To maintain a balance between the dynamic scope compression and the colour constancy, Rahman et al. extended the algorithm of SSR to multiple scales and proposed an MSR algorithm. The intensity of the enlightenment has very little effect on the reflection image, which makes it simple to acknowledge picture dehazing using the RETINEX algorithm. Equation (1) and (2) depict the formula of single scale retinex and large-scale retinex respectively.

$$R_i\left(x, y\right) = \log F_i(x, y) - \log [G(x, y)\ F_i(x, y)] \quad (1)$$

where

$F_i(x, y)$ is the input image on the i-th colour channel
$R_i(x, y)$ is the output of the RETINEX image on the i-th channel,
G is the normalized surround function
(x, y) is the position of a single pixel,

$$R_{MSR_i}\left(x,\ y\right)\ =\sum\nolimits_{K=1}^{N} w_k\{logF_i\left(x,\ y\right)\ -\ log\ [G_k\left(x,\ y\right)*F_i\left(x,\ y\right)]\}\ (2)$$

where

$F_i\ (x, y)$ is the input image on the i-th colour channel
R_{MSRi} is the multiscale retinex image
w_k is the weight of each scale
N is the number of scales
DCP is a pre-processing technique followed by the RETINEX algorithm combined with HAAR

The wavelet transform RETINEX is a post-processing technique. Algorithms depend on RETINEX are used widely in many fields for the enhancement of the image. It is connected for the evacuation of shadow and in addition mist. To get the properties of reflection of the objects from the light influence light effect on the picture is the key idea of the procedure.

The concept depends on the fact that during visual information transmission, the human vision system performs some information processing to expel the uncertainty related to the light source's intensity and irradiation, and only information reflecting the nature of the article, such as the reflection coefficient. DCP is followed by the RETINEX & DWT technique.

As per the conditions, RETINEX can accomplish

- Sharpening
- Compensation for the obscuring presented by a picture development process
- Colour constancy preparing
- Consistency improvement of yield as enlightenment changes dynamic range pressure.

Discrete Wavelet Transforms (DWT)

DWT is utilized to decompose an image. It is more proficient than the DCT transform method. A two-level DWT disintegrates picture into low and high-frequency segments. In this work, HAAR Wavelet Transform is utilized to expel fog from videos. Here first media is separated into outlines. At that point luminance segment of each edge is picked and DWT is connected to it that comes about into various subgroups which are again divided into various parts. We measure the covariance matrix for every piece. Inverse DWT is used to find watermarked luminance segment of the frames.

After applying the Pre-processing algorithm followed by RETINEX & HAAR wavelet transform the fog can be removed from videos captured by moving cars.

The WT is a nearby change of space and recurrence and has focal points of multi-scale examination and multi-determination qualities for image contrast enhancement. The wavelet transform upgrade the details of videos in every pixel. The algorithm depicted here uses HAAR Wavelet transform. The wavelet-based technique is used to calculate the low and high-frequency segments.

Inverse DWT

The inverse wavelet transform is used in the reconstruction of the resulting video. In this procedure, the converted picture is joined with the IDWT algorithm to get a combined picture and afterward all the procedure is reproduced.

Figure 10. Workflow of RETINEX combined with HAAR wavelet method

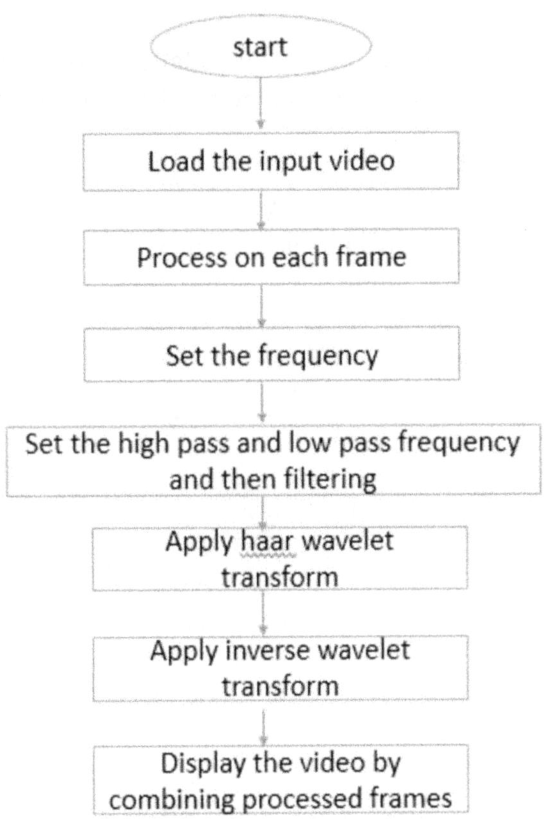

MATLAB Simulation Program

Contents

The retinex and DWT algorithm is implemented by the following sequence of steps

1. Input the haze video
2. Separate the three RGB channels are separated
3. Application of fast Fourier transform to all the channels
4. Put the above fog-video into logarithm domain
5. Acquire the outcome video R (x, y) by utilizing the above detailing previous steps
6. Utilize a straight stretch to make the above R (x, y) video size to be like the original video.
7. Use the wavelet transform method decompose R (x, y) image into two layers.
8. Increase the high frequency and suppress low frequency of the image from the previous step.
9. Output the outcome fog evacuated video

After applying the Pre-processing algorithm followed by RETINEX & HAAR wavelet transform the fog can be removed from videos captured by moving cars.

MATLAB Code

Algorithm 9. Retinex with Wavelet implementation

```
while the video sequence still has frames do
frame = readFrame(obj)
[height,width, ]=size(frame)
[Y,X]=meshgrid(1:height,1:width)
c=50000000
Fnok = exp(-((X.2) + (Y.2))./(c.2))
K= 1/(sum(sum(Fnok)))
F = K.*Fnok
All the coloured channels are separated
IR = double(frame(:,:,1))
IG = double(frame(:,:,2))
IB = double(frame(:,:,3))
FF = fftshift(fft2(F'))
Application of fft to all the colour channels separately
```

```
IFR = fftshift(fft2(IR));
IFR=FF.*IFR
IFR=real(ifft2(ifftshift(IFR)))
IFG = fftshift(fft2(IG));
IFG=FF.*IFG
IFG=real(ifft2(ifftshift(IFG)))
IFB = fftshift(fft2(IB));
IFB=FF.*IFB
IFB=real(ifft2(ifftshift(IFB)))
RR = log10(double(IR))-log10(IFR)
RG = log10(double(IG))-log10(IFG)
RB = log10(double(IB))-log10(IFB)
OUT(:,:,1)=uint8(255*RR/(max(max([RR RG RB]))))
OUT(:,:,2)=uint8(255*RG/(max(max([RR RG RB]))))
OUT(:,:,3)=uint8(255*RB/(max(max([RR RG RB]))))
Haar wavelet is applied to each colored channel
[a h v d]=dwt2(OUT(:,:,1),'haar')
R1(:,:,1)=idwt2(a,2*h, 2*v,2*d,'haar')
[a h v d]=dwt2(OUT(:,:,2),'haar')
R1(:,:,2)=idwt2(a,2*h, 2*v,2*d,'haar')
[a, h v d]=dwt2(OUT(:,:,3),'haar')
R1(:,:,3)=idwt2(a,2*h, 2*v,2*d,'haar')
im3 = rgb2gray(double(R1))
CNR3 = ComputeVM(im3, 5, 5)
set(handles.edit3, 'String', CNR3)
image(uint8(R1), 'Parent', ax)
set(ax, 'Visible', 'off')
pause(1/obj.FrameRate)
diff = toc(tStart) - tElapsed
tElapsed = toc(tStart)
display diff
end while
```

Results of RETINEX With DWT

See Figure 11.

Figure 11. Video implementation of RETINEX
CNR for Original Video = 5.065 Visibility (CNR): 35.4554 Delay: 2.41s

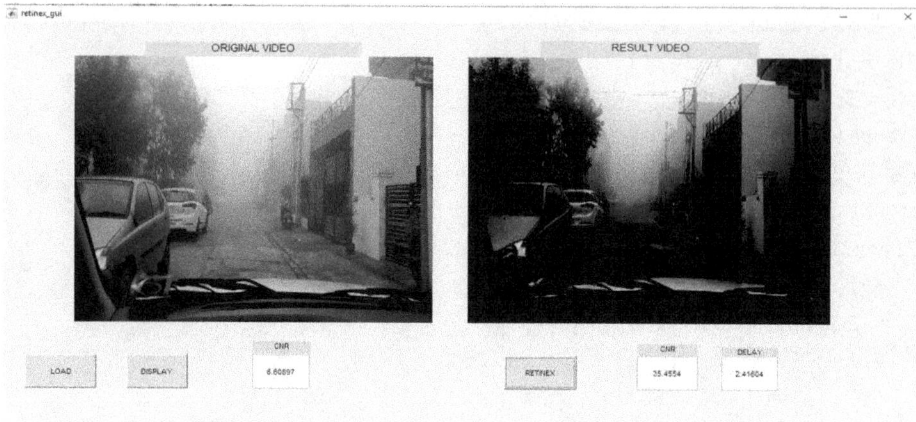

FURTHER RESEARCH DIRECTIONS

The major limitation of this work is the delay that is introduced due to the complexity of the algorithms. This can be improved by devising algorithms which reduce the delay and give quick results to the driver. Also, the processors used in real time driverless cars are much faster as compared to those in PCs and are more suitable for delay sensitive applications. Also, the automatic slowing down of the car during fog according to the delay of the algorithms which is deployed for fog removal will make the work more suitable for real-time applications.

The chapter does not cover the details of how the algorithms will be implemented in the dashcams of the cars. The new generation smart cameras integrate the whole visual system inside one system that can be used anywhere, in any industry where image processing needs to be applied. There is no requirement for separate space for keeping the computing equipment because the computer is housed within the smart camera. In industrial image processing, real-time processing is the opposite of offline processing. In offline processing applications, you record images (many of them) and process them later. The chapter focuses on offline processing as videos recorded while driving the car in foggy weather are used for experimentation. In real time, the positioning and angle of the camera are likely to influence the videos. The work in this chapter can be extended to actual implementation using smart cams on the dashboard of cars.

The proposed techniques can likewise be stretched out for applications like night vision, underwater image restoration and so forth.

CONCLUSION

Smart cities have come up with the intent to better the lifestyle of its residents. The basic idea is to apply practical science, engineering, and innovation in the service of people. Many data analytics, data storage, and communication technology-based mechanisms are being built for efficient utilization and management of a city's resources. However, the most indispensable asset of a city is its citizens. Therefore, the focus needs to be on the security and safety of citizens. Numerous lives are lost annually especially on roads. There is still a need for improvement in the way technology is used in traffic management systems.

Bad weather leads to the degradation in the brightness of the video thus influencing the execution of numerous PC vision applications, such as tracking, video surveillance, transportation and robot navigation This chapter compared three important algorithms DCP, CLAHE, and Retinex for fog removal. The algorithms were analysed based on their computational complexity and Contrast to Noise Ratio (CNR). The algorithms are simulated on videos captured in moving cars using MATLAB2015a

Comparison of the Algorithms

By analysing the results, it is observed that CLAHE is a better algorithm as compared to the other two in terms of delay. Though dark channel gives more delay it is used as a pre-processing algorithm. The Retinex algorithm is better than dark channel prior in terms of delay but not so good when compared with CLAHE. The Retinex algorithm is better than dark channel prior in terms of the visibility (CNR). The dark channel algorithm gives better results during night time followed by CLAHE and Retinex Thus, a combination of the dark channel prior and CLAHE was used to arrive at desired results. Table 1 compares the performance of algorithms discussed in the chapter.

A mathematical demonstration of how the system can suggest driver drive at a certain speed to obtain defogged image reliably is given below

Consider for instance the driver is driving at 30Km/h.

The delay of the algorithm is 0.6 s.

Distance travelled by the driver in 1 sec is 8.33m

Now, distance travelled in 1 + 0.6 s should be 8.33m

Table 1. Comparison of algorithms for fog removal

Parameters / Algorithms	DCP	CLAHE	Combination of DCP and CLAHE	RETINEX
Delay	2.6535s	0.60976s	3.14s	2.41s
Visibility Metric (CNR)	12.9581	43.4573	50.5163	35.4554
Complexity	More Complex	Least Complex	Most Complex	Less complex than DCP but more than CLAHE
Clarity	Least clear (Used for pre-processing purposes)	Clear	Most clear	Brightness increases
Performance at night	Best performance (CNR between 30 40)	Not good (CNR between 18 to 20)	Good performance (CNR between 20-25)	Not good (CNR between 11 to 15)

Thus, the speed of the car should be 18Km/h

Hence, our system can be used to suggest a safe speed for driving to receive the output in a reasonable amount of time. The test results show that the methods in this chapter can effectively remove fog in video streams captured in moving cars

Thus, algorithms discussed throughout the chapter will aid in the development of smart transportation systems for 'smart cities'. These algorithms will further aid in the development of traffic surveillance systems during bad weather. However, even if the best technology is implemented it is of prime importance that the driver of the vehicle must take all the precautionary measures and drive at the safe speed given by the system and diligently follow the traffic guidelines.

REFERENCES

Anwar, M. I., Khosla, A., & Singh, G. (2017). *Visibility enhancement with single image fog removal scheme using a post-processing technique.* In *Signal Processing and Integrated Networks (SPIN), 2017 4th International Conference on* (pp. 280-285). IEEE. 10.1109/SPIN.2017.8049960

Chen, C., Do, M. N., & Wang, J. (2016). Robust image and video dehazing with visual artifact Suppression via gradient residual minimization. *Proc.Eur. Conf. Comput. Vis*, 576–591. 10.1007/978-3-319-46475-6_36

Chen, G., Zhou, H., & Yan, J. (2007). A novel method for moving object detection in foggy day. *8th ACIS international conference on software engineering. Artificial intelligence, networking, and parallel/distributed computing*, 53–58.

Computer Vision System Toolbox. (n.d.). MATLAB. Retrieved from https://www.mathworks.com/products/computer-vision.html

Garima Yadav Saurabh Maheshwari Anjali Agarwal. (2014). *Contrast Limited Adaptive Histogram Equalization Based Enhancement For Real Time Video System.* Rajasthan: ICACCI.

Guo, Tang, & Cai. (2013). Fusion strategy for single image DE hazing. *Int. J. Digit. Content Technol. its Appl, 7*, 19–28.

He, K., Sun, J., & Tang, X. (2011). Single Image Haze Removal Using Dark Channel Prior. *IEEE Transactions on Pattern Analysis and Machine Intelligence*, *33*(12). PMID:20820075

Jobson, D. J., & Woodell, G. A. (1995). *Properties of a center/surround retinex, part 2: Surround design.* NASA Tech. Memo. 110188.

Jobson, Rahman, & Woodell. (n.d.). *A Multiscale Retinex for Bridging the Gap Between Color Images and the Human Observation of Scenes.* Academic Press.

Karthika, P. B., & Sindhu, R. (2017). A simple haze removal algorithm to remove haze from videos. In *Computing Methodologies and Communication (ICCMC), 2017 International Conference on* (pp. 461-464). IEEE. 10.1109/ICCMC.2017.8282732

Nayar, S. K., & Narasimhan, S. G. (1999). Vision in bad weather. *Proc. IEEE Conf. Computer Vision*, 2, 820–827. 10.1109/ICCV.1999.790306

Parthasarathy, S., & Sankaran, P. (2012). A RETINEX based haze removal method. In *Industrial and Information Systems (ICIIS), 2012 7th IEEE International Conference on* (pp. 1-6). IEEE.

Rahman, Jobson, & Woodell. (2004). Retinex processing for automatic image enhancement. *Journal of Electronic Imaging, 13*(1), 100-111.

Singh & Sood. (2017). A Review on Influence of Fog on Road Crash. *International Journal of Engineering Research & Technology, 6*(7). Retrieved from http://www.ijert.org

Sulami, M., Glatzer, I., Fattal, R., & Werman, M. (2014). Automatic recovery of the atmospheric light in hazy images. *Proc. IEEE Int. Conf. Comput. Photogr*, 1–11.

Sun, C.-C., Heng-Chi, L. M.-H. S., & Huang, Y.-H. (2013). *Single Image Fog Removal Algorithm based on an Improved Dark Channel Prior Method.* Depart. of EE, NFU, Taiwan & Depart. of EE, NYUST, Taiwan.

Sun, C.-C., Lai, H.-C., Sheu, M.-H., & Huang, Y.-H. (2016). Single image fog removal algorithm based on an improved dark channel prior method. In *Intelligent Signal Processing and Communication Systems (ISPACS), 2016 International Symposium on* (pp. 1-4). IEEE.

Tan, R. T. (2008). Visibility in bad weather from a single image. *Proc. IEEE Conf. Comput. Vis. Pattern Recognition*, 1–8.

Tarel, J.-P., & Hautière, N. (2009). Fast visibility restoration from a single color or gray level image. *Proc. IEEE Int. Conf.Comput. Vision*, 2201–2208. 10.1109/ICCV.2009.5459251

Wang, W., Li, W., Guan, Q., & Qi, M. (2015, October). Multiscale single image dehazing based on adaptive wavelet fusion. *Mathematical Problems in Engineering, 2015*, 1–14. doi:10.1155/2015/813507

Xie, B., Guo, F., & Cai, Z. (2010). Improved single image dehazing using dark channel prior and multi-scale retinex. In *Intelligent System Design and Engineering Application (ISDEA), 2010 International Conference on* (*vol. 1*, pp. 848-851). IEEE. 10.1109/ISDEA.2010.141

Xu, Z., Liu, X., & Chen, X. (2009). *Fog Removal from Video Sequences Using Contrast Limited Adaptive Histogram Equalization. In Information Science and Technology College Dalian Maritime University.* IEEE.

Zhao, Xiao, Yu, & Xu. (2015). Single image fog removal based on local extrema. *IEEE/CAA Journal of Automatica Sinica, 2*(2), 158-165.

Chapter 4
Green Communication for Cognitive Cities

Indu Bala
Lovely Professional University, India

Ghanshyam Singh
University of Johannesburg, South Africa

ABSTRACT

To meet the explosive data rate requirements of future cognitive cities, mobile and wireless technologies are evolving at a very fast pace. With the faster adoption of bandwidth-hungry applications in our daily lives, the energy consumption in wireless networks has increased. Higher energy consumption has imposed some serious health concerns such as exposure to harmful radiation and increase in carbon dioxide emissions into the environment. To overcome the ecological and health issues and to meet the end user's high data rate requirements for the successful deployment of the cognitive cities, the most important technology is green communication. Taking into consideration the importance of green communication technologies, this chapter presents an intense survey on energy-efficient technologies developed for green communication. Some key research challenges pertaining to the use and deployment of these technologies for green communication are discussed. In the end of this chapter, the worldwide standardization efforts for cognitive cities are also highlighted.

DOI: 10.4018/978-1-5225-8085-0.ch004

INTRODUCTION

The most recent World Urbanization Prospects report has revealed that 55% of the world's population is living in urban areas and it is expected to increase up to 70% by 2050. The factors responsible are: better education and job opportunities and great access to amenities like public transportation, sports, and cultural events etc. (Cisco, 2015). These advantages result in growth, and with growth comes strain on existing public services, infrastructure, resources and public safety. The basic need for urban public safety is also one of the biggest forces driving the adoption of "cognitive city" – the ultimate solution to the urban challenges through technological means. Thus, the term cognitive city refers to a city that collects, analyze and distribute information smartly by means of technologies (Yaqoob et. al, 2017). The foundation of such cities is a contemporary smart infrastructure made up of millions of devices/ sensors connected through communication networks that collect and process data from these devices. This data is leveraged by various systems and platforms to make decisions or to initiate activities, and to address the needs of users via services and applications. Thus, cognitive city need the seamless Internet connectivity not only to the computers and smart devices carried by people, but also to the "things" with embedded sensors, actuators and networking capabilities to manage environmental, social, economic, and public health issues in future (Talari et. al, 2017). Cognitive city deployment is more than refining the old existing system technologically by simply adding electronic sensors, remote surveillance, and control to enhance existing city services (Wang, Huang, and Zou, 2016). Some of the major applications of cognitive cities have been enlisted in Table 1.

ROLE OF ICT IN COGNITIVE CITY

The information & communication technologies (ICT) plays a very vital role in cognitive cities in collecting and aggregating information and data from the field to improvise the city functionalities in terms of resources utilization, better services, and improved lifestyles, as illustrated in Fig. 1. The main key functions to achieve the goals and to maximize the performance of cognitive city are (ITU-T Focus Group, 2014):

- **Information Sharing and Knowledge Updation:** Even though, the conventional cities are well equipped to tackle emergency situations like natural disaster. However, the lack of information sharing outgrows the problem. It is anticipated that with immediate and accurate information

Table 1. Cognitive city applications

S. No.	Application	Examples
1.	Smart Energy	Smart Energy meters Integration of renewables and decentralized energy Network monitoring and control
2.	Smart Transportation	Traffic monitoring and control Real time solutions for pedestrian traffic Services for drivers Integrated public transportation Vehicle-to-everything Driver behavior management
3.	Smart Buildings	Light and temperature control Energy efficient buildings Energy generation using Renewable resources Security Big data management
4.	Smart Water	Pressure management Smart water meters Water conservation and efficient distribution Remote control and prediction management
5.	Smart Health	Real time health care solutions Smart hospitals Remote health monitoring Electronic record management of patient health data Disease controlling and mitigation Adequate sanitation
6.	Smart Education	Interactive teaching aids and learning environment Online world class digital content access Massive open online courses (MOOCs)
7.	Smart Security	Video and CCTV surveillance Seamless connectivity during natural calamity
8.	Smart Waste	Waste management Waste water treatment City cleaning Waste tracking

availability in future cognitive cities, one can gain deep insight to the problems and appropriate actions may be taken in advanced before they escalate.

- **Data Forecasting:** To deal with situation like natural disaster requires extensive data collection to study patterns, identification of risk areas etc. ICT helps in managing this information more efficiently and thus cognitive cities can improve its preparation and response capability.
- **Data Integration:** The timely access of the relevant information such as early warning systems, ensures better understanding of the city's vulnerabilities and strengths.

Figure 1. Components of cognitive city and position of ICTs (Abedlhak and Asmaa, 2016)

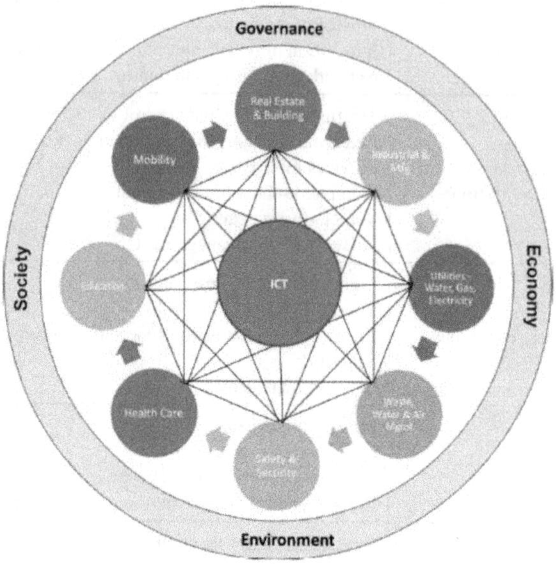

From past few years, the demand of smart electronic gadgets with ubiquitous wireless applications has triggered a sharp increase in energy consumption and greenhouse gas emission (Wang, Huang, and Zou, 2016). It is anticipated that by the end of 2020, the total number of connected devices will reach to 50 billion and therefore, the capacity requirement will also boost 1000 times to the existing capacity (Cisco, 2015). At that time, to provide error-free connectivity to these devices will surely become a big challenge due to interference issues. The ICT is expected to play a very important role here in terms of providing an error free connectivity in these cities. In present scenario, various technologies such as Bluetooth, Zigbee, WiFi, WiMAX, LTE and LTE-A are immediate solutions available for cognitive cities (Yaqoob et. al, 2017). However, these technologies are not the ultimate solution due to the limited transmission range, capacity and higher energy consumption (Refer Table 2).

To realize the idea of cognitive cities, cities must be empowered with new services and applications such as smart grids, e-health, smart infrastructure, smart transport etc. (Theoleyre et. al, 2015). The energy requirements for these billion new devices to provide anytime, anywhere connectivity would contribute significantly towards greenhouse gas emissions (Hamaguchi et. al, 2012). To reduce energy consumption in communication domain, it is important to incorporate energy efficient green communication techniques to achieve significant energy savings. However,

Table 2. Comparison of existing technologies for cognitive cities

S. No.	Technology	Standard/ Project	Frequency Range	Range	Data rate	Advantages	Disadvantages
1.	WiFi	IEEE 802.11 (a/b/g/n)	2.4/3.6/4.9/5/5.9 GHz	100 m	1-54 Mbps	(a) Subscriber Mobility (b) Less Wires	(a) Less Coverage (b) Non reliable connection (c) High Signal attenuation
2.	WiMAX	3GPP	3.5 GHz	50 km	75 Mbps	(a) Wide Coverage (b) High Speed Internet Access	(a) High installation cost
3.	LTE	3GPP	2.5/5/10 GHz	30 km	300 Mbps (Downlink) 75 Mbps (Uplink)	(a) High Spectral Efficiency (b) Backward Compatibility with existing technologies. (c) Future Proofing (d) Less internet connection lagging	(a) High installation cost
4.	LTE-A	3GPP	2.5/5/10/15/20 GHz	30km	1 Gbps Downlink) 500 Mbps (Uplink)	(a) High Spectral Efficiency (b) Suitable for Voice applications	(a) High installation cost
5.	Bluetooth	IEEE 802.15.1	2.4 GHz	1-100m	1 Mbps	(a) Low Cost (b) Easy Installation	(a) Suitable for short range communication only
6.	ZigBee	IEEE 802.15.4	2.4 GHz	10-20 m	250 Kbps	(a) Low Cost (b) Energy Efficient (c) Collision Avoidance	(a) Less Speed

the implementation of green communications for cognitive cities is itself a very challenging task as the connected devices must support diverse range of applications; traffic requirements; different quality of service requirements & deployment scenarios; equipment and hardware compatibility; various energy efficiency tradeoffs; and heterogeneous communication and networking protocols, technologies, & network architectures (Khan, Rehmani, and Rachedi, 2016). In this chapter, various energy efficient green communication technologies for cognitive cities are discussed. The major key challenges in the implementation of these technologies are highlighted. Moreover, the worldwide standardization activities in context with cognitive cities are also emphasized.

GREEN COMMUNICATION: THE NEW COMMUNICATION PARADIGM

As discussed earlier, due to the fast adoption of personal digital assistance (PDAs) such as smart phones, smart glasses, smart watches, personal wearable communication and healthcare devices; have made us to move towards the smart society. By strengthening the mobile and wireless networks, two or more devices can communicate with each other to exchange information and display necessary information anytime, anywhere. The recent development in wireless communication technology has created the imbalance of the resource utilization. Moreover, the faster adoption of wireless applications in our daily lives has increased the volume of the internet traffic and power consumption by these networks manifolds. Thus, it has become the dire need of the hour to think over the new communication paradigms to address these challenges. Green communication technologies are evolving at very fast pace to address these challenges by balancing the resources, information sharing, routing adaption and spectrum efficiency. Thus, green communication has become an emerging research area for future mobile and wireless networks. Many energy efficient green technologies (Refer Fig. 2) have been proposed to reduce the network power consumption, and the detailed overview of these technologies is given in this section.

Spectrum Sharing

Traditionally, the spectrum is allocated for different services for long-term use while restriction on maximum transmission power to avoid interference between users. Since, most of the spectrum has already been allocated to different services; it has become extremely challenging to accommodate new emerging applications developed for cognitive cities (Bala, Bhamrah, and Singh, 2017). However, the most

Figure 2. Energy efficient green technologies for cognitive cities

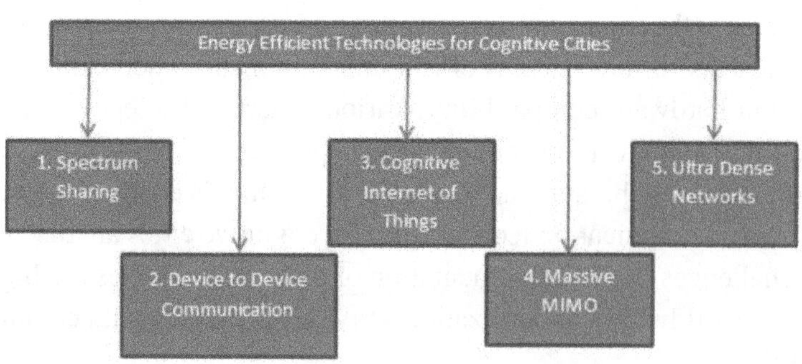

recent Federal Communication Commission (FCC) report has revealed that most of licensed frequency bands are being used sporadically and has left no spectrum to deploy new services/applications (Bala, Bhamrah, and Singh, 2015). The limited or no spectrum availability is considered as a biggest bottleneck for the successful deployment of the cognitive cities in future where, billions of the smart devices/sensors are expected to connect to the cloud for seamless connectivity. To deal with this conflict, the regulatory authorities have proposed an idea of spectrum sharing (Mitola, 2000). Based on the network parameters, spectrum sharing models can be classified as shown in Figure 3.

Architecture Based Spectrum Sharing

Based on the architecture, spectrum sharing may be classified as:

Figure 3. Spectrum sharing models

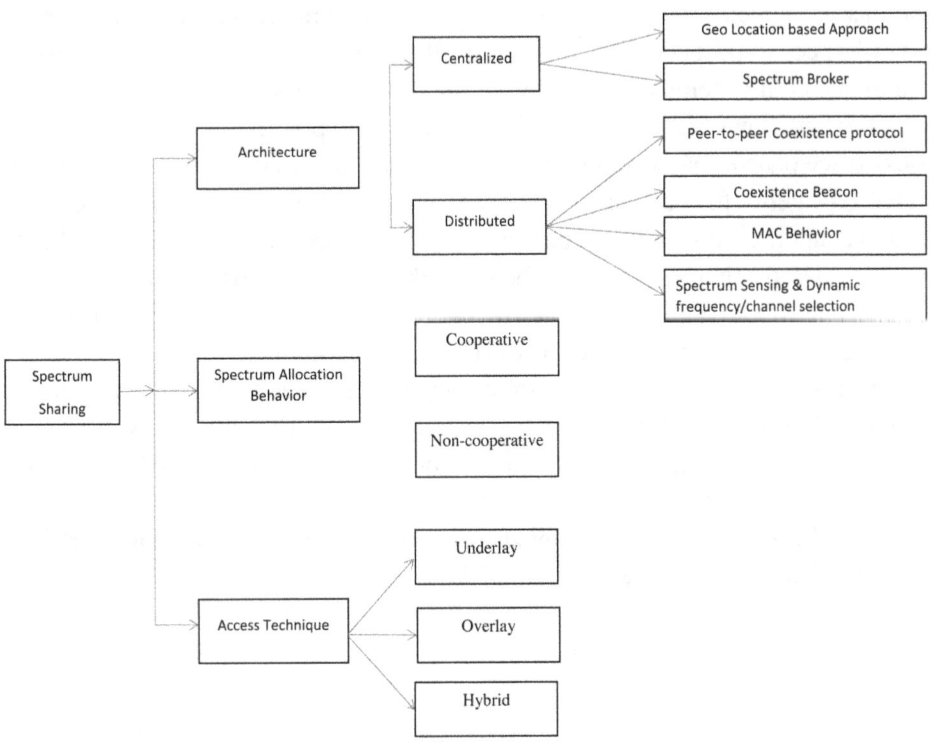

Centralized Spectrum Sharing

In the centralized spectrum sharing scenario, a centralized entity known as central controller coordinates the entire procedure of spectrum sensing and spectrum allocation (Sridhara, Chandra, Tripathi, 2008). All communications are through a central entity and the spectrum sharing decisions like transmission frame duration and transmission power levels are decided by a central controller. It monitors spectrum occupancy level, estimate bandwidth requirements by all secondary users and selects frequency band accordingly to maximize the network efficiency. The decision of central controller is broadcasted among all secondary users in the network (Fette, 2013).

Distributed Spectrum Sharing

In this spectrum sharing scenario, unlicensed users communicate in peer-to-peer manner without central coordinator (Zhao and Sadler, 2007). The secondary users take spectrum access decision independently and thus require greater computational capabilities and power backup. Various solutions have been proposed to evade the commercial agreements between the equipment vendors and network operators, such as (1) the peer-to-peer coexistence protocol allows direct message exchange between communication systems via some well-defined interface (Ahn, Lee, and Lee, 2016). Moreover, it allows high spectral efficiency with less power consumption. (2) The coexistence beacon-based solutions allow licensed system transmits a well-known signal repeatedly to indicate its presence. The unlicensed users can exploit this data to adapt their transmission parameters for fair spectrum sharing. (3) In MAC behavior-based schemes use listen-before-talk approach in which the unlicensed user such as Wi-Fi keeps searching the silent period. On finding channel inactive, unlicensed system gains channel access (Eu et. al, 2008).

Spectrum sensing and dynamic frequency/channel selection (DFS/DCS) based solution allows systems dynamically change its operating frequency. Different types of detectors such as energy detector, or feature detector is used to measure the channel occupancy. The spectrum sharing scenario, DFS/DCS continuously identifies the idle or partially occupied channels to minimize the sharing overheads (Nawaz, Marcenaro, and Regazzoni, 2017).

Spectrum Access Behavior

Based on the spectrum access behavior, CR system may be classified as:

Cooperative Spectrum Sharing

In cooperative spectrum sharing scheme, all secondary users cooperate with each other either through a common base station. The sensing information is shared among them to reduce sensing time, sensing errors and to improve the degree of fairness. The secondary users cooperate among themselves to share the frequency spectrum and to maximize spectral efficiency (Liang, Soon, and Hanzo, 2017).

Non-Cooperative Spectrum Sharing

This spectrum sharing scheme does not allow the secondary users to exchange their sensing data among themselves. This kind of selfish spectrum sharing approach is advantageous for small networks and provides less communication overhead. However, it reduces spectral efficiency of the communication networks. In this mode of spectrum sharing, the spectrum sensing information is used to share the spectrum. The major bottleneck of this approach is that the probability of false alarm is significantly high and therefore does not use natural resources optimally.

Spectrum Access Schemes

Based on the spectrum access technology by the secondary users, spectrum sharing can be classified as

Spectrum Underlay Scheme

In spectrum underlay scheme, both primary and secondary users share a frequency band. The interference temperature model is used by secondary user to minimize the interference to the licensed primary user. The interference temperature may be defined as the interference limit set at primary user's receiver up to which it can tolerate interferences from other users without affecting its quality of communication (Bhattarai, Hong, and Shetty, 2013). Code Division Multiple Access (CDMA), Ultra Wide Band (UWB) technology etc. are generally used by secondary users to exploit the wide range of spectrum. However, due to transmission power constraints, secondary unlicensed users can achieve short-range communication only.

Spectrum Overlay Scheme

The spectrum overlay scheme is also known to as Opportunistic Spectrum Access (OSA). In this scheme, secondary users are allowed to use the licensed spectrum band whenever primary users are not using it. Thus, secondary users can transmit over these unoccupied licensed channels with full transmission power to enhance their channel capacity (Le, Chin, and Chen, 2017).

Hybrid Spectrum Sharing Scheme

Hybrid spectrum sharing scheme allows the secondary users to switch between the underlay and overlay spectrum sharing scheme based on the sensing results. When primary user is detected absent from the licensed band, secondary user can transmit with maximum allowed transmission power in overlay mode (Rebato et. al, 2017). Whereas, on detecting primary user present in a licensed band, secondary user can switch to the underlay mode of communication and scales up its transmission power subjected to the interference constraints.

Device to Device Communication

To accommodate the demand of data intensive applications for future cognitive cities, Device-to-Device (D2D) communication is another communication paradigm for next generation networks. In traditional cellular communication system, all communications are through base station even if both the users are very near to each other (Asadi, Wang, and Mancuso, 2014). On contrary to this, D2D communication allows direct communication between two mobile users without involving base station. It not only reduces the power consumption, but also improves spectral efficiency, energy efficiency and delay. It was firstly implemented for cellular communication networks by Qualcomm's FlashLinQ (Hamaguchi et. al, 2012). The 3GPP is also scrutinizing D2D communications for Proximity Services.

Communication Modes

Based on the spectrum usage, D2D communication is classified as

1. In band D2D communication
   a. Underlay in band D2D communication
   b. Overlay in band D2D communication
2. Out band D2D communication
   a. Centralized Mode
   b. Autonomous Mode

Underlay in band D2D communication share cellular spectrum for D2D communication along with cellular users. Thus, interference to the cellular users is inevitable (Shen and Lok, 2014). To overcome this bottleneck, another mode of D2D communication is Overlay in band D2D communication in which some of the radio resources are reserved for D2D communication only (Khan, Rehmani, and Rachedi, 2016). Keeping in view the spectrum scarcity problem, out band

D2D communication has also gained attention these days. It allows coordination between radio resources through base station (BS) in centralized mode and by users themselves in autonomous mode (Swetha and Murthy, 2017). It may be noted that the user equipment must have two wireless interfaces such as LTE and WiFi simultaneously can use out band D2D communication (Bala, Bhamrah, Singh, 2017). Significant literature is available on in band D2D communication. However, from past few years out band D2D communication has also gained lot of popularity. In Table 3, a brief comparison of both schemes is presented.

D2D communication technology is immature technology and there are several issues that need to be address before making it available for commercial use. 3GPP has already started working on it and proposed it as a viable solution for public safety network. Qualcomm is also working in this direction to use the technology in cellular network for social networking and content sharing. However, there are some bottlenecks to implement D2D cellular networks. For example, the modulation format, CSI acquisition, interference management, throughput power tradeoff, architecture etc.

Table 3. Comparison of different D2D communication schemes

Type	Advantages	Disadvantages
In band D2D	Underlay D2D communication, spatial diversity techniques are used to enhance the spectral efficiency.	Resource wastage cannot be avoided in overlay D2D.
	Any cellular device can use in band D2D communication.	The interference management is very challenging job.
	It fulfills QoS requirements as the cellular spectrum can be fully controlled by the BS.	A user cannot have simultaneous cellular and D2D transmissions.
Out band D2D	There is no interference between cellular and D2D users.	The interference in unlicensed spectrum cannot be controlled by BS.
	Unlike overlay in band D2D communication, no dedicated cellular resources are required.	Cellular devices need two radio interfaces such as LTE and WiFi to use out band D2D communications.
	The resource allocation is easy because the scheduler does not require information about operating frequency, time, and location of the users.	The power management between two wireless interfaces is crucial to control the power consumption of the device.
	Simultaneous D2D and cellular communication is feasible.	The packets encoding and decoding is required as the protocols employed for different radio interfaces are not same.

Cognitive Internet of Things (CIoT)

It is believed that Internet of Things (IoT) is the most efficient and effective technology to realize the concept of cognitive cities in terms of the assisted living, improved learning, industry automation and manufacturing, smart transportation, e-health facilities, real-time industrial process monitoring etc. IoT enabled objects are interconnected by means of wired and wireless technologies (Bala, Bhamrah, and Singh, 2015). The communication between these IoT enabled sensors/ objects and cloud servers is largely dependent on wireless techniques used. Since the available wireless solutions like Bluetooth and Zigbee have limited range and to allocate the fixed bandwidth to the billions of the sensors/ devices in future cognitive cities is very difficult to the existing spectrum scarcity problem. Thus, the paradigm is shifting towards cognitive radio based IoT (Wu, 2014). The concept is also gaining the momentum due to following reasons:

It is expected that IoT devices/objects will increase tremendously in future, it would be very difficult to allocate separate frequency bands to individual device/ service/application due to the cost involved in purchasing license using traditional spectrum assignment policy. Moreover, traditional approaches do not support spectrum sharing among multiple users also. Thus, cognitive radio technology is evolving at a very fast pace as it allows spectrum sharing among users (Koshizuka and Sakamura, 2010).

There are various communication technologies available for IoT from wired to wireless such as cellular communication, Bluetooth, Zigbee etc. The major limitations with these are infrastructure requirements and limited range of operation. Thus, IEEE 802.22 i.e. the wireless regional area network standard is becoming more popular due to its high range and suitability for short to long range applications.

CIoT based devices/ products can tackle the interference issue intelligently by using dynamic spectrum access approach for channel access.

It is anticipated that in future cognitive cities, people will use smart IoT devices without worrying about the huge data storage and information access from servers. Cognitive capabilities embedded in IoT based products will autonomously identify the storage space. To support user mobility along with seamless connectivity is the most anticipated feature of future IoT-based structures. To meet this requirement, the idea of cloud services is already in consideration for cognitive cities where the cloud server will autonomously take care of the data storage spaces by sensing the spectrum intelligently.

Modes of Communication

The modes of communication for CIoT are

1. Intra-Cognitive Communication (INTRACC) mode
2. Inter-Cognitive Communication (INTERCC) mode

In INTRACC mode, all cognitive objects have similar cognitive capabilities whereas in INTERCC mode, cognitive objects have different cognitive capabilities (Khan, Rehmani, and Rachedi, 2017).

Though, CR-based IoT is the most upcoming technology however, there are lots of challenges that still need to address. For example, for CIoT, spectrum sensing is indispensable. Similar to the cognitive radios, IoT objects monitor frequency spectrum in a dynamic environment to know the presence of a active licensed users in a band. Since spectrum sensing needs time and consumes battery energy also, novel energy-efficient algorithms need to design. Moreover, the standardization efforts in IoT are also in their early stages. Thus, the overall benefits can be achieved if standardization activities are carried out jointly around the globe.

Massive MIMO

Massive MIMO is another enabling technology to develop future energy-efficient broadband networks to use frequency spectrum efficiently (Arauio et. al, 2016). It uses large antenna arrays at base stations to serve multiple terminals simultaneously, as illustrated in Figure 4.

The main advantages of the technology are:

- Massive MIMO technology uses large number of antennas that could send more independent data streams simultaneously to serve multiple terminals.

Figure 4. Massive MIMO system

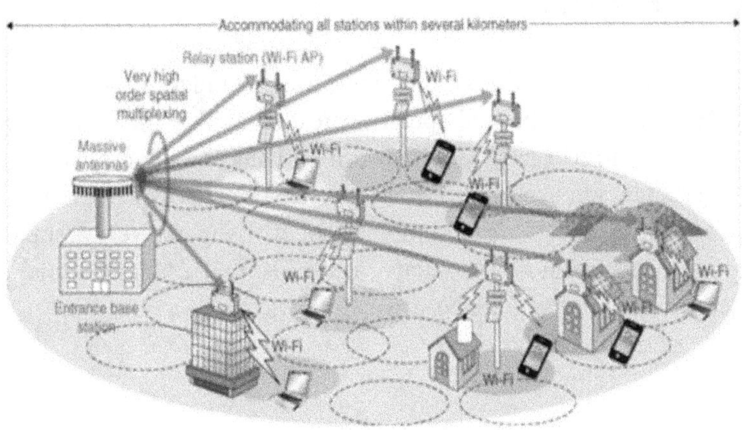

- It uses aggressive spatial multiplexing and coherent superposition of wavefronts to improve system capacity.
- Massive MIMO systems can be designed with components with low-power requirements.
- It enables a significant reduction of latency on the air interface.
- Since, more antennas are deployed; radio signals can propagate over more distinct paths. That makes these systems are highly reliable.
- Massive MIMO systems can focus energy spatially towards terminals. It helps in minimizing interference as the base stations can intentionally stop transmitting in directions where signals could interfere with each other or with other systems.

Modes of Operation

Massive MIMO requires different amounts of CSI at the base station and at the terminals which can be obtained either by estimating received pilot signals, or through feedback from the receiver to the transmitter, or both (Rusek, 2013). Based on this, there are two modes of operations for massive MIMO, these are:

1. **Time-Division Duplexing (TDD) Mode:** In this mode, the base station uses uplink pilot signals to learn about the uplink channel. Since, the channel is reciprocal after learning from uplink channel, base station exploit this information to estimate the downlink channel.
2. **Frequency-Division Duplexing (FDD) Mode:** In this mode, base station sends the pilot signals and terminals learn about the downlink channel from these. The estimated channel state information (CSI) is communicated back to the base station over a control channel (Jiang, Sun, and Chiu, 2016).

Ultra Dense Networks

Basic notion of the cognitive cities is to provide users seamless connectivity anytime, anywhere. However, with limited spectrum resources, it is difficult to enjoy the luxury of full bandwidth by individual subscriber. Thus, the paradigm is also shifting towards the frequency reuse by shrinking the size of cells. Based on this idea, the Ultra-Dense Network (UDN) technology is also evolving at a very fast pace where the number of cells are more than active users (Kamel, Hamouda, and Youssef, 2016). The salient features of UDNs are as follows:

- Every user is surrounded with multiple low power small cells. The coverage area of these cells is few meters only.
- Since, the density of small cells is huge in UDN. There are chances that some of the cell sites may be inactive at a time. Thus, the concept of the Idle mode is introduced in which the inactive cells are turned off to partially or fully avoid interference to other active cells.
- Due to small distance between adjacent cells, adjacent channel interference is unavoidable. Therefore, robust interference mitigation techniques are required to avoid interference among neighboring cells.
- Traditional cellular networks use spectrum reuse concept with certain cluster size whereas in UDN novel frequency reuse concept needs to introduce.
- Since the separation between adjacent cells is very small in UDN, it support Line of Sight (LOS) communication. Thus, the Rician fading channel is most appropriate propagation model for UDN due to the presence of dominant LOS component in the received signal.

Various types of small cells such as macrocells, microcells, femto cells etc. are deployed in UDN to offer provide high rate connections to the subscribers. Thus, interference management is a big challenge in UDNs. To mitigate interference in highly densed network, inter-cell coordination increase signaling overheads manifolds. The interference management amongst interfering cells takes place in the time domain, frequency domain, power domain, space domain, or hybrid domain. For example, in the frequency domain, orthogonal frequency channels can be used by static or dynamic allocation. In time domain, blank sub-frames can be inserted to avoid interference between adjacent frames. The advances in MIMO systems allow spatial interference management. Moreover, the power control is another method to manage inter-cell interference.

Sleep Mode Capability of UDN

The introduction of sleep mode capabilities in dense small cell networks is another method to avoid the interference. In heavy cells deployment scenario, the chances of finding idle cells is always high. Thus, by turning them off, the interference to the neighboring cells can be avoided partially or completely. This is called sleep mode of UDN cell (Xu et. al, 2018). Fig. 5 shows the three different techniques for idle mode operation of small cells.

Three different methods are used to control the sleep mode of UDN cells, namely

Figure 5. Sleep mode techniques in UDNs

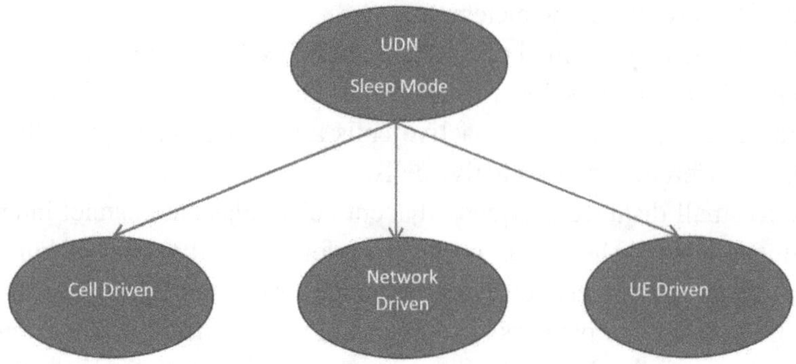

1. Cell driven: In cell driven approach, the small cells use low powered sensors to detect the active users in its coverage area (Zhao and Sadler, 2007). Depending on the presence of active users, cells are activated or deactivated accordingly.
2. Core network driven: In core network driven technique, wake-up control messages sent from the core network to turn on and off the small cell.
3. UE driven: In UE driven approach, wake-up signals are broadcasted by user equipment to activate the small cells in its vicinity.

Challenges in UDN

There are several challenging issues which have not been addressed so far, such as

1. Propagation Modeling: As discussed earlier also, in dense cell deployment scenario, there exists a probability for line of sight communication. Thus, the fading channels can be modeled as Rician fading channel. However, rigorous research is required to consider multi-slope propagation models (Al-Ammouri, Andrews, and Baccelli, 2018).
2. Mobility Management: In highly dense network comprised of small cells, a fast moving user needs multiple handovers. Therefore, user mobility in small cell networks imposes a critical challenge in mobility management. To address these issues, collaborative solutions are required (Demarchou, Psomas, and Krikidis, 2018).
3. Interference Management: Due to highly dense deployment of small cell, UDN supports LOS communication. However, LOS components cause extreme interference between the nearby cells and impose serious concerns to such dense network dispelling the densification gains. The interference management to these highly dense networks is a big challenge and therefore to tackle this

issue novel interference management techniques are required such as, idle mode capabilities of the small cell BSs (Peng, 2016). Another approach is proactive turning off of lightly-loaded cells to reduce interference to the neighboring cells.

4. Backhauling: The backhauling is another bottleneck for the highly dense networks. Many wireless backhauling techniques are available for these networks such as relays links, mmWaves links, and massive-MIMO backhaul links etc. However, the actual performance can be evaluated by knowing the real traffic distributions and user distributions in UDN.
5. Limited Range: In highly dense small cell deployment scenario, it is difficult to expand the range due to adjacent channel interference. Moreover, the future networks will be heterogeneous in nature thus each user may have association with multiple cells using single radio access technology (RAT) or have multiple RAT options within a cell.
6. Small Cell Discovery: Cell detection in the proximity of a given user in a cellular network is very crucial for the optimal operation of a network. However, the efficient detection of a cell is a tedious task as it requires reuse of synchronization signals among neighboring cells, which are in the interference range of each other. Optimum cell discovery in terms of time and energy-efficiency is an open problem in UDN that needs to address.

Challenges for Next Generation Network Deployment

This section discusses the challenges for the deployment for next generation communication networks for cognitive cities.

Interference Management

Due to the extraordinary proliferation of wireless devices, the interference problem among users is unavoidable that causes frequent communication errors. As a matter of fact, it impedes the deployment of sensors, embedded systems and WLANs etc. in cognitive cities. The existing interference models and mechanisms are not adequate to resolve the interference problem completely for cognitive cities due to the large number of connected devices (Khan, Rehmani, and Reisslein, 2016). Thus, novel interference management models and mechanisms are required to monitor and control the interference smartly.

High Energy Consumption

To realize the cognitive cities idea into reality, many advanced communication technologies are available today such as WiMAX and LTE-A etc. These technologies offers tremendous uploading and downloading speed at a cost of high energy consumption. It is anticipated that future cognitive cities, the devices will be battery operated thus new power management schemes/algorithms need to be designed for prolong battery life. The paradigm is also shifting towards the optimum utilization of renewable energy resources to meet the needs of the devices.

Scalable Wireless Solutions

As discussed earlier, it is anticipated that billions of devices will be connected to the Internet in future cognitive cities. Thus, there exists a dire need of scalable wireless solutions. Although, the existing state-of-the-art low powered wireless technologies, such as RFID, ZigBee, Bluetooth and other WPAN technologies are immediate solution available for the cognitive cities. However, their capabilities are limited in terms of a transmission range, throughput, and transmission range as these were designed to offer high throughput for short range communication (Li et. al, 2018). Therefore, significant attention is required for scalable communication technologies to support cognitive city environment.

Interoperability Support among Heterogeneous Wireless Networks

Future cognitive cities will be amalgamation of various wireless network technologies such as WiFi, WiMAX, MANETs and wireless mesh networks. However, the interoperability is a serious issue among these heterogeneous wireless networks. The interworking of the diverse wireless technologies leads to several challenges pertaining to the network architecture, resource management, mobility management, quality of service (QoS) provisioning, and security (Barbu and Fratu, 2011). Thus, considerable attention is required to address these challenges in the future.

Cyber Security Challenges

Cognitive cities are susceptible to the cyber security threats and criminal misuses. The Cognitive city data is generated in vulnerable environments and stored to the backend Cloud. Consequently, data security and storage facilities are indispensable in order to preserve evidence to conduct cyber-crime investigations. Thus, cognitive city must be capable to protect sensitive data/information for malware, unintentional damage and natural disasters (Zhang et. al, 2017).

IEEE Standardization Activities for Cognitive Cities

In order to meet the increased data rate requirement of futuristic cognitive cities, lot of research is happening around the world to cater the needs of smart citizens. To realize the vision of anytime, anywhere connectivity for future cognitive cities, various governing bodies are working on the standardization of the infrastructure and networking essentials to redesign and upgrade the existing communication networks. The detail of such activities is provided in Table 4.

CONCLUSION

The notion of smart city appeared in order to build future sustainable cities that benefit their inhabitants in a context of exponential urban growth, resource scarcity, demographic explosion, and strong environmental constraints. The applications and processes of smart cities make use of massive data and information collection, processing, and sharing. Therefore, the error-free connectivity requirement in smart cities has attracted the attention of the IT industry toward communication technologies. The coexistence of a high number of intelligent devices in cognitive cities has brought several problems in terms of connectivity that can impede the realization of existing communication technologies. In this chapter, efforts have been made from the perspective of communication and networking technologies available for cognitive cities. A discussion on state-of-the-art enabling communication technologies used in cognitive cities is presented to facilitate the reader in comprehending the recent efforts in this direction. Furthermore, several challenges are discussed in context

Table 4. IEEE standardization activities for cognitive cities

S. No.	Standard	Description
1.	IEEE 802 series	Wired and wireless networking
2.	IEEE Std 802.22 series	Wireless regional area networks (WRAN)
3.	IEEE Std 1451 series	Addressing sensors
4.	IEEE Std 1775 series	Powerline communication equipment
5.	IEEE Std 1801 series	Low power chip design
6.	IEEE Std 1900 series	Dynamic spectrum access
7.	IEEE Std 1901 series	Broadband over powerline networks
8.	IEEE Std 1903 series	Next generation system overlay networks
9.	IEEE Std 2302 series	Intercloud and its related testbed
10.	IEEE Std 2700 series	Sensors
11.	IEEE Std 1278 series	Distributed interactive simulation
12.	IEEE Std 1451 series	Smart sensors
13.	IEEE Std 1828 series	Systems with virtual components
14.	IEEE Std 1907.1 series	Real time mobile video communications

with cognitive cities that need open discussion. It is concluded that to make use of the existing communication technologies to provide error-free connectivity in future cognitive cities is not a workable solution as these technologies are designed only for a limited number of devices to support specific range of communication. Therefore, much attention must be paid to enable unimpaired connectivity to the cognitive cities.

REFERENCES

Abdelhak, A., & Asmaa, E. H. (2016). *The Role of Communication Technologies in Building Future Smart Cities. In Smart Cities Technologies* (pp. 55–76). Intech.

Agiwal, M., Roy, A., & Saxena, N. (2016). Next generation 5G wireless networks: A comprehensive survey. *IEEE Communications Surveys and Tutorials*, *18*(3), 1617–1655. doi:10.1109/COMST.2016.2532458

Ahn, J. H., Lee, S. H., & Lee, T. (2016). Anti-Collision Protocol for Coexistence of RFID and NFC P2P Communications. *IEEE Communications Letters*, *20*(11), 2185–2188. doi:10.1109/LCOMM.2016.2602865

AlAmmouri, A., Andrews, J. G., & Baccelli, F. (2018). SINR and Throughput of Dense Cellular Networks With Stretched Exponential Path Loss. *IEEE Transactions on Wireless Communications*, *17*(2), 1147–1160. doi:10.1109/TWC.2017.2776905

Araújo, D. C., Maksymyuk, T., Almeida, A. L. F., Maciel, T., Mota, J. C. M., & Jo, M. (2016). Massive MIMO: Survey and Future Research Topics. *IET Communications*, *10*(15), 1938–1946. doi:10.1049/iet-com.2015.1091

Asadi, A., Wang, Q., & Mancuso, V. (2014). A Survey on Device-to-Device Communication in Cellular Networks. *IEEE Communications Surveys and Tutorials*, *16*(4), 1801–1819. doi:10.1109/COMST.2014.2319555

Bala, I., Bhamrah. M.S., & Singh, G. (2015). Adaptive Power Control Scheme for Cognitive Radio Systems Based on Receiver Sensitivity. Computational Advancement in Communication Circuits and Systems, 69-79.

Bala, I., Bhamrah, M. S., & Singh, G. (2017). Capacity in Fading Environment Based on Soft Sensing Information Under Spectrum Sharing Constraints. *Wireless Networks*, *23*(2), 519–531. doi:10.100711276-015-1172-0

Barbu, O., & Fratu, O. (2011). *An enabler of interoperability in heterogeneous wireless networks. In Proceedings of Wireless Communication, Vehicular Technology, Information Theory and Aerospace & Electronic Systems Technology* (pp. 1–5). Chennai: Wireless VITAE.

Bhattarai, S., Hong, L., & Shetty, S. (2013). Optimum transmit beamforming scheme for underlay cognitive radio networks. In *Proceedings of International Symposium on Wireless Personal Multimedia Communications (WPMC)* (pp. 1-6). Academic Press.

Cisco. (2015). *Visual Networking Index.* White paper, Feb. 2015. Retrieved from www.Cisco.com

Demarchou, E., Psomas, C., & Krikidis, I. (2018). Mobility Management in Ultra-Dense Networks: Handover Skipping Techniques. *IEEE Access: Practical Innovations, Open Solutions*, *6*, 11921–11930. doi:10.1109/ACCESS.2018.2810318

Eu, Z. A., Winston, K., Seah, G., & Tan, H. P. (2008). A study of MAC schemes for wireless sensor networks powered by ambient energy harvesting. *Proceedings of 4th Annual International Conference on Wireless Internet (WICON '08)*, 9.

Feng, D., Lu, L., Yuan-Wu, L., Li, G. Y., Feng, G., & Li, S. (2013). Device to-Device Communications Underlaying Cellular Networks. *IEEE Transactions on Communications*, *61*(8), 3541–3551. doi:10.1109/TCOMM.2013.071013.120787

Fette, B. (2013). Fourteen Years of Cognitive Radio Development. In *Proceedings of International Military Communications Conference* (pp. 1166-1175). Academic Press.

Hamaguchi, K., Ma, Y., Takada, M., Nishijima, T., & Shimura, T. (2012). Telecommunications systems in smart cities. *Hitachi Rev*, *61*(3), 152–158.

ITU-T Focus Group on Smart Sustainable Cities (FG-SSC). (2014). *An Overview of Smart Sustainable Cities and the Role of Information and Communication Technologies.* Focus Group Technical Report.

Jiang, J., Sun, H., & Chiu, W. Y. (2016). *Energy efficient massive MIMO system design for smart grid communications. In Proceedings of IEEE Communications Workshops* (pp. 337–341). ICC.

Kamel, M., Hamouda, W., & Youssef, A. (2016). Ultra-Dense Networks: A Survey. *IEEE Communications Surveys and Tutorials*, *18*(4), 2522–2545. doi:10.1109/COMST.2016.2571730

Khan, A. A., Rehmani, M. H., & Rachedi, A. (2016). When Cognitive Radio Meets the Internet of Things? In *Proceedings of International Wireless Communications and Mobile Computing Conference (IWCMC)* (pp. 469-474). Academic Press. 10.1109/IWCMC.2016.7577103

Khan, A. A., Rehmani, M. H., & Rachedi, A. (2017). Cognitive-Radio-Based Internet of Things: Applications, Architectures, Spectrum Related Functionalities, and Future Research Directions. *IEEE Wireless Communications*, *24*(3), 17–25. doi:10.1109/MWC.2017.1600404

Khan, A. A., Rehmani, M. H., & Reisslein, M. (2016). Cognitive Radio for Smart Grids: Survey of Architectures, Spectrum Sensing Mechanisms, and Networking Protocols. *IEEE Communications Surveys and Tutorials*, *18*(1), 860–898. doi:10.1109/COMST.2015.2481722

Koshizuka, N., & Sakamura, K. (2010). Ubiquitous ID: Standards for Ubiquitous computing and the Internet of Things. *IEEE Pervasive Computing*, *9*(4), 98–101. doi:10.1109/MPRV.2010.87

Le, T. N., Chin, W. L., & Chen, H. H. (2017). Standardization and Security for Smart Grid Communications Based on Cognitive Radio Technologies—A Comprehensive Survey. *IEEE Communications Surveys and Tutorials*, *19*(1), 423–445. doi:10.1109/COMST.2016.2613892

Li, Z., Kang, J., Yu, R., Ye, D., Deng, Q., & Zhang, Y. (2018). Consortium Blockchain for Secure Energy Trading in Industrial Internet of Things. *IEEE Transactions on Industrial Informatics*, *14*(8), 3690–3700.

Liang, W., Ng, S. X., & Hanzo, L. (2017). Cooperative Overlay Spectrum Access in Cognitive Radio Networks. *IEEE Communications Surveys and Tutorials*, *19*(3), 1924–1944. doi:10.1109/COMST.2017.2690866

Miah, M. S., Schukat, M., & Barrett, E. (2018). An enhanced sum rate in the cluster based cognitive radio relay network using the sequential approach for the future Internet of Things. *Human-centric Computing and Information Sciences*, *8*(16), 1–27.

Mitola, J., III. (2000). *Cognitive Radio an Integrated Agent Architecture for Software Defined Radio* (Doctoral dissertation). KTH Royal Institute of Technology, Stockholm, Sweden.

Nawaz, T., Marcenaro, L., & Regazzoni, C. S. (2017). Defense against jamming attacks in wide-band radios using cyclic spectral analysis and compressed sensing. In *Proceedings of Ninth International Conference on Ubiquitous and Future Networks (ICUFN-09)* (pp. 874-879). Academic Press.

Peng, K. H. (2016). Ultra Dense Network: Challenges, Enabling Technologies and New Trends. *China Communications*, *13*(2), 30–40.

Rebato, M., Boccardi, F., Mezzavilla, M., Rangan, S., & Zorzi, M. (2017). Hybrid Spectrum Sharing in mmWave Cellular Networks. *IEEE Transactions on Cognitive Communications and Networking*, *3*(2), 155–168. doi:10.1109/TCCN.2017.2707551

Rusek. (2013). Scaling Up MIMO: Opportunities and Challenges with Very Large Arrays. *IEEE Signal processing Magazine*, *30*, 40–60.

Shen, S., & Lok, T. (2014). Power control for underlay in-band spectrum sharing in device-to-device communications. In Proceedings of IEEE Wireless Communications and Networking Conference (WCNC-14) (pp. 1350-1355). IEEE.

Sridhara, K., Chandra, A., & Tripathi, P. S. M. (2008). Spectrum Challenges and Solutions by Cognitive Radio: An Overview. *Wireless Personal Communications*, *45*(3), 281–291. doi:10.100711277-008-9465-6

Swetha, G. D., & Murthy, G. R. (2017). Selective overlay mode operation for D2D communication in dense 5G cellular networks. In *Proceedings of IEEE Symposium on Computers and Communications (ISCC)* (pp. 704-709). IEEE.

Talari, S., Khah, M. S., Siano, P., Loia, V., Tommasetti, A., & Catalão, J. P. S. (2017). A Review of Smart Cities Based on the Internet of Things Concept. *Energies*, *10*(4), 421. doi:10.3390/en10040421

Theoleyre, F., Watteyne, T., Bianchi, G., & Tuna, V. C., Gungor, & Pang, A.C. (2015). Networking and communications for smart cities. *Computer Communications*, *58*(1-3).

Usman, A., Gebremariam, A., Raza, U., & Granelli, F. (2015). A Software- Defined Device-to- device Communication Architecture for Public Safety Applications in 5G Networks. *IEEE Access: Practical Innovations, Open Solutions*, *3*, 1649–1654. doi:10.1109/ACCESS.2015.2479855

Wang, X., Huang, C., & Zou, Z. (2016). The analysis of energy consumption and greenhouse gas emissions of a large-scale commercial building in Shanghai, China. *Advances in Mechanical Engineering*, *8*(2), 1–8. doi:10.1177/1687814016628395

Wu, Q., Ding, G., Xu, Y., Feng, S., Du, Z., Wang, J., & Long, K. (2014). Cognitive Internet of Things: A New Paradigm Beyond Connection. *Internet of Things.*, *1*(2), 129–143. doi:10.1109/JIOT.2014.2311513

Xu, H., Hematian, Y. A., Griffith, D., & Golmie, N. (2018). *Performance Evaluation of Energy Efficiency with Sleep Mode in Ultra Dense Networks. In Proceedings of Computing, Networking and Communications* (pp. 747–751). Maui, HI: ICNC.

Yaqoob, I., Targio Hashem, I. A., Mehmood, Y., & Gani, A. (2017). Enabling Communication Technologies for Smart Cities. *IEEE Communications Magazine*, *55*(1), 112–120. doi:10.1109/MCOM.2017.1600232CM

Zhang, J. N., Yang, K., Liang, X., Ren, J., & Shen, X. S. (2017). Security and privacy in smart city applications: Challenges and solutions. *IEEE Communications Magazine*, *55*(1), 122–129. doi:10.1109/MCOM.2017.1600267CM

Zhang, K., Ni, J., Yang, K., Liang, X., Ren, J., & Shen, X. S. (2017). Security and Privacy in Smart City Applications: Challenges and Solutions. *IEEE Communications Magazine*, *55*(1), 122–129. doi:10.1109/MCOM.2017.1600267CM

Zhao, Q., & Sadler, B. M. (2007). A Survey of Dynamic Spectrum Access: Signal Processing, Networking and Regulatory Policy. *IEEE Signal Processing Magazine*, *55*(5), 2294–2309.

Chapter 5
Impact of Dust for Solar PV in Indian Scenario:
Experimental Analysis

Rashmi Chawla
J.C. Bose University of Science and Technology YMCA, India

Poonam Singhal
Deenbandhu Chhotu Ram University of Science and Technology, India

Amit K. Garg
Deenbandhu Chhotu Ram University of Science and Technology, India

ABSTRACT

The facile conversion of sunlight into electricity of a solar-energy system is predominantly associated with the sun's irradiance as well as myriad environmental/climatic factors and photoactive material employed. However, the impact of other external factors associated with solar-rich geographical location (India) can be more influential on system performance. Among these, dust is oft-times an overlooked or understated issue that can be a major stumbling block in a solar panel's output performance. This chapter provides an insight of dust's impact on photovoltaic modules and evaluates mitigation of power loss and other performance parameters due to dust accumulation. In addition, this chapter analyzes dust's impact on the real-time data collected for 46 inverters with total 114819.30 kWh productions in a month with an average of 4416.13 kWh/day. The research further reviews key contributions to the understanding and performance effects of dust on solar module and presents an inclusive literature survey/assessment.

DOI: 10.4018/978-1-5225-8085-0.ch005

INTRODUCTION

With the increase in population and thrive for smart city, the self-sustainable energy demand globally has stipulated over 500 million tera-joules/year (reported by International Energy Agency, 2015). India has bagged third position after China and USA in energy consumption. This reveals that the energy consumption in the nation is on a significant rise and as predicted, will grow the fastest among the major economies by 2040 (as illustrated in Figure 1) ("Physical Progress (Achievements)". About 67% of the country's electricity is procured from conventional sources such as fossil fuels, mostly coal. Moreover, India accounts for 11% of global coal consumption making it world's second largest coal consumer (("Achieving 2022 target four years ahed of schedule"), "India hits 20 GW solar capacity milestone", 2018). This demonstrates the massive dependence on non-renewable energy sources such as coal, which is culpable of CO2 emissions. The coal based power plants contribute about 35% to the total all-India CO2 effusions ("India hits 20 GW solar capacity milestone", 2018).

To meet the extensive energy demand and to move towards cognitive cities, there has been a remarkable transition from non-renewable to sustainable energy resources. Amongst the several renewables such as wind, hydro, solar is the most boosted sector with the annual potential of 1,575-49,837 exa-joules ("India hits 20 GW solar capacity milestone", 2018). However, it is good to know that India is steadily switching to renewables, their share in India's installed capacity being 40%. Moreover, 18% of India's energy production is reliant on renewable sources (Pal Chandra, 2018). Earlier, solar energy used to be expensive, but now the solar and wind tariffs are declining. Since 2010, there has been a noteworthy decline in tariffs for solar energy owing to reduction in cost of PV modules, increase in scale of projects etc. In 2017, solar tariff dropped to historic low of Rs.2.44/kWh from Rs15/kWh in 2010 (Economic Times, 2017).

By 2027, it is estimated that the renewable energy capacity would become 275 GW ("Economic Times"). Over a few years, renewable energy could become as affordable as conventional sources. Knowing the fact, that clean power is now cheaper, would widen its use throughout the country, in particular for the 24 crore Indians still without electricity. The total available solar energy in one hour, if harnessed completely, could meet the global energy needs for a year. Being a tropical country, India receives tremendous amount of sunlight throughout the year (illustrated in Fig.2 (a)), 300 days of sunlight contributing to 5000 trillion kWh ("Economic Times" and Bhushan, 2018). Solar Photovoltaics Cells (PVC) provides a means of harnessing sunlight and converting it into usable electricity and is expected to yield maximum outputs. Although solar PV technology is approaching maturity from its first to

Figure 1. World energy outlook 2017, IEA

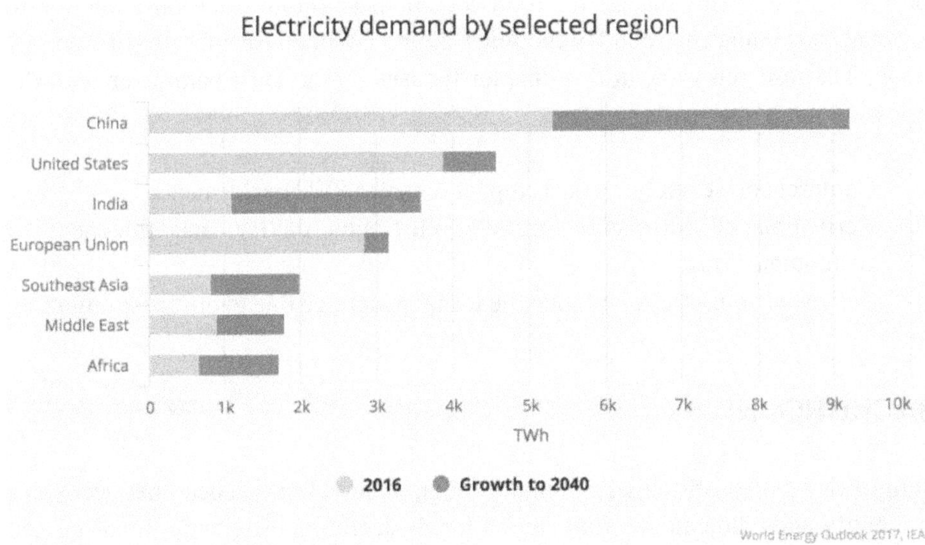

fourth generation, it cannot be denied that solar panels have quite low reported efficiency. Due to perpetual outdoor exposure, many environmental factors hinder panel's capability to yield ideal outputs. One of such imperative factor influencing panel performance is dust. With the passage of time, solar panels develop a thin layer of dust on their surface. Sources like pollution, vehicular movement, rising temperatures, changing weather patterns, nature and many others result in dust generation. Large parts of North India as reported are turned into Dust Bowl owing to depleting groundwater, unsustainable agriculture and deforestation (Nattakarn, 2017). Fig.2 (b) illustrates dust storm in north region on May 14, 2018.

Accounting for the fact that, dust on solar module results in decline in power output of the solar panels, knowledge on its impact and cleaning mechanisms becomes essential. The research is categorized in three major parts: (1) Impact of dust (2) Correlation of dust thickness and efficiency to stipulate cleaning when necessary (3) Cleaning Mechanism. The research work in past (Karim, 2017; Bergin, 2017; Maghami, 2016; Sulaiman, 2014; Ortiz-Conde, 2014; Kumar, 2013; Mani, 2010; Garcia et. al, 2011), shows dust as the least acknowledged, yet the most crucial parameter responsible for deteriorating panel performance. To date, only few research works have exploited correlation between dust thickness and efficiency to manifest right time cleaning of solar PV module. The accurate correlation however is a function of various on field parameters. To find out this value, mostly the method of trial-and-error is used.

As is evident, a smart city needs self-sustainable energy harvesting mechanism for which solar energy has proved to be a copious solution. If a robust solution to correlate dust impact on solar PV module is found it will give a boost to solar energy usage. The research work in this chapter focuses on the Dust parameter with the following main objectives:

- Comprehensive study on dust impact on solar PVC performance.
- Derivation of correlation between efficiency of PV and thickness of accumulated dust.
- Review of solutions to mitigate the losses generated due to dust accumulation.

BACKGROUND

Being aware of the substantial potential of solar photovoltaics, it becomes necessary to identify and eradicate the root causes for its declining efficiency. As discussed above, dust is a culpable factor in hindering ideal performance of PVCs. A multitude of studies evaluating the impact of dust on PV modules have been reported in the past.

Figure 2. (a) Solar map of India depicting annual average global horizontal irradiance (b) The above map depicts aerosols, inclusive of dust, over northern India on May 14, 2018, around the time of the second dust storm

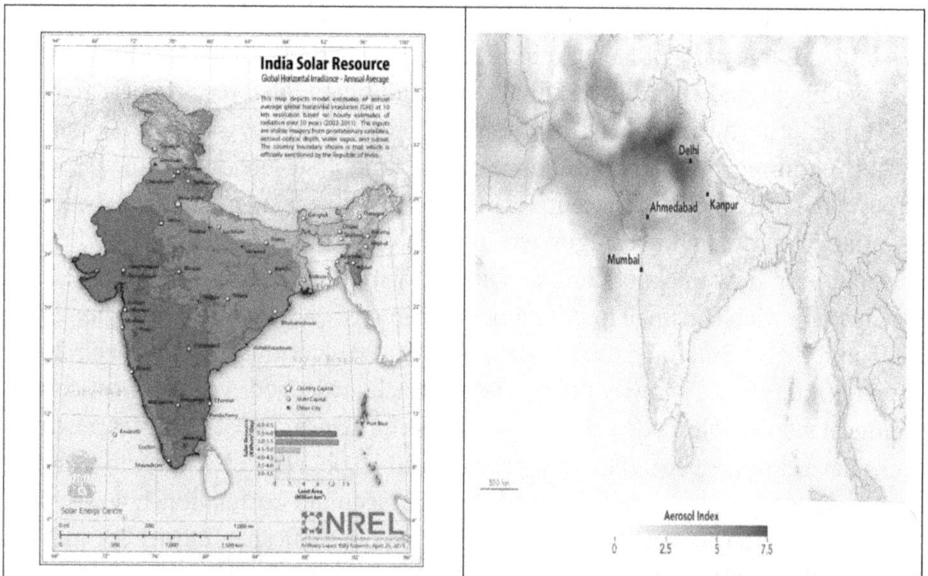

(a) (b)

**For a more accurate representation see the electronic version.*

Influence of dust has been a persistent topic of research since the last few decades. Many reviewed literature on the topic, while others quantified the losses due to dust in many forms. They showed that dust is not only responsible for affecting a single PV parameter, but governs myriad of them. This section provides a qualitative as well as a quantitative understanding of how dust is a significant parameter impacting PV performance. A comprehensive review focusing on the research studies conducted in the last two decades is presented here.

LITERATURE REVIEW: DUST IMPACT ON PHOTOVOLTAIC PERFORMANCE

See Table 1.

ABBREVIATIONS

- Pc-Si: polycrystalline silicon
- mc-Si: monocrystalline silicon
- a-Si: amorphous silicon
- P_MAX: maximum power rating
- V_MAX: maximum voltage rating
- I_MAX:maximum current rating
- V_OC: open-circuit voltage
- I_SC: short-circuit current
- V_MP: voltage at maximum power
- I_MP: current at maximum power
- FF: fill factor
- η: efficiency
- α: temperature coefficient of current
- β: temperature coefficient of voltage
- MPP: maximum power point
- PMPP, MPPT: maximum power point tracking
- T: time
- y:year
- m:month

Table 1.

Author and Region	PV Type	Study Features and Findings	Methodology Adopted	Condition	Noteworthy Deduction	Shortcomings/Suggestions
Hegazy, 2001 Egypt	Glass plates	The reduction in normal transmittance depends essentially on dust deposition along with tilt angle, period of exposure as well as climatic conditions.	Experimental methodology	Different tilt angles	Developed an empirical correlation for calculating the reduction in transmittance of glass plate for a fixed tilt angle after a certain number of days of exposure.	The correlation needs to be generalized for other regions, and desert belts.
Al Hasan and Ghoneim, 2005 Kuwait	Two PV modules	The short circuit current and output power decrease due to dust accumulation. The efficiency of PV module decreased by 33% for each 1 g/ m^2 of dust deposited.	Plotted I-V characteristics curve for dusty and clean module.	Considered sand dust. Tilt angle 30°	I_{sc} decreased less sharply for dust accumulation >1.5 g/ m^2 because at higher amounts of dust, particles assemble on top of each other and therefore do not cover large area.	
Eliminir et al, 2006 Egypt	100 glass samples	Dust deposition density decreased from 15.84 g/m^2 at 0° to 4.48 g/m^2 at 90° and transmittance decreased from 15.84% to 4.48% respectively.	Experimental setup	Different tilt (β) and azimuth angles(γ)		Optimum tilt angle for least dust deposition not specified. Detailed wind direction data not specified.
Mani and Pillai India (2010)	Flat panel and CPV	Evaluated the status of research on influence of dust on PV performance and analyzed dust accumulation mechanisms and mitigation approaches.	Review		Identified challenges for further research.	
Vivar et al. Spain and Australia (2010)	Concentrated PV system	Investigated the effects of soiling, losses could reach up to 26%.	Experimental analysis at different sites	Conducted 3 different tests.	Concentrated PV systems are more sensitive as compared to flat panels, output current losses being 14%.	

continued on following page

Table 1. Continued

Author and Region	PV Type	Study Features and Findings	Methodology Adopted	Condition	Noteworthy Deduction	Shortcomings/Suggestions
Kaldellis et al. Athens, Greece (2011)	Two pairs of PV panels pc-Si (max power of each pair 102 W_p and collectors' area 988 x 448 mm^2).	Deposition of dust particles deteriorated PV performance. Reduction in efficiency of module by natural dust was found to be smallest.	Experimental measurements in urban environment.	Studied the impact of 3 pollutants: limestone, red soil, fly ash.	Theoretical model to predict dust deposition effect. Reduction in PV energy performance strongly depends on particle composition and source.	Laid more attention on artificial dust.
Cabanillas et al. Mexico (2011)	Amorphous and crystalline Si module	Studied the electrical efficiency effects of dust accumulation on PV panels.	Experimental studies	Considered natural dust.	Output power of crystalline module declined by 4-7% whereas that of amorphous declined by 8-13%	The mathematical model developed was aimed to ascertain maximum potential as a function of solar irradiance and temperature. Dust density/thickness must also have been involved in this formulation.
Mohamed et al. Southern Libya (2012)	Four PV modules	2-2.5% performance loss due to dust was recorded.				Environmental conditions during test not reported.
Maghami et al. Southeast Asia (2013)	1 KW solar array Area= 125 × 62.5 mm^2	Showed that tracking flat PV is more appropriate than fixed flat PV.		Tropical climate		Precise results.
Kalagirou et al. Cyprus (2013)	Monocrystalline polycrystalline, amorphous Si module	Investigated impact of soiling on the efficiency of three types of modules. Discovered that amorphous modules were less influences than the rest two.	On-site PV characterization	Considered artificial and natural dust.	Winter rains are apposite to keep module surfaces cleaned. In summers, t PVs require cleaning every 2–3 weeks. Reduction of module performance after a dust event proceeded by heavy rain is approx. 13%.	Quantitative analysis of dust not presented.

continued on following page

Table 1. Continued

Author and Region	PV Type	Study Features and Findings	Methodology Adopted	Condition	Noteworthy Deduction	Shortcomings/Suggestions
Adinoyi et al. Saudi Arabia (2013)	PV system consisted of 4 modules: Mod-1 and 2(mc-Si): 85 W (were used in pair.)	Reported 50% power reduction due to dust. Single dust storm could reduce output by 20%.	Characterization of samples using EDS, SEM	Power output was daily measured.	Cleaning was recommended once every two weeks.	Dust density and properties not specified.
Pavan et al. Italy (2013)	Pc- Si PV module Nominal power= 4.4 kW	Efficiency reduction ranges from 1-5% by dust deposition.	BNN and regression polynomial methods		Clean-up process can be scheduled once the percentage loss of power exceeds an admissible threshold.	Dust deposition mechanisms and their effects on performance of PV require more investigation.
Appels et al. Belgium (2013)	PV modules	Reported a persistent power loss of 3% to 4%	SEM analysis	Suitable tilt angle 35°. Periods of uniform rainfall.	Rain showed a little cleaning outcome on smaller dust particles (2-10 µm), whereas on bigger particles (60 µm) the cleaning effect was evident.	
Zaki Ahmed (2014)	Solar array	Studied effect of type of dust pollutant, and found that six of them namely ash, calcium, limestone, soil, sand and silica have more noteworthy effect on PV				The study mostly concentrated on artificial dust. Hence, there's a need to pay attention on natural dust also.
M. Mani IISc, Bangalore (2014)	PV Module dimensions= 675 x 485 mm^2 Cell dimension	Accumulated dust did not influence open circuit voltage, but impacted short circuit current.	Studied I-V characteristics	Indoor and outdoor tests		Characterization of dust must be clearer.
Sayyah A (2014)	Flat panel PV	PV power loss was found to be the highest for horizontal surface, while least for vertical, because sedimentation is the governing process.	Review		Dust removal was intensified with high tilt angle.	Cannot be assumed that particle removal would be maximized by inclination.
Said, Husam, and Walwil (2014)	PV glass cover was used. Cell type= mc-Si Number of cells=6	Fundamental studies on dust fouling effects. The spectral transmittance reduction was found to be 35% and overall to be 20% with the corresponding dust deposition around 5 g / m^2	characterization of samples by various techniques		Reviewed about dust particle shape and considered it generally spherical. Most of the particles have a diameter<2 µm with presence of few large particles 10 µm which had greater impact in scattering the irradiance.	

continued on following page

Table 1. Continued

Author and Region	PV Type	Study Features and Findings	Methodology Adopted	Condition	Noteworthy Deduction	Shortcomings/Suggestions
Semaoui et al. Algeria (2015)	The PV generator installed in roof of PV laboratory included 31 mc PV modules.	Average transmittance of a glass plate tilted 32° to the horizontal reduced by 8% over a day.	Improved Hottel model of solar irradiation	Desert area, no rainfall		Data regarding dust amount and properties missing.
Klugmann-Radziemska Northern Poland (2015)	Three mc-Si PV modules with nominal power of 70 W, 75 W and 100 W were used.	Exhibited a linear relationship between reduction in conversion efficiency of PV module and thickness of dust layer. The maximum daily efficiency loss was reported to be 0.8%	I-V characterization, SEM analysis.	Natural and artificial dust deposition. Tilt angle 37°	Dust physical properties depend upon location and environment.	Optimum dust thickness resulting in worst CE not specified.
Maghami et al. (2016)	PV module	Classified shading into two types: hard shading and soft shading. Hard shading occurs when solid blocks the incoming sunlight and causes decrease in PV voltage. Smog/dust in environment results in soft shading and affects PV current.	Review		Two interdependent parameters affect characterization of soiling accumulation, dust property and the environment.	Measure of decrease in voltage/ current due to a particular amount of pollutant should have been mentioned.
Hasan et al. (2016)	PV module	Found that PV power loss is linearly proportional to soiling mass.				Not applicable when the surface is highly soiled due to flattening of transmission loss.
Figgis BW Doha, Qatar (2016)	PV module	Dominant dust particle size is 10-30μm. Mean particle size was found to be 18μm and 90% of mass comprised of particles <30μm.	In-situ testing	Tilt angle of module 22°.		Dust particle size significantly depleting panel efficiency should have been investigated.
Al-Shabban (2016)	PV module	Fine sized particles spread and cover more uniformly over module and decrease solar radiation approaching the cells.				Quantitative decrease in solar irradiation should be reviewed.
Mohsin et al. Jordan (2016)	PV panel	Performance ratio of cleaned pane increased by 8% and energy yield by 8.7% as compared to the panel with no cleaning mechanism.	Designing and implementation of automatic cleaning/ cooling system	Cleaned panel's temperature maintained below 30°C using water.		Exposure time not specified.

continued on following page

Table 1. Continued

Author and Region	PV Type	Study Features and Findings	Methodology Adopted	Condition	Noteworthy Deduction	Shortcomings/Suggestions
Lu et al. China (2016)	PV panels installed on building roof	Effects of dust on PV panels were investigated using computational fluid dynamics simulation. Developed a model to estimate PV η reduction ratio by dust pollution. It was concluded that dust deposition rate first showed a rise and thereafter declined with increase of size of dust particle.	CFD simulation		Empirical model can be utilized to estimate cleaning frequency and reduction in PV efficiency due to dust.	Dust amounts on the windward and leeward roof faces were not reported.
Paudyal et al. Kathmandu, Nepal. (2016)	Pc-Si solar module, installed on building roof.	Efficiency of dusty PV module decreased by 29.76% w.r.t cleaned module.	Regression model	Natural dust, with density: 9.6711 g / m^2 Tilt angle:27°	Accumulated dust is mainly present at the bottom of module, possessing high risk of hot spots, which can finally lead to module's permanent damage.	The study was performed in rainless period, hence the results may vary for other periods.
Guan et al. China (2017)	PV module nominal η=14.98% Dimensions=1650 x 990 x 40 mm^3	Analyzed effect of dust deposition coefficient on module temperature, transmittance and power output. Relative transmittance of PV module decreased by 20% in 8 days of natural dust accumulation.	In-situ investigation. Test were performed under no-load as well as load conditions.	Tilt angle 30°	Natural dust was considered for experiments. The dust deposition density not only influenced the transmittance of panel, but also deteriorated the module temperature.	
Figgis et al. Doha, Qatar (2017)	PV module	Dust accumulation on PV takes place through the processes of particle accumulation, rebound and resuspension. Reviewed literature on particle mechanics pertinent to desert conditions.	Review	Desert environment	Explained PV soiling behavior and identified possible strategies for its mitigation.	Suggested that further research should be performed on elevated PV modules (with faster air flow), significance of particle rebound in the soiling process, effect of azimuth orientation of module.
Hammad et al. MENA region (2017)	The system is grid–connected, consisting of 28 pc-Si modules of nominal power of 285 W_P (STP285–24/ V_d) wired in two strings with a nominal system power of 7.98 kW_P.	Proposed models to imitate effects of dust and surrounding temperature on photovoltaic performance. Average efficiency reduction due to dust are 0.768%/day and 0.607%/day using MLR and ANN models respectively.	Artificial Neural Network(ANN) and Multivariate linear regression(MLR) models.		Estimated cleaning frequency as 12-15 days which agreed with other research works in MENA region.	Limited to a specific location only, a generalized approach is needed to procure better results.

continued on following page

Table 1. Continued

Author and Region	PV Type	Study Features and Findings	Methodology Adopted	Condition	Noteworthy Deduction	Shortcomings/Suggestions
Abderrezek and Fathi Algeria (2017)	Mc-Si PV module with PMPP=20W	Investigated the effect of dust on thermal and electrical behavior of solar modules . Efficiency drops by 17.76% and 9.52% in indoor and outdoor tests. Dust type plays a key role in thermal properties of panel.	Experimentation, energy2D software tool	Test were conducted indoor as well as outdoor. Considered several types of dust: ash, salt, soil.	Showed that light transmission varies with dust type. Soil and sh and overheat the glazing of module, while salt reduces the temperature of panel.	Dust thickness corresponding to the lowered efficiency not reviewed.
Gholami et al. Iran (2017)	Green soda lime silica window glasses were used as the samples.	Investigated parameters influencing dust such as tilt, azimuth angle, wind direction. Transmission coefficient dropped by 15-24.83% for different tilt angles.	Experimental setup	Different geographical directions	Correlation between transmission reduction and dust accumulation density. The equation is generalized and applicable to different regions and weather conditions.	A long term experiment should be done in order to obtain more accurate correlation. The study did not provide any information on the dust particles' size and materials, hence further research is needed in this area.
Athar Hussain et al. India (2017)	PV module	Analyzed the characteristics and topography of dust particles. The minimum power value of 3.88 W was noted during accumulation of rice husk.	SEM analysis of dust samples.	Data for dust samples of various weights with change in power loss at several solar irradiation levels of 650,750,850 W/ m^2 was collected.	The effect of dust particles on reduction of power has been assessed by measuring the electrical performance index such as voltage, current and power which rarely appeared in previous research works.	
Picotti et al. (2017)	Concentrated solar power(CSP) Plant	Loss of reflectivity as a result of soiling decreases the whole productivity of the plant by limiting the harvested energy.			Organized previous soiling reviews as 'dust life cycle' which involves generation, deposition, adhesion and removal. The objective was to provide a helping hand to solar field workers to find a balance between dust losses and cleaning expenditures.	
Tanesab et al. Australia and Indonesia (2017)	Location ROTA: PV 1 (a-Si):	Evaluated seasonal effect of dust on PV degradation. Asserted that Deterioration is more influenced by dust as compared to non-dust related parameters.	Experimental methodology, curve scanning, SEM analysis, image processing, EDS analysis, XRD	Various seasons.	More cleaning is necessary for PV mounted at lower latitude and those employed in tropical climate.	

IMPACT OF DUST ON SOLAR PHOTO-VOLTAIC PERFORMANCE

Dust is a natural phenomenon which is triggered by various environmental factors viz. air pollution, wind, rainfall etc. Dust particles impinge and stick onto the surface, obstructing the sunlight reaching the solar cells. Soiling reduces the transmittance of the cover glazing of PV modules. Guan et. al, 2017 revealed that the relative transmittance of PV module declined by 20% in 8 days of natural dust accumulation. The dust deposition indirectly impacts PV module temperature, as it causes a decline in incident radiation and results in increase in the thermal resistance of the PV modules. Accumulated dust causes a reduction in the overall energy produced by the module. Jiang, Lu, and Sun, 2011 reported that the reduction in overall power outcome from 0% to 26% when dust density rose from 0 to 22 g/m^2. Mastekbayeva and Kumar, 2000 conducted experiments in India on PV glass, and showed that power output reduced from 87 to 75% after one month of exposure. Researchers inferred that a short span outdoor subjection of the panels' in the presence of air pollution considerably deteriorates the outcome energy yield of PV panels. For instance, Sarver et al, 2013 reported a decrement of 6.5% in energy production on exposure of 2 months.

The amount of accumulated dust varies with numerous factors such as geographical location, direction of wind, type of panel surface, tilt angle, exposure time, dust properties etc. Desert regions, which have a great potential for PV installation are more prone to dust storms, and experience little rainfall. Hence, the situation is more problematic in these areas. The surface finish has a major role in determining the dust deposition process, such as a rough surface allows more dust to settle. Furthermore, accumulation of dust is dependent on orientation of the panel. A horizontal surface has more tendency to accumulate dust. Mani and Pillai, 2010, in their research work reported that dust related degradation in solar performance is worst in tropical regions where PV arrays are employed with lower tilt angles. Moreover, dust deposition is accountable to the characteristics of dust such as size, shape, physical and chemical properties. Al Shabban et. al in 2016 reviewed that fine sized particles spread and cover more uniformly over module and decrease solar radiation approaching the cells. A thermal study of various types of dust on the modules showed that materials such as ash and soil are more prone to overheat the glazing of module than others (Abderrezek and Fathi, 2017).

The size of dust particles is mostly < 10 μm (diameter) depending on area and environment. In 2014, Sanaz Ghazi, Sayigh, and Kenneth analyzed the pattern of dust distribution in various parts of the world and reported that the Middle East and North Africa have the worst dust accumulation zones in the world. In experiments conducted by Paudyal and Shakya in 2016, it was reported that the efficiency of dusty PV module decreased by 29.76% w.r.t cleaned module. Said, Husam, and Walwil in 2014 found the overall transmittance reduction of PV glass cover to be 20% with the corresponding dust deposition around 5 g / m^2.

The above graph in Fig.3(left) illustrates an exponential decrease in I_{sc} / I_m with respect to the rise in dust deposition density for various particle sizes. This manifests that an increase in dust density results into reduction of short circuit current of PV as compared to maximum current. Here,

I_{sc} =short-circuit current of PV module,

I_m =maximum current of PV module.

The above graph in Fig.3(right) also shows a series of exponentially decreasing curves(corresponding to various dust particle sizes) which indicate an increase in dust deposition density causes a decrease in the output power of dusty panel as compared to clean panel. Here, P= output power delivered by dusty PV module,

P_{clean} =output power of clean module. Solar intensity=196

W / m^2 is assumed constant in both the plots.

Figure 3. Short-circuit current (left) and output power (right) for different particle sizes as a function of dust deposition density (El-Shobokshy and Hussein, 1993)

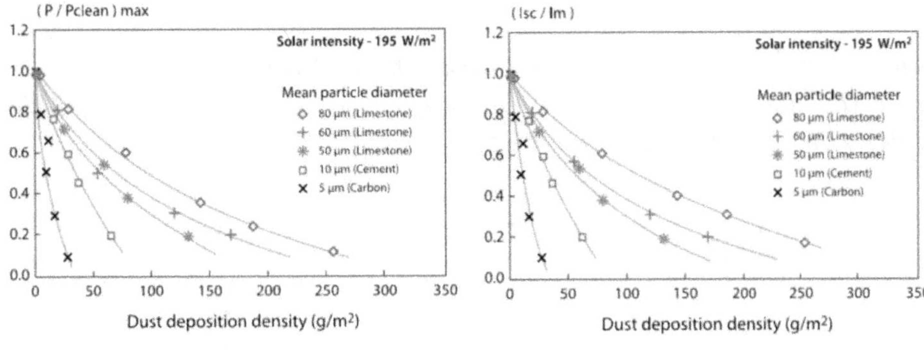

Figure 4. Decrease in solar intensity (left) and fill factor (right) for different particle sizes as a function of dust deposition density (surface density in g/ m^2) (El-Shobokshy and Hussein, 1993)

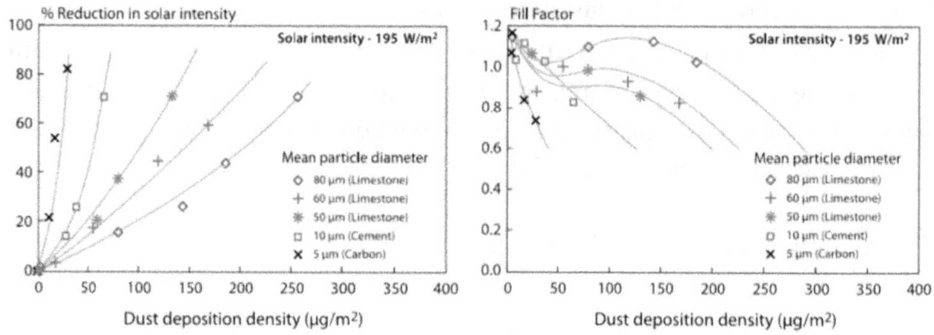

The above graph in Fig.4 (left) illustrates the exponentially increasing curves (corresponding to various particle sizes) between reduction in solar intensity and dust deposition density. This reveals that as dust density on solar panel increases, the solar intensity reaching the panel surface declines. Fig.4 (right) illustrates the inverse relationship between dust density and fill factor of solar panel. Fill factor is defined as the ratio of maximum power of solar panel to its theoretical power. Hence, an increase in dust density on solar panel causes a reduction in its fill factor, or the maximum power delivered by the panel. The dust impact and relevance of India-The Dust bowl can be very well figured out in Fig.5.

The chemical composition of the aerosol Fig. 5(a) changes the overall refractive index of atmosphere abd directly affects how it interacts with solar radiation. The variation in refractive index further determine scattering and absorption of light photons by solar photovoltaic module.

Figure 5. Impact of Dust (a) Aerosol global effect (b) Haze blanket at northern region (India) ("NASA"). (c) Pictorial view of heavy dust storm, on May 2, 2018 in Noida, India (Hindustan Times, 2015).

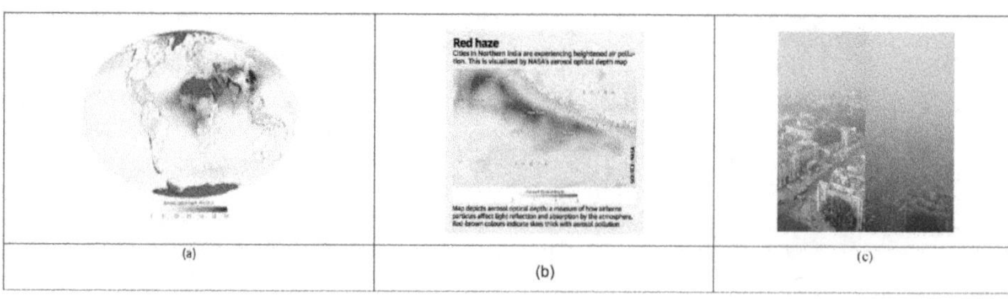

(a) (b) (c)

Correlation Between Thickness of Accumulated Dust and Difference in Efficiencies of PV Module

Experimental data was collected for a period of 25 days in the month of February 2017. Data was collected for 46 inverters, and output energy was recorded in kWh. The first column of table 2 shows average per inverter against each day. Assuming the inverter is 100% efficient, its output can be considered equivalent to PV output. Maximum efficiency of PV was assumed to be 20%. The maximum output was found to be 115 kWh on day-23 and 24.

Input of PV was calculated as:

Efficiency (η) = output delivered by the panel/input to the panel

0.20=115/input

Therefore, Input=575 kWh (assumed constant throughout the month)

The second column of this table shows the efficiency of PV corresponding to each day. The third column was obtained by subtracting η from its maximum value i.e. 20%.

The duration of 25 days was divided into five equal periods, each consisting of 5 days. The second column of table 3 was obtained by taking the mean of Δ η of days in that period (here, 5 days). Experimental values of thicknesses were used here.

CALCULATIONS

The motive is to derive a relationship between Δ η and t. Here, we have analyzed the linear regression process, quadratic regression and symmetrical sigmoidal as shown from Fig.5-Fig.7

This is the final equation which exhibits the correlation between thickness of accumulated dust (t .) and difference in efficiencies ($\Delta\eta$) of PV module. Goodness measure represents and validates various measures calculated by curve fitting tool to quantify how well the data is modelled. Here,

R^2 = 1-(sum of squares of residual)/ ($Y_{fit} - Y_{mean}$. value) ^{2}

R^2 corresponds to value '1' for perfect fit and '0' to bad fit.

aR^2 is the adjusted value of R^2 compensated downward to fit the curve.

F is the F-statistic and is given by ratio of variance by curve fitting model by resual variance.

Table 2. Difference in efficiencies of PV module

Day	Output (in kWh) Average per Inverter	η (%)	Δ η (%) Where, Δ $\eta = \eta_{max-}\eta_i$ (Efficiency Difference Between Clean and Dusty Panel)
1	101	17.5652	2.434783
2	87	15.1304	4.869565
3	92	16	4
4	62	10.7826	9.217391
5	54	9.3913	10.6087
6	84	14.6087	5.391304
7	110	19.1304	0.869565
8	99	17.2174	2.782609
9	100	17.3913	2.608696
10	108	18.7826	1.217391
11	106	18.4348	1.565217
12	87	15.1304	4.869565
13	102	17.7391	2.26087
14	105	18.2609	1.73913
15	104	18.087	1.913043
16	85	14.7826	5.217391
17	66	11.4783	8.521739
18	101	17.5652	2.434783
19	99	17.2174	2.782609
20	93	16.1739	3.826087
21	91	15.8261	4.173913
22	106	18.4348	1.565217
23	115	20	0
24	115	20	0
25	112	19.4783	0.52173

Table 3. Method of least squares for two variables

Period	Δ η (%)	Average of Thickness of Dust Collected t(mm)	t^2.	Δ η*t
Day 1-5	6.52174	0.00085	7.3E-07	0.00557
Day 6-10	3.30435	0.00045	2.06E-07	0.0015
Day 11-15	2.46957	0.00035	1.24E-07	0.00087
Day 16-20	4.52174	0.00055	3.03E-07	0.00249
Day 21-25	2.34783	0.00024	5.76E-08	0.00056
Σ(Total)	Σ Δ η= 19.1652	Σ t . =0.00245	Σ t^2 . =1.42E-06	Σ Δ η*t= 0.01099

SE is the standard error and is defined as root mean square error of the residuals. P - value compares variance attributed to the model with the variance of residual. RMS is root mean square.

Fig.7 employs linear regression equation given by eq (1) and output analysis as evaluated on curve fitting tool by eq (2)

$$y = mx + c \tag{1}$$

$$y = 7343.185*x + 0.2495717 \tag{2}$$

Figure 6. (a) Linear regression applied on data in Table-2 where x axis represents thickness (t) and y axis (Δ η) (b) Represents calculated goodness measure parameter for linear regression

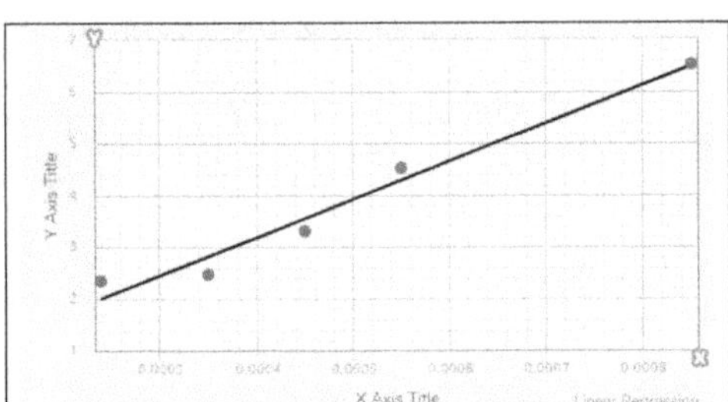

(a)

R^2		0.9707
aR^2		0.9609
P		0.002149
SE		0.3431
F		99.35
Coefficients		
m	7343.185	± 736.7
c	0.2495717	± 0.3909
Equation		
y = 7343.185x + 0.2495717		

(b)

Figure 7. (a) Quadratic regression applied on data in Table-2 where x axis represents thickness (t) and y axis ($\Delta\ \eta$) (b) Represents calculated goodness measure parameter for linear regression

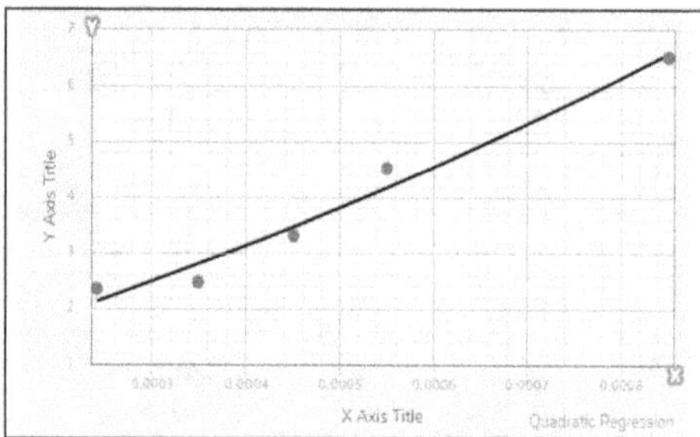

R^2		0.9743
aR^2		0.9486
P		0.02569
SE		0.3934
F		37.93
Coefficients		
a	0.867005	± 1.245
b	4716.827	± 5014
c	2359439	± 4440000
Equation		
$y = 0.867005 + 4716.827x + 2359439x^2$		

(a) (b)

Fig.8 displays quadratic regression equation given by eq (3) and eq (4) evaluates output alysis as by curve fitting tool.

$$y = a + bx + cx^2 \tag{3}$$

$$y = 0.867005 + 4716.827*x + 2359439*x^2 \tag{4}$$

Fig.8 employs symmetrical sigmoidal equation given by eq (5) and output analysis asvaluated on curve fitting tool by eq (6). The comparative analysis in Table 4 and Fig.9 suggest symmetrical sigmoidal correlation as to be the best one in terms of lowest calculated standard error and fitness ratio.

Figure 8. (a) Symmetrical sigmoidal applied on data in Table 3 where x axis represents thickness (t) and y axis ($\Delta\ \eta$) (b) Represents calculated goodness measure parameter for linear regression

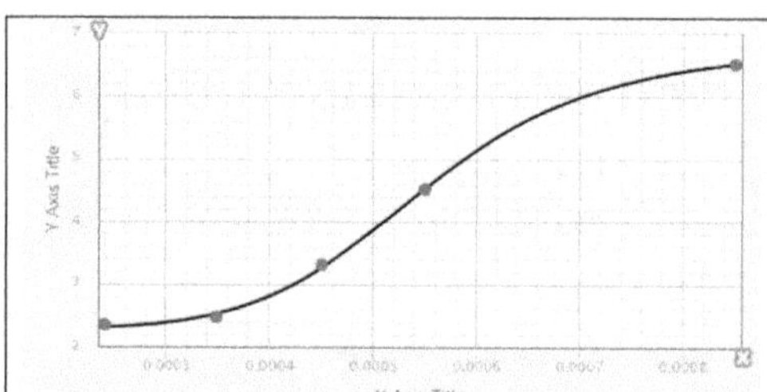

R^2		0.9995
aR^2		0.9974
P		0.03417
SE		0.07909
F		481.3
Coefficients		
a	2.290761	± 0.08045
b	6.377645	± 0.7599
c	0.000550064	± 0.000009488
d	6.782991	± 0.1505
Equation		
$y = 6.782991 + (2.290761 - 6.782991)/(1 + (x/0.000550064)^{6.377645})$		

(a) (b)

$$y = d + \frac{a-d}{1+\left(\frac{x}{c}\right)^b}. \tag{5}$$

y = 6.782991 + (2.290761 - 6.782991)/ (1 + (x/0.000550064)^6.377645) (6)

SOLUTIONS AND RECOMMENDATIONS

The problem of dust accumulation on solar panels affect their performance parameters significantly, hence it is essential to adopt appropriate techniques for mitigation of the losses caused due to dust. Cleaning of solar panel to clear out the dust from panel surface, is one solution to reduce these losses and maintain panel efficiency. The graph in Fig.10 is formulated from the experimental data of 46 inverters and indicates a sudden increase in PV efficiency after cleaning of modules. The next section presents a more elaborative view of PV module cleaning. Figure 10 illustrates the cleaning effect on PV efficiency.

Cleaning Mechanisms

This section lists the noteworthy points regarding panel cleaning mechanisms (as illustrated in Fig.11).

As is evident from the earlier sections, dust on solar panels reduces their performance capability, and therefore is an issue of concern. With the aim to enhance

Table 4. Comparison between the models

Parameter	Linear Regression	Quadratic Regression	Symmetrical Sigmoidal
R^2.	0.9707	0.9743	0.9995
aR^2	0.9609	0.9486	0.9974
P	0.002149	0.02569	0.03417
SE	0.3431	0.3934	0.07909
F/100	0.9935	0.3793	0.4813

Figure 9. Comparison chart for various parameters of the three models

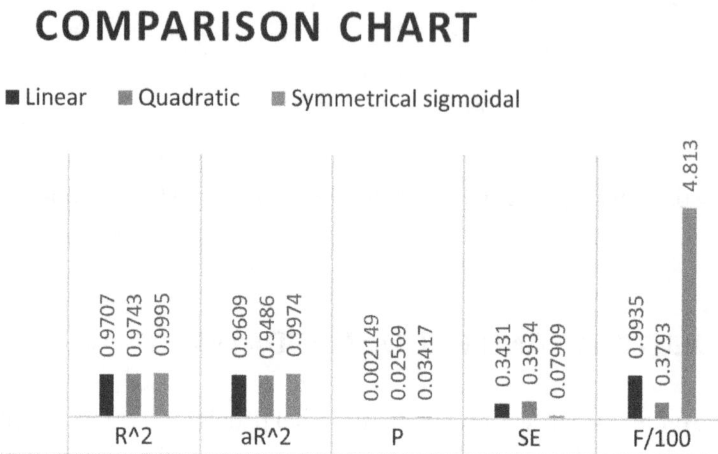

	R^2	aR^2	P	SE	F/100
Linear	0.9707	0.9609	0.002149	0.3431	0.9935
Quadratic	0.9743	0.9486	0.02569	0.3934	0.3793
Symmetrical sigmoidal	0.9995	0.9974	0.03417	0.07909	4.813

panel efficiency, various cleaning methods have been proposed/adopted in the past, as summarized below:

- The mechanical means consists of moving parts and consumes substantial amount of power. Cleaning panel by brush/broom is driven by energy. Air blowing and water spraying are efficient methods, they reduce temperature and avoid excessive heat. However, blowing requires high energy consumption. Cleaning by wiping can result into scratches on the surface of solar panel.
- Cleaning by manual method requires huge manpower and is not a feasible solution for large solar arrays.
- Cleaning by electrostatic method requires electrical power for operation. Electrostatic method removes dusts, mainly lunar dusts outside from electric curtain using electrostatics' standing and travelling wave. Standing wave removes the dust by waving it upward and downward, whereas travelling wave waves dust horizontally.
- Coating is one of the capable, reliable and low cost option. It does not consume any power and no scratches are produced by its usage. There are two types of coatings, namely hydrophilic and hydrophobic. The hydrophilic coatings are made of low surface energy materials and diminish dust through photocatalytic reaction. On the other hand, hydrophobic coatings are those made of high surface energy materials, and clean solar panel by rolling water

Figure 10. Graphical representation of per day PV output where cleaning days are highlighted

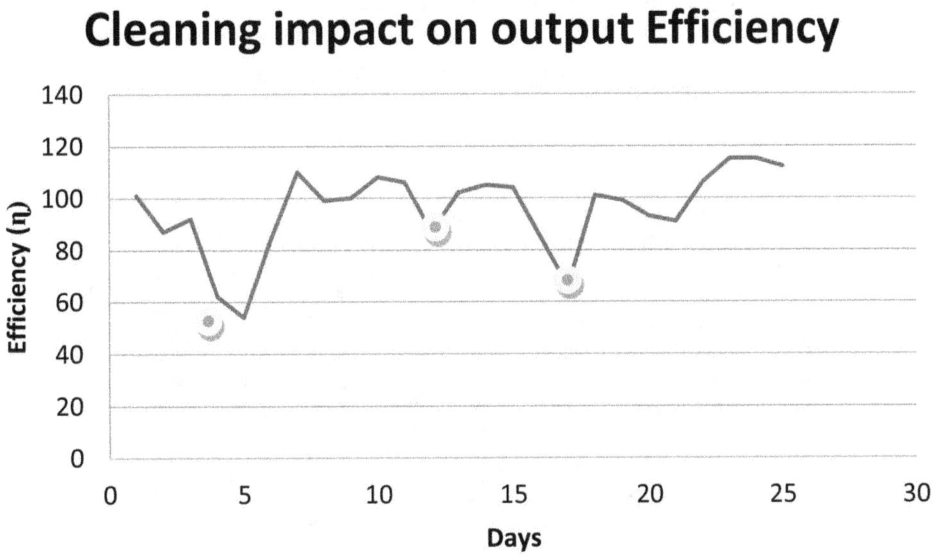

Figure 11. Cleaning techniques incorporated for PV efficiency enhancement

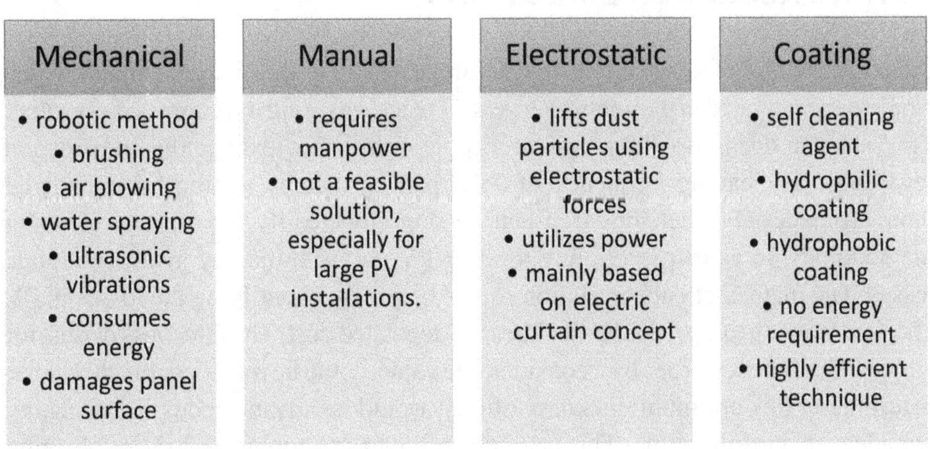

droplets to carry away dirt from surfaces (Syafiq et. al, 2014). For example, 3M anti-soiling coating is a hydrophilic coating which acts as a self-cleaning agent, prevents dust particles from sticking onto the panel surface, hence conserves panel efficiency (as shown in Fig.12).

Figure 12. Monthly average performance of PV modules in Arizona USA showing energy increases for modules coated with hydrophilic 3M anti-soiling coating as compared to uncoated modules (Brown et. al, 2012)

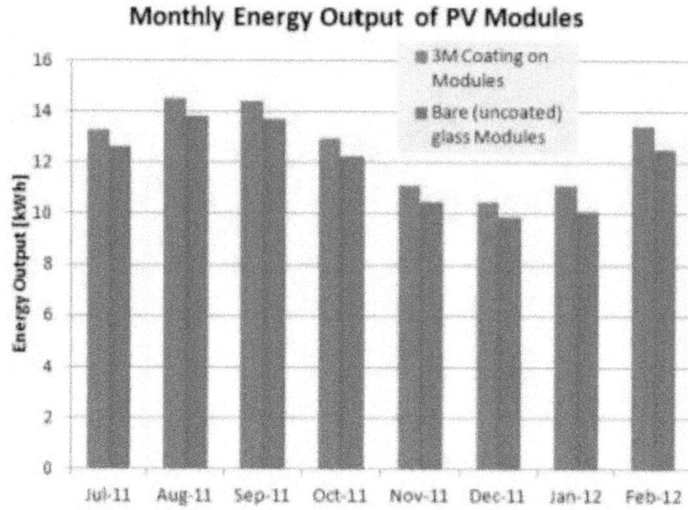

FUTURE RESEARCH DIRECTIONS

Solar energy has a foreseeable bright future, catering to the energy demand which would be accelerated in the upcoming years. Hence, ensuring its progress is essential. This chapter opens up future research opportunities to extract the information regarding the cleaning frequency of solar panels, thereby aiming for enhancing their efficiency. Further, research can be done to identify appropriate cleaning methodology so as to prevent any kind of PV damage, thereby avoid economic losses. The mathematical correlation can be helpful in quantifying the losses in PV efficiency owing to a particular thickness of deposited dust. The thickness of dust for which PV efficiency drops by a considerable value, could be recorded. Furthermore, determining this optimum thickness of dust would be advantageous in intimating the adequate cleaning time. This can be regarded as mature approach for cleaning of PV modules, thereby reducing time and effort, as well as avoiding PV damage.

CONCLUSION

The smart city has become the top priority for any country owing to worldwide socioeconomic and technological developments. In this tech driven data age interest to grow cognitive city has become the basic need owing accommodation of rapid urbanization. Smart city need self-sustainable energy solution. The chapter reviewed influence of dust on PV modules to give a wider knowledge of this critical issue in the world of solar photovoltaics. The inclusive literature review presented in this chapter summarized the noteworthy information regarding dust impact for the last two decades viz. (2000-2017). An in-depth study was done to find the impact of dust in various regions of the world, on various types of PV and appraisal on the current research was provided. It was concluded that dust critically influenced the panel performance parameters and is responsible for deteriorating panel efficiency. In addition, the study revealed that the amount of dust accumulated on module surface is a function of various parameters such as geographical location, wind direction, exposure time, dust properties and characteristics, etc. An attempt was made to derive the correlation between difference in efficiencies of PV module and dust thickness for the experimental data recorded for a month. The results were found to be in best match with the symmetrical sigmoidal model, with the least standard error of 0.07909 using curve fitting tool. The correlation developed can be utilized for intimating the optimum thickness of dust which is responsible for degrading the panel efficiency by a significant value. The thickness, for which difference in efficiency ($\Delta\eta$) will be maximum, could be referred to as the optimum thickness corresponding to lowest efficiency. This can be of immense use for predicting the appropriate time for the cleaning process. Rather than frequent cleaning sessions, cleaning would only be done when that thickness level is reached. In the last section, we suggested solutions to cure the dust problem and discussed numerous cleaning mechanisms to mitigate losses due to dust. Further research in the field of best suitable cleaning method is suggested.

ACKNOWLEDGMENT

We are grateful to Dr. Naresh Thakral, Chef Executive Officer, Renewable Energy for his constant support and guidance throughout and IIT Delhi for resourceful data.

REFERENCES

Abderrezek, & Fathi, M. (2017). Experimental study of the dust effect on photovoltaic panels' energy yield. *Solar Energy*, *142*, 308–320. doi:10.1016/j.solener.2016.12.040

Adinoyi, M. J., & Said, S. A. M. (2013). Effect of dust accumulation on the power outputs of solar photovoltaic modules. *Renewable Energy*, *60*, 633–636. doi:10.1016/j.renene.2013.06.014

Al-Hasan, A. Y. (2016). A new correlation between photovoltaic panel's efficiency and amount of sand dust accumulated on their surface. Academic Press.

Al-Hasan, A. Y., & Ghoneim, A. A. (2005). A new correlation between photovoltaic panel's efficiency and amount of sand dust accumulated on their surface. *International Journal of Sustainable Energy*, *24*(4), 187–197. doi:10.1080/14786450500291834

Al-Shabaan. (2016). *Effects of dust grain size and density on the monocrystalline PV output power* (Vol. 6). Academic Press.

Appels, R., Lefevre, B., Herteleer, B., Goverde, H., Beerten, A., Paesen, R., ... Poortmans, J. (2013). Effect of soiling on photovoltaic modules. *Solar Energy*, *96*, 283–291. doi:10.1016/j.solener.2013.07.017

Bergin, M. H. (2017). *Large Reductions in Solar Energy Production Due to Dust and Particulate Air Pollution*. Academic Press.

Bhushan. (2018). *India's very own Dust Bowl*. Retrieved from https://www.downtoearth.org.in/blog/india-s-very-own-dust-bowl-60913

Brown. (2012). Soiling Test Methods and their Use in Predicting Performance of Photovoltaic Modules with and without 3M Anti-Soiling Coating. Renewable Energy Division, 3M Company.

Cabanillas, R., & Munguía, H. (2011). Dust accumulation effect on efficiency of Si photovoltaic modules. *Journal of Renewable and Sustainable Energy*, *3*(4), 043114. doi:10.1063/1.3622609

Chandra. (2018, March 29). Solar Energy a Path to India's Prosperity. *Journal of Institution of Engineers India Ser. C.*

Darwish, Z. A., Kazem, H. A., Sopian, K., Al-Goul, M. A., & Alawadhi, H. (2015). Effect of dust pollutant type on photovoltaic performance. *Renewable & Sustainable Energy Reviews*, *41*, 735–744. doi:10.1016/j.rser.2014.08.068

El-Shobokshy, & Hussein, F. M. (1993). Degradation of Photovoltaic Cell Performance Due to Dust Deposition on to Its Surface. *Renewable Energy*, *3*(6-7), 585–590. doi:10.1016/0960-1481(93)90064-N

Elminir, H. K., Ghitas, A. E., Hamid, R. H., El-Hussainy, F., Beheary, M. M., & Abdel-Moneim, K. M. (2006). Effect of dust on the transparent cover of solar collectors. *Energy Conversion and Management*, *47*(18-19), 3192–3203. doi:10.1016/j.enconman.2006.02.014

Figgis, B., Ennaoui, A., Ahzi, S., & Rémond, Y. (2017). Review of PV soiling particle mechanics in desert environments. *Renewable & Sustainable Energy Reviews*, *76*, 872–881. doi:10.1016/j.rser.2017.03.100

Figgis, B. W. (2016). Solar test facility in Doha. Academic Press.

Garcia, M., Marroyo, L., Lorenzo, D., & Perez, M. (2011). Soiling and other optical losses in solar-tracking PV plants innavarra. *Progress in Photovoltaics: Research and Applications*, *19*(2), 211–217. doi:10.1002/pip.1004

Garcia, M., Marroyo, L., Lorenzo, D., & Perez, M. (2011). Soiling and other optical losses in solar-tracking PV plants innavarra. *Progress in Photovoltaics: Research and Applications*, *19*(2), 211–217. doi:10.1002/pip.1004

Ghazi, Sayigh, A., & Ip, K. (2014). Dust effect on flat surfaces-a review paper. *Renewable & Sustainable Energy Reviews*, *33*, 742–751. doi:10.1016/j.rser.2014.02.016

Gholami, A., Saboonchi, A., & Alemrajabi, A. A. (2017). Experimental study of factors affecting dust accumulation and their effects on the transmission coefficient of glass for solar applications. *Renewable Energy*, *112*, 466–473. doi:10.1016/j.renene.2017.05.050

Guan, Zhang, Xiao, & Yan. (2017). In-situ investigation of the effect of dust deposition on the performance of polycrystalline silicon photovoltaic modules. *Renew Energy, 101*, 1273–84.

Hammad, Al-Abed, M., Al-Ghandoor, A., Al-Sardeah, A., & Al-Bashir, A. (2018). Modelling and analysis of dust and temperature effects on photovoltaic systems' performance and optimal cleaning frequency: Jordan case study. *Renewable & Sustainable Energy Reviews*, *82*, 2218–2234. doi:10.1016/j.rser.2017.08.070

Hegazy, A. A. (2001). Effect of dust accumulation on solar transmittance through glass covers of plate-type collectors. *Renewable Energy*, *22*(4), 525–540. doi:10.1016/S0960-1481(00)00093-8

Hindustan Times. (2015, November 6). The stubble trouble: Gasping for air as smog reigns. *Hindustan Times*.

Hussain. (2017). An experimental study on effect of dust on power loss in solar photovoltaic module. Renewables: Wind, Water, and Solar, 4(9).

IEA. (2015). *Key world energy statistics*. Retrieved from https://www.iea.org/weo2017

India's solar capacity to cross 20GW in next 15 months: Piyush Goyal. (2017, April 6). *Economic Times*.

Jiang, H., Lu, L., & Sun, K. (2011). Experimental investigation of the impact of airborne dust deposition on the performance of solar photovoltaic (PV) modules. *Atmospheric Environment*, *45*(25), 4299–4304. doi:10.1016/j.atmosenv.2011.04.084

Kaldellis, J., & Kapsali, M. (2011). Simulating the dust effect on the energy performance of photovoltaic generators based on experimental measurements. *Energy*, *36*(8), 5154–5161. doi:10.1016/j.energy.2011.06.018

Kalogirou, Agathokleous, R., & Panayiotou, G. (2013). On-site PV characterization and the effect of soiling on their performance. *Energy*, *51*, 439–446. doi:10.1016/j.energy.2012.12.018

Klugmann, & .... (2015). Degradation of electrical performance of a 326 crystalline photovoltaic module due to dust deposition in northern Poland. *Renewable Energy*, *78*, 418–426. doi:10.1016/j.renene.2015.01.018

Lu, H., Lu, L., & Wang, Y. (2016). Numerical investigation of dust pollution on a solar photovoltaic (PV) system mounted on an isolated building. *Applied Energy*, *180*, 27–36. doi:10.1016/j.apenergy.2016.07.030

Maghami, Hizam, H., Gomes, C., Radzi, M. A., Rezadad, M. I., & Hajighorbani, S. (2016). Power loss due to soiling on solar panel: A review. *Renewable & Sustainable Energy Reviews*, *59*, 1307–1316. doi:10.1016/j.rser.2016.01.044

Maghami, M., Hizam, H., Gomes, C., Hajighorbani, S., & Rezaei, N. (2015). Evaluation of the Southeast Asian Haze on Solar Generation Performance. *PLoS One*, *10*(8), e0135118. doi:10.1371/journal.pone.0135118 PMID:26275303

Maghami, M. R., Hizam, H., Gomes, C., Radzi, M. A., Rezadad, M. I., & Hajighorbani, S. (2016). Power loss due to soiling on solar panel: A review. *Renewable & Sustainable Energy Reviews*, *59*, 1307–1316. doi:10.1016/j.rser.2016.01.044

Mani. (2010). *Impact of dust on solar photovoltaic (PV) performance: Research status, challenges and recommendations*. Academic Press.

Mani, M., & Pillai, R. (2010). Impact of dust on solar photovoltaic (PV) performance: Research status, challenges and recommendations. *Renewable & Sustainable Energy Reviews*, *14*(9), 3124–3131. doi:10.1016/j.rser.2010.07.065

Mastekbayeva, G. A. (2000). Effect of dust on the transmittance of low density polyethylene glazing in a tropical climate. Sol. Energy, 68, 135-141.

Menoufi, K. (2017). Dust Accumulation on the Surface of Photovoltaic Panels. *Introducing the Photovoltaic Soiling Index*, *9*, 963.

Mohamed, & .... (2012). Effect of dust accumulation on performance of photovoltaic solar modules in Sahara environment. *Journal of Basic and Applied Scientific Research*, *2*(11), 11030–11036.

Mohsin, L. (2016). Optimized cleaning and cooling for photovoltaic modules based on the output performance. *Thermal Science*, *145*.

Ortiz-Conde. (2014). *A review of diode and solar cell equivalent circuit model lumped parameter extraction procedures*. Academic Press.

Paudyal, & Shakya, S. R. (2016). Dust accumulation effects on efficiency of solar PV modules for off grid purpose: A case study of Kathmandu. *Solar Energy*, *135*, 103–110. doi:10.1016/j.solener.2016.05.046

Pavan, A. M. (2013). A comparison between BNN and regression polynomial methods for the evaluation of the effect of soiling in large scale photovoltaic plants. *Applied Energy*, *108*, 392–401. doi:10.1016/j.apenergy.2013.03.023

Physical Progress (Achievements). (2018). Ministry of New & Renewable Energy.

Picotti, G. (2017). Soiling of solar collectors– Modelling approaches for airborne dust and its interactions with surfaces. *Renewable & Sustainable Energy Reviews*.

Rao (2014). Influence of dust deposition on photovoltaic panel performance. *Energy Procedia, 54*, 690-700.

Said, S. A., & Walwil, H. M. (2014). Fundamental studies on dust fouling effects on PV module performance. *Solar Energy*, *107*, 328–337. doi:10.1016/j.solener.2014.05.048

Sakarapunthip. (2017). *Effects of dust accumulation and module cleaning on performance ratio of solar rooftop system and solar power plants*. Academic Press.

Sarver, Al-Qaraghuli, A., & Kazmerski, L. L. (2013). A comprehensive review of the impact of dust on the use of solar energy: History, investigations, results, literature, and mitigation approaches. *Renewable & Sustainable Energy Reviews*, *22*, 698–733. doi:10.1016/j.rser.2012.12.065

Sayyah, A., Horenstein, M. N., & Mazumder, M. K. (2014). Energy yield loss caused by dust deposition on photovoltaic panels. *Solar Energy*, *107*, 576–604. doi:10.1016/j.solener.2014.05.030

Semaoui, S., Arab, A. H., Boudjelthia, E. K., Bacha, S., & Zeraia, H. (2015). Dust effect on optical transmittance of photovoltaic module glazing in a desert region. *Energy Procedia*, *74*, 1347–1357. doi:10.1016/j.egypro.2015.07.781

Sulaiman, S. A. (2014). *Influence of Dirt Accumulation on Performance of PV Panels*. Academic Press.

Suresh Kumar, E. (2013). Soiling and Dust Impact on the Efficiency and the Maximum Power Point in the Photovoltaic Modules (Vol. 2). Academic Press.

Syafiq, A., Pandey, A. K., Adzman, N. N., & Rahim, N. A. (2018). Advances in approaches and methods for self-cleaning of solar photovoltaic panels. *Solar Energy*, *162*, 597–619. doi:10.1016/j.solener.2017.12.023

Tanesab, Parlevliet, D., Whale, J., & Urmee, T. (2017). Seasonal effect of dust on the degradation of PV modules performance deployed in different climate areas. *Renewable Energy*, *111*, 105–115. doi:10.1016/j.renene.2017.03.091

Vivar, M., Herrero, R., Antón, I., Martínez-Moreno, F., Moretón, R., Sala, G., ... Smeltink, J. (2010). Effect of soiling in CPV systems. *Solar Energy*, *84*(7), 1327–1335. doi:10.1016/j.solener.2010.03.031

Chapter 6
Innovative Ideas to Manage Urban Traffic Congestion in Cognitive Cities

Subhankar Das
Duy Tan University, Vietnam

Anand Nayyar
Duy Tan University, Vietnam

ABSTRACT

The cognitive city transport plays a vital role in setting the targets of monetary similarity and aggressiveness, social harmony, and reasonable significant development. As per the principle of urban development and versatility, the urban communities and their formative experts, which are dynamic in their considerations, are supporting neighborhood governments in handling portability issues, which shape basic examples over all subjective and dynamic populated urban areas. Interest for traveler and cargo transport is continually expanding in urban communities and towns, adding to movement clog, contamination, and car crashes. Understanding the movement clog is a typical test for governments and transport experts in all these urban formative specialists. As per them, putting the current versatility inclines on a maintainable cutting-edge way requires an aggregate rejig of urban portability.

DOI: 10.4018/978-1-5225-8085-0.ch006

INTRODUCTION

Confined chiefs need to convey modern comprehensive and coordinated transport arrangements that streamline the utilization of all vehicle modes in the urban system for the two travellers and merchandise. Government subsidized research on urban transport critical thinking is coordinated to help nearby chiefs in conveying maintainable and cutting-edge arrangements and offering very powerful condition cordial green arrangements.

Transportation is a main thrust behind advancement what's more, the prosperity surprisingly around the globe. Current life requests developing portability. Often it is anchored through consistently expanding utilization of private autos. The subsequent loads on the vehicle framework, that is as of now intensely extended, are increasing. In spite of major consumptions on new street frameworks, movement blockage keeps on rising. Past additions in street security and natural enhancements are diminishing. These issues can be tackled basically by building more streets or by depending on past methodologies. Creative endeavours are plainly required on a wide front. Imaginative Transportation Method is a transport framework that apply data, correspondences, and the control innovations to move forward the activity of transport systems. Creative Transportation Technique devices depend on three center highlights data, interchanges and combination that encourage administrators and explorers improve and composed choices. The point of this chapter is to make mindful of the individuals with respect to ideas of Innovative Transportation Strategies.

Fundamental level grass root examine is sponsored by illustrative useful activities and displaying activities to quicken different arrangements in understanding rush hour gridlock misfortunes in urban areas and towns.

Existing street foundations are getting to be unequipped for taking care of the fast activity development in metro urban communities. Advance development of foundation in these urban communities is having impediment of land accessibility. Keeping in light of these, it has end up basic to build the proficiency of these street frameworks by making them and vehicle moving over them as insightful. The creative transportation procedure can decrease the deferral at the movement crossing point and henceforth lingering time. This outcomes in aside from explorers' efficient, vehicular discharge decrease. The street movement limit can be expanded numerous folds and voyagers can be guided fittingly to achieve their goals. In the present examination, the advantages of Innovative Transportation Technique in handling the urban movement have been condensed. Additionally, the outline work to utilize this developing innovation on urban streets in India is depicted in short.

Programmes like Transport Research and Innovation Portal (TRIP) & New and Innovative Concepts for Helping European Transport Sustainability (NICHES) are the flag bearer ones in this fields. They bring out the very fundamental policies acceptable in all aspect for the cognitive cities and their urban development authorities. They are not only the bench marking programmes but also very feasible and practically implacable projects.

This section attempts to feature the commitment of research and advancement to the incorporated and practical urban transport frameworks that help financial thriving alongside social improvement for urban zones. Its general point is to team up the coordination of research exercises for scholarly establishments, industry, transport administrators and specialists in regards to key urban transport advancements that need wide arrangement. In addition to this, this part needs to help urban transport leaders and specialists alongside approach agents to discover advancements that could be connected in their urban areas, by giving some detail improvements of the most encouraging imaginative urban transport ideas around world. It incorporates the aftereffects of the down to earth inquire about did, i.e. a nitty gritty portrayal of the NICHES Innovative Urban Transport Concepts. These are creative urban transport estimates that have turned out to be fruitful in (for the most part) European urban communities. Specialties has investigated twelve creative ideas with the end goal to empower their take-up in various urban settings.

Also, this chapter will try to provide a snippet of all the new futuristic innovations in making the transport in urban areas a success around the world. It will give certain prototype ideas for the transport problems around all cognitive cities.

Mission of the Chapter

The chapter aims in providing all the basic fundamental innovations around the world in urban traffic control in all possible way. This will highlight all the futuristic prototype innovations which are either tested or on verge of implementation.

Concerns of the Chapter

The most important concern is to provide a research platform for all the future candidates & policy makers to get maximum reach in urban traffic control in mostly developed countries and to take a clue from it in implementing it in third world countries.

BACKGROUND

In 1990s, as the computerized transformation accumulated pace, some social reporters guessed that it would prompt the passing of the city. Individuals' land area would turn out to be less vital, the contention went, as they came to associate for the most part in the internet. Two decades on, the inverse has happened: human creatures keep on living especially in the physical domain, also, early this century passing the defining moment of something beyond individuals living in urban than rustic regions. The UN predicts that by 2050 the world's urban populace will be as large as the world's aggregate populace in 2002. However, what will the urban areas of tomorrow resemble? Individuals keep on being attracted to urban areas by the financial, social and imaginative open doors they offer; substantial urban communities are more gainful than country territories, delivering more licenses furthermore, yielding higher profits for capital. McKinsey estimates that the world's best 100 urban areas will represent 35% of worldwide GDP development amongst now and 2025. Be that as it may, urbanization likewise shows significant difficulties. The world's quickest developing urban communities have seen issues changing to development and industrialization, stifling under the weight of contamination, clog and urban destitution. In the created world, urban sprawl can prompt individual levels of asset utilization far surpassing those in the creating scene. Urban settings amplify worldwide dangers, for example, atmosphere change, water and nourishment security and asset deficiencies, yet in addition give a system to tending to them. On the off chance that the fate of urban areas can't be one of unsustainable extension, it ought to rather be one of vigorous development. This chapter emphasizes on how urban communities are making creative answers for an assortment of issues. A significant number of these arrangements are versatile, replicable and can be adjusted to an assortment of particular urban conditions. Some are conceivable just because of new advancements while others apply innovation to thoughts that are as old as the city itself. This is explained in below points as discussed in the abstract.

Additionally, this chapter addresses the challenges of urban mobility and their possible solutions.

Hindrances in Cognitive City Transport

Expanding interest for traveller and cargo transport in urban regions is a test for governments and transport specialists. Clog in urban communities and towns is adding to air contamination and to auto collisions and is hampering availability. Additionally, the hindering effects on the urban condition itself, blockage unfavourably influences monetary aggressiveness social union and feasible development.

Activity for Daily Traffic Problems

Urban advancement and high dependence on the auto and truck have incited stop up in urban territories and towns. Blockage and the running with issues of road security and biological defilement show genuine troubles in preparing for improved accessibility and diminishing the characteristic impact of transport. The need is to make urban transport structures that solves particular issues for compactness and the money related and social enthusiasm for fast and capable improvement of items and people, safely and cost-enough. A fused portion of capable urban flexibility is restricting natural and prosperity impacts, particularly in thickly populated domains.

Dynamic Culture and Mobility

Breaking the cycle of growing urban stop up and running with effects on economy, society and the earth requires a modification in attitude by the two boss and transport customers. Another culture in urban adaptability is required remembering the true objective to pass on composed and viable transport masterminding, and customers need to modify their manners and direct as for convenience.

Great transport masterminding incorporates more than creating and expanding transport establishment. It requires a framework in perspective of the guidelines of temperate change, with perfect usage of all vehicle modes in the adjacent framework. A joined method to manage urban adaptability requires an overall vision on urban transport game plan and organizing reinforced by fitting essential administration instruments and frameworks.

Societal and Environmental Changes

With developing auto reliance, a maturing population, and requests for new and adaptable ways of life, urban communities are looked with finding new transport answers for quick and simple development of individuals. Moreover, new advancements make elevated standards for savvy versatility choices, for example, constant activity data for explorers, drivers, armada administrators and system supervisors. New advancements furthermore bring open entryways for organizing data for movement masterminding and electronic ticketing, and sharp cards to energize interoperability between open transport modes.

The troubles of an all-inclusive temperature support, uncommon imperativeness sources and extending essentialness costs are on the most noteworthy purpose of European, national and adjacent methodology designs. In this remarkable condition, green game plans are relied upon to diminish the environmental impact of transport in urban zones. A critical concern is to find ways and means to sensibly lessen transport spreads in light of the fact that urban action is accountable for 40% of CO2 surges and 70% of various releases from road transport.

CONTEMPORARY RESEARCH ADVANCEMENT ON CITY TRANSPORT

The key noteworthiness of urban transport in achieving the general money related, social and regular target. Approach reports and critical financing is obliged research and advancement improvement for urban transport systems. Urban transport is by and by course of action plan with dedicated sponsoring undertakings to encourage urban zones and towns move towards a sensible and facilitated approach to manage urban transport.

Research is consistent for the bleeding edge urban vehicle with cut down radiations and higher imperativeness capability. Inventive approaches to manage city coordination are being created that wire new way of thoughts and reliable flexibility organizations. Urban domains are getting the opportunity to be 'labs for mechanical and various levelled progression in transport.

COORDINATED AND VIABLE TRANSPORT PLANNING

The methodology objective of diminishing dependence on the private auto is urgent in achieving cleaner and more successful transport in urban domains. The structure conditions have been made and fiscal spurring powers provided for help close-by specialists in making Sustainable Urban Mobility Plans. At adjacent level, sensible compactness requires a whole deal vision together with strong political will, and the assistance of creative courses of action in perspective of sound research. Consistently, financed inquire about has added to broadening the extent of measures that oversee transport ask for while upgrading elective and sensible suitable offers, particularly with no attempt at being subtle transport.

Down to earth transport masterminding requires a consolidated method to manage regulate transport ask for and to give benefits that redesign the general urban structure and condition. In like manner, a fused heap of measures ought to be made that consolidates thought of the normal and social impacts of transport, and that impacts convincing usage of open to space. Research has incited the progression of mechanical assemblies, models and field shows to help boss in grasping facilitated frameworks and creative courses of action, understanding the social and environmental consequences of methodology decisions and furthermore the costs and points of interest.

Finally, viable transport orchestrating is an issue for stunned organization and requires helpful energies among commonplace and national associations. Moreover, early enthusiasm of accomplices in close-by essential authority adds to more broad affirmation of new adaptability measures. Diverse research exercises have made imaginative ways to deal with develop made trade between system makers, accomplices and regular society.

Urban Traffic Management

Expanding transport request is making a noteworthy test in rush hour gridlock administration in urban zones all through Europe. Chiefs have available to them an extensive variety of innovation arrangements that have risen up out of late innovative work, particularly in Intelligent Transport Systems (ITS). These frameworks are presently being utilized to enhance utilization of street foundation and to oversee urban activity streams by adjusting street use by private autos, open transport and cargo vehicles, streamlining vitality utilization, and decreasing blockage and transport discharges.

Best Qualitative Transportation Availability

Open transport is profoundly productive regarding space per explorer and for versatility in thickly populated zones. One transport can convey the same number of individuals as roughly 30 autos, while possessing the street space of only three. Along these lines, moderate and family inviting open transport arrangements are basic in decreasing auto reliance. Financed inquire about on open transport goes for creating and presenting more proficient vehicles and foundation, and coordinating administrations into the urban transport organize. Coordinated to invigorating utilization of open transport, explore has been completed on foundation plan, (for example, availability of stations), operational and support frameworks, and on the general population picture of the general population transport administrations. Research on multi-purpose open transport is coordinated to consistent exchange starting with one mode then onto the next by, for example, shrewd ticketing frameworks and travel organizers. Other imaginative arrangements incorporate consolidated open transport, and auto and bike sharing administrations.

New advances are being produced to enhance open transport utilize, for example, PRT (Personal Rapid Transit), BRT (Bus Rapid Transit) and GRT (Group Rapid Transit), and new administrations, for example, request responsive open transport. These advances and administrations are being tried in the CIVITAS Initiative, in the CityMobil ventures and in the Intelligent Energy-Europe (IEE) program.

Sustainability Mobility Concepts

Research has demonstrated that feasible portability ideas incorporate arrangements that animate reasonable auto utilize, (for example, auto sharing and auto pooling) and utilization of adaptable open transport administrations, for example, transports on request.

MOBILITY NEEDS OF URBAN CITIZENS

To be a viable other option to the auto, open urban transport should be effectively available, solid, sheltered and secure. In accomplishing these goals, the extraordinary needs of client bunches in various urban communities should be considered. European approach and research are devoted to creating imaginative answers for pull in more travellers to aggregate methods of transport.

Increase in Accessibility for Transportation in Urban Areas

Basic access to capable transport organizations – regardless of whether road, rail or waterborne – is essential to the individual fulfilment of people in urban areas. Executing technique to improve accessibility requires innovative responses for clear fiscal and social preventions to movability and to upgrade access to reliable organizations in all legitimate transport modes. Procedure revolves around upgrading the movability of the weakened, the elderly, people running with little youths, and people running with overpowering things. This fuses upgrading access to metro stations, link auto stops and transport stops. EU-financed explore is attempting new progressions and thoughts to make open transport workplaces more accessible for people with diminished flexibility.

Trustworthy Accountable Information

An essential segment in making open transport more accessible and along these lines an all the more engaging other alternative to the auto is palatable, strong and forward information. All explorers – laborers, adjacent tenants, visitors – require steady information to settle on choices on when and how-to development, for instance, plans, travel choices, presentation guides, interfacing organizations, and alerts to delays. In Europe, investigation is being coordinated to make customer organized information advances and perfect channels to address the issues of different customer social events. One such measure being attempted in European urban territories is altered journey outlines. Travel information center's in a couple of urban networks are by and by outfitting locals and visitors with progressing travel information and transport orchestrating organizations both very close and by methods for the web. Research adventures have focused on working up normal measures to ensure that information stages are straightforwardly open through various types of PDAs. This engages successful pooling of information, which can be gotten to by transport chairmen, establishment chiefs, information center's, and customers beforehand and in the midst of a journey. EU-financed investigate is moreover making transport information systems to meet the specific needs of the disabled.

Safety of City Transportation

In the course of the most recent ten years, transport has turned out to be more secure on account of enhanced wellbeing principles and advances, and better transport administration. By and by, security in urban zones is influenced by a blend of variables including expanding transport request, incorporation of transport into local locations and expanding lack of room in downtown areas.

INNOVATIVE AND GREEN URBAN TRANSPORT SOLUTIONS

Greener urban transport includes to diminishing transport's impact natural change, the earth and human prosperity. While these issues occur at close-by level, their belongings connect past city limits. In like manner, greening urban transport requires a total effort at all levels - close-by, common and national independently. Setting a methodology framework at national level for the market take-up of cleaner vehicles. also, elective forces reinforce urban transport course of action while with respect to adjacent abilities and commitments.

New Vehicles and Fuel Technologies

New advancements are being made to enable a useful move from fossil-made a beeline for decarbonized transport. Promising decisions are directly being investigated to set up them closer available to be purchased to general society entry.

- Electricity, hydrogen, and biofuels

Carmakers like Tesla, Mahindra are thinking of autos worked by sun-oriented power and battery for which transportation will be sans contamination. Hydrogen fuelled inflatables can be attempted in future as well. Biofuels are additionally a source which will decrease the reliance on biofuels and therefore increment the odds of securing our condition alongside taking care of movement issues.

- Natural gas as Compressed Natural Gas (CNG)

CNG is combustible gas fuel that has been compacted to under 1% of its volume (at standard barometrical weight), making typically unscented, inauspicious and vaporous. CNG is temperate to convey and store making it ideal for different naval force sizes: dismiss trucks, transports, transports, taxis and considerable trucks.

- Liquefied Natural Gas (LNG), or Gas-To-Liquid (GTL)

LNG (Liquefied Natural Gas) is flammable gas that has been melted for transport and capacity. Methane has a low thickness and is in this manner exorbitant to transport and store. High weight gas pipelines can be utilized to transport gas ashore or for short sea intersections.

- Liquefied Petroleum Gas (LPG).

Despite elective fuel advancements, the change to cleaner transport in urban regions is empowered by shorter travel partitions, affecting lower to expand vehicles achievable choices, for instance, cars with electric driving force, electric bikes and individual transporters.

Innovation in Urban Logistics and Transportation

Streamlining freight transport coordination is central to reducing urban blockage and vehicle releases. Regardless, to ensure capable 'last-mile' movement, the entire load coordination stay ought to be progressed, including joins between long-independent, among urban and urban transport. The goal is to diminish 'last mile movement' to the briefest possible course since transport in urban objectives has been gave off an impression of being the most inefficient bit of the ascertained chain.

Solutions on Market Price

As opposed to the focal points, the expense of the natural impacts of transport is generally not borne by the vehicle customer. Procedure intervention is relied upon to ensure that customers consider these outside costs in settling on transport decisions. The Action Plan on Urban Mobility progresses mask of external costs in the urban setting. Transport White Paper. Research is being done on the ampleness and efficiency of esteeming courses of action in light of the customer pays run the show. Concentrates consolidate use issues, for instance, open value, social outcomes, cost recovery, availability of ITS gadgets, and how urban assessing methodologies can be joined with other green zone game-plans. Well beyond in the customer pays standard is the coordination of biological impacts into securing decisions with the change of principles and order on Green Public Procurement (GPP).

Essentialness Efficient Road Transport Vehicles requires that imperativeness use and natural impacts of a vehicle over its life cycle are melded in purchase decisions. The order covers vehicles purchased by open and private substances giving open transport organizations. Such necessities are required to update exhibit take-up of earth pleasant vehicles, while ask about is studying the impact of this new open securing plot on transport structure improvement.

INVESTIGATIVE RESEARCH GUIDELINES AND OVERVIEW

It perceives its part in supporting urban transport specialists to address the perplexing challenges exhibited by the growing enthusiasm for voyager and payload transport. A procedure structure to encourage urban networks and towns in Europe to move to more viable and facilitated urban flexibility is displayed in the Green Paper Towards a New Culture of Urban Mobility, the Action Plan on Urban Mobility and the White Paper on Transport. The game plan hopes to confine the consequences of obstruct and sullying in the urban condition, and to construct desires for regular solaces, budgetary force, and social connection. The realistic execution of the plans is in the hands of neighbourhood governments for acclimation to the adjacent setting.

Research and Innovation are crucial in making and completing measures to make more viable urban versatility. The examination has added to upgrading urban transport orchestrating and development organization systems, making open transport additionally appealing and accessible to all customers and developing more achievable vehicles.

UPDATED EASY CITY TRANSPORTATION SERVICES

From advancing the cost or accommodation of an adventure to hailing cabs and ride-sharing, innovation is getting to be vital to day by day travel. The capacity to get to a 'pay per ride' travel with one tap on a telephone application has upset worldwide portability. 22% of individuals have embraced shared administrations as their most prevalent other option to having their own auto. Of these, auto clubs like ZipCar are overwhelming urban communities; 13% of urban natives currently consistently ride-share and up to half of recent college grads in urban areas like London utilize e-hailing administrations, for example, Uber and Gett in any event once per month. The appearance of new innovation has opened up more individuals' personalities to various portability arrangements. In Europe, for instance, the utilization of carpooling extended from 22% to 29% somewhere in the range of 2013 and 2016, while proficient auto sharing expanded from 4% to 7%. Auto makers and portability players need to get ready for this to keep on top of things. Generally, car brands have rushed to react to this pattern, with any semblance of GM, VW and Toyota all putting resources into ride-hailing applications. We've additionally observed a few brands, including BMW, Ford and Renault, put millions in auto sharing ventures. By 2025, eager interests in associated mass travel frameworks and non-mechanized arrangements will expand the urban versatility go and enhance the movement encounter. Car brands should offer consistent administrations in urban regions to meet forceful administrative approaches and interest for versatility sharing.

Urban Lift-Sharing Facility (ULSF)

Urban Lift-sharing Services are organizing organizations that join people going a comparable route, expecting to ask individuals to share private vehicles for particular journeys. Creative formal lift-sharing designs use incline setting developments like organizing programming, the web and optional call networks for excursion planning, and are accessible to all. Such designs can take the old idea of lift-sharing to another level and to all the more promptly accomplish the fundamental least measure of customers to impact them to work. Urban Lift-sharing Services have accomplished a create status and end up being reliable. They got positive notoriety, for example as response to rising gas costs. Regardless, the possibility of Urban Lift-sharing Services is up 'til now not by and large spread across over Europe, notwithstanding the way that masters see no tremendous obstacles to trade the arrangement to various countries. The progression and execution of Lift-sharing Services has tolerably low costs and segment blocks.

Advantages

- Helps clients to accomplish impressive cost funds on gas and stopping.
- Increases versatility decisions and availability in zones that are not all around served by open transport, with moderately low start-up and running expenses;
- Has the possibility to decrease the quantity of parking spots at taking an interest organizations;
- Reduces the requirement for a private auto;

- Has the possibility to decrease clog, vitality utilization, air contamination and CO2 discharges;
- Can enhance the personal satisfaction through less driving pressure and enhanced amiability;
- Improves work-life balance for lift sharers that will probably leave the workplace on time.

Public Bicycles Service (PBS)

Public Bicycles are imaginative plans of rental or free bicycles in urban areas. They differentiate from customary, generally entertainment arranged bicycle rental organizations as they give brisk and straightforward access and can be used for step by step flexibility with one-way use being possible. Open Bicycles can be seen as a part of general society transport system and offer the customer an exceedingly versatile travel decision for inner urban excursions. Well working Public Bicycle designs exist in different setups and the thinking is getting up to speed in various countries. Truth be told direct yet robbery fragile plans have quite recently been executed since the late sixties. The present current thoughts are significantly more mind boggling and have diversified in progressive arrangement, plans of activity, and the advancement associated towards "smart bikes" (rental process through splendid card or wireless). The transferability of Public Bicycle wants to urban zones with legitimate structure conditions has been shown all things considered.

Advantages

- It gives a quick, advantageous and adaptable inward urban transport alternative;
- It can be an "entryway opener" to expand the acknowledgment of cycling as a profitable urban transport mode in urban areas that still do not have a decent level of bike utilize;
- It additionally bodes well in urban communities that have a decent level of cycling as it adds a significant component to existing portability administrations;
- It builds portability decisions, with low costs contrasted with other open transport measures;
- It empowers multi-purpose going regarding open transport;
- It reinforces the nearby way of life as it turns out to be a piece of the urban scene.

Call-a-Bus Services (CBS)

Call-a-transport Services ask for responsive open transport plans that modify their calendar and timetable to suit a particular transport ask. Customers need to call and request a stretch out beyond time. Claims to fame bases on picked innovative plans that give (nearly) approach to-portal benefits, that have midway or totally flexible motivation and that usage creative progressive measures to give their help of the all-inclusive community. Call-a-vehicle Services are particularly sensible for medium to low thickness zones. This consolidates country zones and what's more neighbourhoods and their surroundings in common zones, where traditional mass travel can't satisfy the prerequisites of the voyagers in an expense efficient way. They can in like manner be a recommendable option amidst low demand. Call-a-vehicle organizations have not bigly influenced the general transport outlines yet. The musing is spreading and a broad assortment of plans exists all completed world.

Advantages

- Improve the openness in territories where or now and again when ordinary administrations can't do this attractively;
- Tackle social rejection of individuals that don't approach an auto;
- Offer potential cost decreases when supplanting traditional administrations in regions or times of low interest;
- Can increment open transport use through an enhanced administration quality (flexible and advantageous way to-entryway transport);
- Potentially decrease the requirement for a private auto utilize

INNOVATIVE APPROACHES IN CARGO MOVEMENT AROUND CITY

Cargo movement around the city in day time as well as in night may create problems. The space is managed by below points which will give an idea what can be proposed.

Space Management for Urban Delivery (SMUD)

Space Management for Urban Delivery is an Innovative Concept that makes proficient utilization of framework in urban territories thinking about the particular needs of urban merchandise conveyance. The administration of foundation use as far as time and space is a key issue for urban transport organizers. Data and correspondence advancements, together with mechanical access entryways or variable message signs, turn out to be more affordable and offer an assortment of complex new access plans custom-made to singular foundations of conveyance regions. One conceivable arrangement is the establishment of 'multiuse paths' which can be utilized for various purposes (e.g. stopping, stacking and emptying, transport path, free ride) for the duration of the day.

Another arrangement is the reservation of a devoted 'conveyance territory' for stacking and emptying at particular occasions of the day, where conveyance hardware and support from committed staff is offered to the vehicle administrators.

Advantages

- Brings a decrease in movement time for administrators between 12 to 15% while applying the multi-utilize path idea;
- Reduces clog and postponements for cargo administrators because of expanded street space for conveyances at specific occasions;
- Reduces vitality utilization as the cargo vehicle faces less clog issues or invests less energy discoveries a parking spot for conveyances;
- Enables a more pleasant sharing of significant space/asset;
- Reflects a decrease in generally illicit stopping exercises for both traveler and merchandise vehicles;
- Reduces 'twofold stopping' (stopping in the second path) and prompts enhanced flow.

Intra-City Night Delivery Alternative Solutions (INAS)

Elective Solutions for Home Delivery is a creative methodology for efficiently arranging "last mile" forms, i.e. the conveyance of packages to the final clients. Transport administrators have a business enthusiasm to deal with the "last mile" efficiently, for the most part from a cost point of view. Contrasted with conventional doorstep conveyances, the idea thinks about elective conveyance areas (e.g. locker boxes), time windows for the conveyance, and in addition elective redelivery techniques, if the proctor isn't at home. At the point when bolstered by an efficient transport arranging and fleet observing framework, and guaranteeing the time windows gave, significant investment funds in urban vehicle km driven can be accomplished. Home conveyance administrations can be found in various structures over the universe. The most widely recognized methodology is the conveyance of bundles by postal administrations. As of late, new home conveyance approaches have showed up because of web based business.

Advantages

- Improves personal satisfaction and activity designs – less blockage in the inward city, more freedom for the client as he doesn't have to meet the conveyance individual.
- Improves the productivity of transport – unsuccessful conveyance endeavours can be dodged, and the last mile expenses can be decreased.
- Provides more choices for private people (open 24 hours, clients can get bundles at whatever point needed).
- Reduces the quantity of rides and stops for the conveyance vans and gives more autonomy for the administrators in arranging visits (less expenses for the administrators) which results in lower vitality utilization.

NEW NON-POLLUTING AND ENERGY EFFICIENT VEHICLES

Module electric vehicles (generally called electric automobiles or EVs) can help keep your town and your world clean. At the point when all is said in done, EVs make less releases that add to ecological change and fumes cloud than customary vehicles. There are two general classes of vehicle radiations: direct and life cycle.

Arrange releases are transmitted through the tailpipe, through vanishing from the fuel structure, and in the midst of the stimulating method. Arrange spreads consolidate debilitate cloud moulding poisons, (for instance, nitrogen oxides), distinctive harms terrible to human prosperity, and ozone hurting substances (GHGs), fundamentally carbon dioxide. Each and every electric vehicle convey zero direct releases, which especially upgrades air quality in urban areas. Module cross breed electric vehicles (PHEVs), which have a gas engine despite an electric motor, make evaporative surges from the fuel system and furthermore tailpipe releases while taking a shot at gas. Nevertheless, in light of the fact that most PHEVs are more viable than proportional customary vehicles, in any case they make less tailpipe surges despite while relying upon fuel.

Life cycle releases fuse all radiations related to fuel and vehicle age, dealing with, scattering, use, and reusing/exchange. For example, for a common gas vehicle, releases are conveyed when oil is removed beginning from the soonest arrange, to fuel, dispersed to stations, and duplicated in vehicles. Like direct releases, life cycle surges consolidate a variety of pernicious defilements and GHGs.

All vehicles convey impressive life cycle releases, and finding out them is flighty. Regardless, EVs normally convey less life cycle releases than customary vehicles in light of the fact that most transmissions are bring down for power age than replicating fuel or diesel. The right proportion of these spreads depends upon your capacity mix, which varies by geographic territory. While the U.S. national midpoints are above, investigate your specific postal area's capacity mix and EV releases on the Alternative Fuels Data Center. EV drivers can moreover confine their life cycle spreads by using power delivered by non-defiling supportable sources like daylight based and wind. Make sense of how to buy maintainable power or present it on your home at the Energy Saver site.

Policy Strategy for Clean Vehicles (PSCV)

Arrangement Strategy for Clean Vehicles expects to give long haul solidness to the partners of the Alternatively Fuelled Vehicles (AFV) market and endeavours to include new buyer gatherings. The premise is frequently a solid political responsibility. The procedure will vary from city to city contingent upon the type(s) of fuel accessible, the type(s) of vehicles, target bunches and so on.

For instance, vaporous energizes require costly foundation and costly vehicles, while flammable gas is a considerably less expensive option than oil and diesel. Ethanol vehicles and foundation are moderately shabby be that as it may, contingent upon the charges, the fuel is costlier than diesel. An essential to prevail with regards to influencing privately owned businesses to pick AVFs is to settle on this decision more sparing. This requires various motivating forces and disincentives for fuel, vehicles, foundation and so on.

Advantages

- Provides the market long haul steadiness for speculations;
- Gives organizations security when putting resources into new innovation;
- Decreases outflows of ozone depleting substances;
- Reduces neighbourhood emanations (sum contingent upon the sort of fuel);
- Reduces reliance on oil;
- Reduces clamour emanations;
- Increases the interest for elective fills, which prompts expanded accessibility;

Biogas in Captive Fleets (BCF)

Biogas produced using wastewater or strong natural material is the cleanest vehicle fuel economically accessible. It barely delivers any risky outflows and next to no ozone harming substances. Delivering and utilizing biogas as a fuel is an approach to diminish vehicle discharges and can likewise be utilized to lessen the waste issue, and enhance the aggressiveness of provincial social orders. Biogas is reasonable for city fleets as it is normally accessible in all urban communities from water treatment plants. A major assortment of vehicles is accessible, including transports, waste vehicles, transporters, autos, vans and lorries. The utilization of biogas for fleets therefore can possibly additionally spread crosswise over numerous nations, yet the idea needs more advancement and scattering.

Advantages

- Provides for a waste transfer and reusing methodology;
- Stimulates country advancement and the utilization of rural feedstock for bio-fuel creation;
- Reduces the reliance on oil;
- Reduces commotion.

Joint Procurement of Clean Vehicles (JPCF)

A standout amongst the most well-known issues while presenting Alternatively Fuelled Vehicles (AVFs) is finding fitting vehicles. AVFs still face a vulnerability of interest and auto merchants thusly frequently go without presenting a model in a nation.

Social affair a significant number of purchasers can beat this issue. Giving the vehicle merchants an ensured interest likewise makes it conceivable to bring down the unit cost. Joint obtainment can be utilized to present new advancements onto the market, and in addition more models, as well as get a lower cost for new innovations. Joint acquirement can be set up inside one nation or crosswise over nations.

Advantages

- Shows the market the interest for clean vehicles;
- Introduces new models of clean vehicles available;
- Gives intrigued purchasers the chance to purchase clean vehicles;
- Improves business through presenting new innovation;
- Can prompt a diminished reliance on oil;
- Decreases ozone depleting substance discharges;
- Improves air quality;
- Reduces commotion.

INNOVATIVE DEMAND MANAGEMENT STRATEGIES

Request administration is a wide term used to allude to any ponder change to the measure of power requested from the system, and incorporates two unmistakable zones:

- **Demand Reaction**: A short run choice to decrease vitality utilization in light of a particular occasion (for instance, stack shortening in light of a system limitation).
- **Energy Productivity**: Alludes to the utilization of less vitality for a similar action or level of yield, or expanding the level of yield from a similar measure of vitality.

Transportation Management Associations (TMA)

Transportation Management Associations (TMA) are open private associations that location neighbourhood transport related issues and work under an assortment of hierarchical structures, contingent on the nature and term of the issues tended to, the mission, the individuals and the financing sources accessible. They propose chosen approaches that encourage the development of individuals and merchandise inside a region, give transport request administration help and give a voice in transport basic leadership, including carpooling and open transport administrations.

Advantages

- Solves concrete and saw issues and proposes down to earth arrangements at a fitting scale;
- Considers and coordinates the points of various partners and makes the vital conditions for an efficient partner inclusion;
- Ensures better monetary participation and financial reserve funds to organizations and workers;
- Creates great transport frameworks and choices to ease driving;
- Contributes to decreasing blockage, stopping issues and to enhancing work availability;
- Increases availability and moderateness of transport choices;
- Ensures adequate travel/travel times and unwavering quality of transport.

Local Taxes or Charges, Ring-Fenced for Transport (LTCRF)

This Concept alludes to the presentation of duties or charges at the neighbourhood level that plan to benefit the urban transport framework overall. They vary from conventional duties or charges, as all incomes gathered through these expenses are straightforwardly reinvested to enhance the neighbourhood transport framework. These charges or expenses are approach instruments requiring great participation between all levels of administration.

The charges or expenses can take different structures, for example, a street client charge, a working environment stopping demand or a blockage charge. This specific idea has been widely examined in the course of the most recent couple of years, yet the quantity of common-sense precedents accessible remains very low. The way that the plans in task are strong and working legitimately may not be sufficient to demonstrate that the Concept is as of now develop. This idea is probably going to be executed by more nearby specialists after some time.

Advantages

- Act as signs to purchasers about the genuine expenses related with specific products or administrations.
- Generate incomes for the change of the general transport framework, empowering a superior modular equalization.
- Reduce movement blockage and volumes in metropolitan zones
- Create a superior personal satisfaction and open acknowledgment of high ecological and societal expenses of ordinary portability
- Elaborate powerful advances to ease driving and guarantee street wellbeing
- Contribute to an effective circulation of merchandise and ventures.

CONCLUSION

This part accentuates on advanced improvements and supportable determination of strategies with the goal that this issue can be understood for all underdeveloped nations. This distribution needs to help urban transport chiefs and experts to discover developments that could be connected in their urban communities, by giving an outline of so of the most encouraging imaginative urban transport ideas. It incorporates the consequences of the functional research completed, i.e. a definite depiction of the NICHES Innovative Urban Transport Concepts. These are imaginative urban transport estimates that have turned out to be fruitful in (for the most part) European urban areas. Specialties has investigated twelve creative ideas with the end goal to support their take-up in various urban settings. The NICHES Innovative Concepts, chosen by European urban transport specialists.

The formative experts recognize its job in supporting urban transport divisions to address the complex difficulties presented by the expanding interest for traveller and cargo transport. An arrangement system to help urban communities and towns to move to more manageable and coordinated urban versatility is very much recorded. The Action Plan depends on Urban Mobility and Transport. The strategy means to limit the results of clog and contamination in the urban condition, and to expand expectations for everyday comforts, financial intensity, and social union. Along these lines the urban movement issues can be tackled in future and for the most part they are breathing a rent of natural air to the officially congested field.

REFERENCES

Access2All. (2010). *EU FP7 project, Mobility schemes ensuring accessibility of public transport for all users*. Retrieved from http://www.access-to-all.eu/

AMTS. (2006). Annual Data booklet. Ahmedabad Bus Operator Association.

APTIE. (2011). *EU FP7 project*. Accessible Public Transport in Europe. Retrieved from http://www.aptie.eu/site

D. Associates. (2005). *A Study of innovations in the Funding and Delivery of Transportation Infrastructure Using Tolls*. Author.

ByPAD. (2008). *EU FP6 project, Bicycle Policy Audit*. Retrieved from http://www.bypad.org

CITyLOG. (2012). *EU FP7 project, Sustainability and efficiency of city logistics*. Retrieved from http://www.city-log.eu

Citymobil Consortium. (2011). *CityMobil – Advanced Transport for the Urban Environment*. Final CityMobil Brochure.

Civitas Initiative. (2012). *Cleaner and Better Transport in Cities*. Retrieved from http://www.civitas.eu

Department for Transport. (2004). *Bus Priority: The Way Ahead*. Department of Transport. Retrieved from http://www.dft.gov.uk/pgr/regional/buses/bpf/busprioritythewayahead12/)

EBSF. (2012). *EU FP7 project*. European Bus System of the Future. Retrieved from http://ebsf.eu/

ERTRAC Research and Innovation Roadmaps. (2011). *Implementation of the ERTRAC Strategic Research Agenda 2010*. Author.

European Commission. (1995). Green Paper, The Citizens' Network - Fulfilling the potential of public passenger transport in Europe. COM (95) 601 final, Brussels.

European Commission. (1998). Communication from the Commission Developing the Citizens' Network. COM (1998) 431 final, Brussels.

European Commission. (2006). Communication from the Commission on the Thematic Strategy on the Urban Environment. COM (2005) 718 final, Brussels.

European Commission. (2007a). Green Paper, Towards a New Culture for Urban Mobility. SEC (2007) 1209 final, Brussels.

European Commission. (2007b). Regulation of the European Parliament and of the Council on Public Passenger Transport Services by Rail and by Road. Regulation (EC) No 1370/2007, Brussels.

European Commission. (2009a). Communication from the Commission, Action Plan on Urban Mobility. SEC (2009) 1211, SEC (2009) 1212, final, Brussels.

European Commission. (2010a). Communication from the Commission, Europe 2020 Flagship Initiative, Innovation Union. SEC (2010) 1161, Brussels.

European Commission. (2010b). Communication from the Commission, European Disability Strategy 2010-2020: A Renewed Commitment to a Barrier-Free Europe. SEC (2010) 1324, Brussels.

European Commission. (2010c). Communication from the Commission, Towards a European road safety area: policy orientations on road safety 2011-2020. COM (2010) 0389 final, Brussels.

European Commission. (2010d): Communication from the, The EU Internal Security Strategy in Action: Five steps towards a more secure Europe, COM (2010) 673 final, Brussels.

European Commission. (2011a). *White Paper, Roadmap to a Single European Transport Area, Towards a competitive and resource efficient transport system*. COM (2011) 144 final, Brussels.

European Commission. (2011d). Proposal for a Regulation of the European Parliament and of the Council, establishing Horizon 2020 – The Framework Programme for Research and Innovation. COM (2011) 809 final, Brussels.

European Commission. (2012a). *Road Transport - A change of gear*. Author.

European Commission. (2012b). CARE - EU road accidents database. Author.

European Commission. (2013a). *Communication Clean Power for Transport: European alternative fuels strategy* (COM(2013) 17).

European Commission. (2013b). *Proposal for a Directive on the deployment of alternative fuels infrastructure* (COM(2013) 18).

European Commission. (2009b). *Directive of the European Parliament and the Council on the Promotion of Clean and Energy Efficient Road Transport Vehicles*. Directive 2009/33/EC, Brussels.

European Green Cars Initiative. (2012). *Public-Private Partnership European Green Cars Initiative*. Retrieved from http://www.green-cars-initiative.eu

IRU. (2009). *Annual Report 2009*. Retrieved from http://www.iru.org/en_bookshop_item?id=214

Louis Berger Group Construction (GIDB). (2006). *A Report submitted to Gujarat Infrastructure Development Board*. Author.

Mediate. (2011). *EU FP7 project - Methodology for describing the accessibility of transport in Europe*. Retrieved from http://www.mediate-project.eu/

National Urban Transport Policy. (2006). Ministry of Urban Development, Government of India.

Polzin, S., & Baltes, M. (2002). Bus Rapid Transit: A Viable Alternative? *Journal of Public Transportation*, *5*(2), 47–69. doi:10.5038/2375-0901.5.2.3

Rao, Bhardwaj, & Subba Rao. (n.d.). *Intelligent Transportation Systems (ITS): A Technology Solution For 21st Century Public Transport System Management*. Academic Press.

Schaller Consulting. (2002). *Bus Rapid Transit For New York City: Prepared for Transportation Alternatives and NYPIRG Straphangers Campaign*. Retrieved from http://www.straphangers.org/pokeyaward/ letter.html

Smart Freight. (2011). *EU FP7 project*. Smart Freight Transport in Urban Areas. Retrieved from http://www.smartfreight.info/

SUMPs Guidelines. (2011). *Developing and implementing a Sustainable Urban Mobility Plan*. Retrieved from http://www.mobilityplans.eu/docs/SUMP_guidelines_web0.pdf

TEMS. (2012). *The EPOMM Modal Split Tool*. Retrieved from http://www.epomm.eu/tems/index.phtml

Transport for London (TfL), (2003). *The case for investing in London's buses: Presenting the results of the London Buses Strategic Review*. Author.

Transport for London (TfL). (2006). *Bus priority at traffic signals keeps London's buses moving Selective Vehicle Detection (SVD)*. Author.

Transportation Research Board. (n.d.). *Bus Rapid Transit Volume 1: Case Studies in Bus Rapid Transit*. TCRP Report 90 Volume 1 BRT Case Studies. Retrieved from gulliver.trb.org/publications/tcrp/tcrp_rpt_90v1. pdf

Transports en Commun d'Agglo. (n.d.). Retrieved from http://www.tcar.fr/presentation/index.asp?rub_code=52&thm_id=317&gpl_id=)

United Nations. (2011). World Urbanization Prospects: the 2011 Revision. New York: Author.

Chapter 7
Intelligent Anomaly Detection Video Surveillance Systems for Smart Cities

Gajendra Singh
Dr. B. R. Ambedkar National Institute of Technology, India

Rajeev Kapoor
Delhi Technological University, India

Arun Khosla
Dr. B. R. Ambedkar National Institute of Technology, India

ABSTRACT

With the growing demands of safety for people and their properties, video surveillance has drawn much attention. These requirements have led to the positioning of cameras almost every corner. Smart video surveillance systems can interpret the situation and automatically recognize abnormal situations, which plays a vital role in intelligence monitoring systems. One vital aspect is to detect and alert generation of suspicious events then to notify operators or users automatically. A long time may pass before an event of interest to take place. In such situations, human attention may get diverted and an event of interest may get missed. In such case, video surveillance systems can effectively improve safety and security for the control and management of public areas or personal life. Independent surveillance systems to replace the traditional (human observer-oriented) systems also can relieve the workload of relative personnel.

DOI: 10.4018/978-1-5225-8085-0.ch007

INTRODUCTION

Anomaly: something that deviates from normal, expected or standard pattern e.g. A camel between a flock. Detection of anomaly finds application everywhere, one of them application area is in video surveillance systems in smart cities and very active research area in computer vision, visual/video surveillance systems in dynamic scenes tries to find, recognize and track specific type objects/patterns from video data feed, and more generally to recognize and define object pattern/behavior. The motive is to develop intelligent video surveillance systems to substitute the traditional visual surveillance systems for smart cities that provide effectiveness and increase capability of operators, who monitor them, as the number of cameras exceeds. Briefly, the aim of video surveillance systems is not only to use cameras in the place of human eyes, but also to achieve the complete surveillance job as automatically as possible (Hu, Tan, Wang, & Maybank, 2004).

Video surveillance systems in real world has an extensive range of possible applications, such as traffic surveillance in cities, a security guard for community buildings, important buildings and expressways, detection of military interest targets, etc. Some of the video surveillance areas involving people or vehicles discussed below.

Access Control in Security Sensitive Areas

In some restricted or security wise sensitive areas such as defense laboratories, important private/governmental units and military bases, only persons with a specific identity are acceptable to enter at that areas. A biometric feature enabled database including legal visitors is built earlier using different biometric methods. When someone wants to get access to that area, the system could automatically acquire the person's features, such as facial appearance, height and walking gait from images taken from surveillance cameras in real time, and then decide whether the person can enter at that place.

Person-Specific Identification in Certain Scenario

Personal identification from distant view point by use of an intelligent surveillance system can provide assistance to the authorities, to catch accused/suspected persons. In such cases, authorities have to build a biometric feature-based database of suspects/accused, and put video surveillance at that locations where the probability of finding accused/suspects are high, e.g., bars, railway stations, casinos, bus stops, etc. The intelligent systems automatically identify and judge whether or not the people in scene are accused/suspects. If yes, alarms are raised with immediate effect. Such type of systems with only face recognition have already been used at public places, but the reliability of these type of system are not up to the mark and generate a large number of false alarms.

Crowd Congestion Management

As our urban areas are growing, persons density is growing rapidly in urban areas and creating congestion at public places due to mismanagement. But by using techniques based on human detection, video surveillance systems can automatically find/compute the density of persons at important public places such as big-stores, stadiums and travel sites, and then deliver congestion analysis to assist in the crowd management. In the same way, visual surveillance systems can monitor highways and intersections of the road network, and further analyze the vehicle flow and the position of road congestion, which are of great use for traffic management.

Anomaly Detection and Alarming

In some environments, it is essential to investigate the activities of people and movement of vehicles to find whether these activities belong to normal or abnormal categories. For example, visual surveillance systems placed in supermarkets and parking lots could examine abnormal behaviors indicative of burglary/shoplifting. there are many ways of giving an alarm. One way is to display burglary or shoplifting on large screens or automatically make public broadcasting whenever any abnormal activity happens. Another way is to immediately contact the authorities.

Interactive Surveillance Using Multiple Cameras Networks

For social safety, collaborative investigation using several cameras at different places can be used to guarantee the safety of an individual or entire community, for example by tracing accused over a widespread area by using the collaboration of several cameras. For traffic management, interactive surveillance using several cameras can help the traffic police to find, track, and catch vehicles that violates traffic rules.

Different Types of Anomalies in Surveillance Scenes

An important part of an anomaly detection techniques is to find the nature of the desired anomaly. Anomalies can be broadly divided into following three categories:

1. Point anomalies
2. Contextual anomalies
3. Collective anomalies.

1. **Point Anomalies:** A particular data point or single pattern is anomalous if it diverges significantly from the rest of the data or its surroundings.

An example of point anomaly is shown in figure 1, here all smiles are blue colour except only one green colour smiley. So, in present scene red colour fish does not fit with neighbourhood or different from other fishes that why red fish is anomalous in this scene.

2. **Contextual Anomalies:** A particular data point is anomalous within a context & it requires a notion of context. This type of anomaly is also referred as conditional anomalies.

An example of point anomaly is shown in figure 2, here is road scene where one-way traffic is only allowed, but one auto rickshaw coming from opposite direction from rest of the vehicles. Here auto rickshaw is normal, but its context is wrong (direction of movement). another example of contextual anomaly is shown in figure 3 below.

Here graph of monthly temperature of a city is plotted, so it is supposed to have high temperature in month of June with respect to other months temperature, but at point t_2 recorded temperature is drastically low in month of June due to some recording device problem. So, this is contextual anomaly, here context is month.

Figure 1. Example of point anomaly

Figure 2. Example of contextual anomaly

Figure 3. Monthly temperature graph of a city

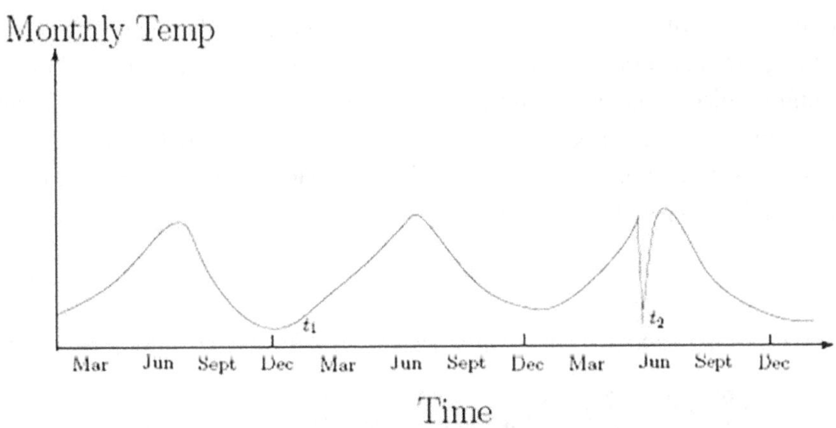

2. **Collective Anomalies:** A collective behaviour of multiple data instances is anomalous while the individual data points are not anomalous by themselves and it requires a relationship among data instances. As an example, a biker is riding his bike, but it suddenly creates problem, and it stops for few seconds and again starts but after few meters it again stops, it repeats it behavior again and again within few meters of driving. So, this type of collective behavior makes whole scene anomalous that's why this type of anomaly is referred as collective anomaly, here both stopping a bike on road at intersections is normal and riding a bike on road is also normal but collectively whole sequence is abnormal.

ANOMALY DETECTION FRAMEWORK

A general framework of an anomaly detection using video surveillance systems is given below. In video surveillance systems, the data source is cameras.

Feature Extraction techniques in literature which employ video surveillance system can be broken down into 1) those methods which tracks target or identify target first, then extract high-level features from the desired targets: e.g. a system which detects and tracks individual vehicles from full motion video, 2) other techniques that are not based on objects tracking, extract low-level features directly at the pixel level from the image sequences. an example of this type of technique is a system that calculates the frequency (rate of change) for each pixel over subsequent images of a video and build a map of motion level in the video frame.

Type of feature extraction method depends on the type of target and scene environment: low level feature extraction at pixel level are used to detect anomaly in crowed scenes, while object-based feature extraction and tracking methods are a much more commonly used for the less crowed scene or less complex scene. This is due to the complexity involved in segmentation of objects in the image, the splitting of an image into several individual objects, when the frame consists of multiple overlapping objects like occluding conditions. However, this difference in feature extraction techniques applied to individuals compared with crowds causes difficulty in defining a single anomaly detection technique that is likewise effective whether an individual or a crowd is in the scene.

Figure 4. General Framework of anomaly detection in intelligent surveillance systems

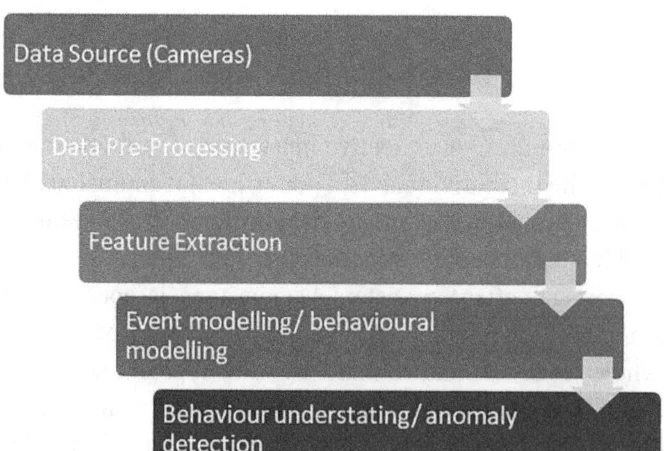

1. **Low-Level Feature Extraction**: This type of feature extraction processes are based on characterizing color, motion, and other image fundamental properties of scene. The low-level feature extraction-based methods have the benefits of being robust in some conditions, like occlusion, which highly affect tracking accuracy in highly dense crowd. Since no objects-based features are extracted from the scene, these methods are able to perform better even with large numbers of objects in the scene. However, these methods provide less detailed data about the scene and has only been used to spot anomalies involving all or many objects in the scene. These primary low-level features can be further processed to extract additional features. Pruteanu-Malinici and Carin (2008) proposes a method for video anomaly detection based on use of infinite hidden Markov models (iHMMs) and review several other approaches, including independent component analysis, invariant-subspace analysis, and shift-invariant wavelets. Low-level features can be directly fed to anomaly detection algorithms as inputs. Several studies (Andersson, Rydell, & Ahlberg, 2009; Zweng & Kampel, 2010; Andrade, Blunsden, & Fisher, 2006) also uses optical flow based approaches for this task. Using this technique, anomaly detection algorithms work on global features, for instance total kinetic energy of the scene. Some multicamera network-based systems match low-level features between different cameras. Ermis et al. (2008) match dwell times of foreground pixels (after motion-based background subtraction), and Li et al. (2009) match the motion of objects in regions from optical flow.
   a. Dong et al. (2009) proposed an approach that combines both low level features and object's features. In this approach, an image is transformed to HSV (hue, saturation, value) color space in which the image color characterizes motion data extracted from visible-spectrum. Different objects are then recognized within the HSV color space image. Thus, recognized objects correspond to regions with similar motion characteristics rather than physical objects. Finally, features like color, shape and texture are extracted from the objects, that is used in the anomaly detection.
   b. Saykol et al. (2010) proposed a method, in this method all movement pixels of video are transformed into a low-resolution grid sections which encodes data about the movement information of corresponding image sections. The grid sections values represent stopping, merging, splitting, or running. Low-level features are frequently used as an initial stage to object-based feature extraction methods. A low-level feature based methods such as background subtraction is used to distinguish between foreground objects/pixels and the background, where foreground objects/pixels are defined as objects/pixels that are of attention, on the other hand background is defined as static objects/pixels that are least of interest or

no use. Features of objects are then pull out from the identified foreground pixels/objects.

2. Objects based Feature Extraction: To extract additional detailed info about an individual object than it is not possible with the low-level feature extraction-based methods, object based feature extraction methods are typically employed. In this type of methods specific features of objects are extracted and it generally depends upon the object, environment, and the probable type of anomaly. Field of view (FOV) and resolution of image are also important factors in the selection of features; higher resolution of image is required for more detailed feature set of objects in the scene, the lower the resolution, the limited the less detailed feature set will be obtained. the position and trajectory of the objects are most common features to be extracted. These two features are important and enough in many areas targeting individuals to find restricted access in a restricted area (Duque, Santos, & Cortez, 2007; Shah, Javed, & Shfique, 2007; Duque, Santos, & Cortez, 2006), unusual paths indicating loitering or confusion (Calderara, Cucchiara, & Prati, 2007; Jan, Piccardi, & Gunes, 2003; Diamantopoulos & Spann, 2005), or specific abnormal behaviors such as falling or running (Hsieh & Hsu, 2007). Position of object and object trajectory are very useful in detection of anomaly in traffic scene to detect dogging (Diamantopoulos & Spann, 2005), or illegal traffic movement (Dong et al., 2010; Albusac et al., 2009). These features can also be used to extract object related features like acceleration, motion energy, and relative distance between different objects (Zhang et al., 2010). Using size and shape of objects as an additional feature allow to distinguish between truck and cars, and are often applied to detection of abnormalities at crosswalks and intersections (Vallejo et al., 2009; Albusac et al., 2009; Varadarajan & Odobez, 2009; Safara & Eftekhari-Moghadam, 2006).

In some research findings, with a high resolution targeting individual objects, limb angles are also used as a features (Zhicheng & Jun, 2008; Zhang & Liu, 2010; Lee, Kim, & Woon, 2010). These limb angles are used to find posture specific action the individual is acting. Lee, Kim, and Woon (2010) proposed a method where limb's entropy is used to find whether the individual's behavior is abnormal or not. Other studies also use to extract features about other specific body parts or regions additionally to limb angles. Zhou and Wu (2009) proposes a method for detection of shoplifting, for that color information is extracted to detect exposed skin on the hands and then hands trajectories are extracted to determine if an individual is shoplifting. Chen et al. (2007) extract the History Motion Image (HMI) of regions of the objects to determine movement of legs and arms of the target. Wu et al. (2009)

extract torso angle of individual of interest to determine whether an individual's behavior is normal or abnormal (running or falling).

Some object-based techniques do not need tracking of objects. Human posture information can be collected by considering features based on extraction of object's boundary. Zhang and Liu (2008) proposes the method based on extraction of edges of an recognized human silhouette to create a posture signature feature, on the other hand Lao et al. (2009) also used the horizontal and vertical projections of a silhouette to get posture information. The object's enclosing rectangle information is one more object-based feature that does not require tracking of objects. Guo and Miao (2007) proposed a method to identify four activities and six postures based on the fraction of black to white pixels in the enfolding rectangle of a human silhouette. Tang, Wang, and Lu (2009) uses the length and width of the enclosing rectangle of the elevator passengers for the detection of crime in an elevator. Xiang and Gong (2005) included the filling ratio of foreground pixels to a feature vector that also contains the centroid location, dimensions, and first-order moments of objects to distinguish between normal and abnormal behaviors of individual persons. In addition to a bounding-rectangular box, Chen, Tsai, and Chan (2010) proposed a method for the detection of drowning behaviors and swimming strokes based on a convex-hull approach.

LEARNING METHODS

This section will give idea about the different learning approaches applied for the modeling of behavior and to detect anomaly. Learning methods are closely linked to the methods for anomaly detection. For instance, a semi-supervised learning technique that includes learning of normal pattern will employ an anomaly detection technique for the detection of such events which shows different nature or do not fit the learned model. Learning methods can be broadly classified into three primary groups:

1. Supervised learning
2. Unsupervised learning
3. Semi-Supervised Learning
4. Reinforcement learning

Figure 5. Broad Classification of Machine Learning Algorithms

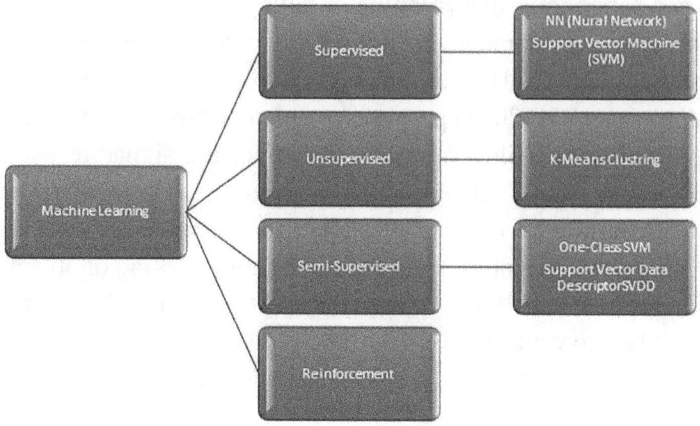

Supervised Learning

Supervised learning denotes to a technique of modeling in which an algorithm learns a model from a representative set of data of features, this date is called as training dataset, in training dataset each element is labeled with its associated true class. In training duration, the model fits/learns associations between the data points and its respective membership class. After the completion of training phase, the supervised learning techniques will be able to determine the class membership of new data points which was not present during the training phase. Supervised learning techniques for anomaly detection can be further subdivided into two categories:

Figure 6. Data required for supervised learning

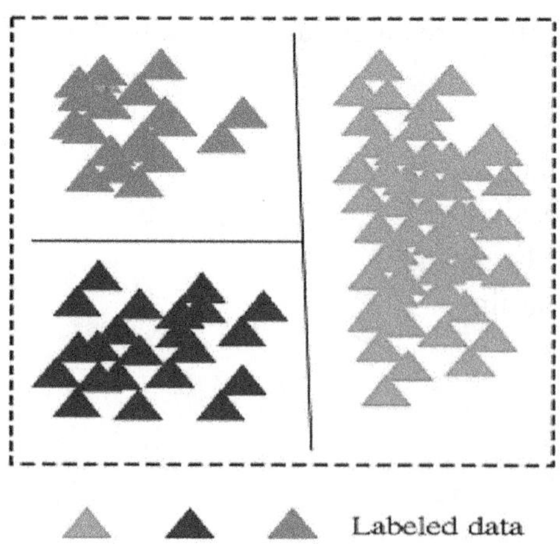

- Training from labelled dataset consisting of both normal and anomalous events.
- For multiclass / multiple events training, training is done on labelled data for each class/event.

First method is a two-class classification problem, where the data membership can belong either of the class (one or two), and second method is for multiclass problem e.g. dataset containing different types of vehicles car, bus, truck, bicycle, bike etc. in both type of supervised learning methods, class labels are required for each categories of class data.

Unsupervised Learning

Unsupervised learning techniques are those in which both type of (normal and abnormal) data is presented for training but there are no class labels. In such cases, a hypothesis is made, based on the fact that normal events are frequently occur while abnormal events rarely occur. These techniques have the advantage of not needing labeling of training dataset but suffer from the downsides associated with the guess that all rarely happening events are abnormal. Unsupervised learning techniques can be further sub divided into two categories:

- **Clustering Based Learning Techniques:** The main aim of these type of techniques is to group or cluster input training data into different categories

Figure 7. Data required for unsupervised learning

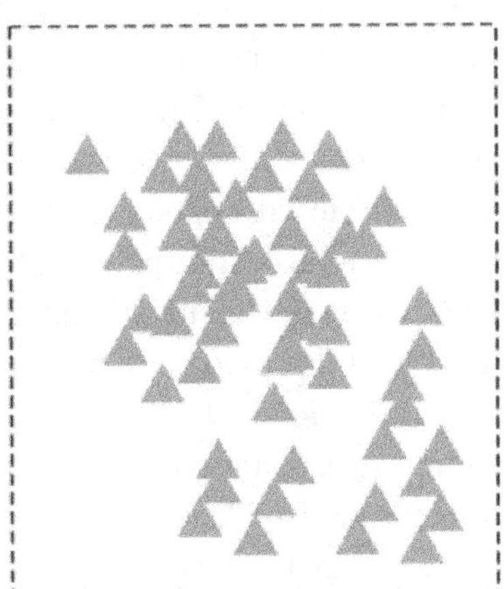

Unlabeled data

or classes using just the features extracted from the training data only and without other additional information. Unlike supervised techniques, the output labels for data points are not previously known in clustering based learning methods. There are various methods to build clustering models, for example by using means, hierarchies, medoids, and many others. Some other common clustering algorithms include k-means, hierarchical clustering, and k-medoids.

- **Association Rule Based Learning Techniques:** These techniques are commonly used to mine and extract patterns and rules from training datasets itself. These association rules describe the relations between different attributes and variables, and also illustrate item sets and patterns which arise in the data frequently. These rules help to discover valuable understandings for any organization or business from their massive data storage. Commonly used algorithms are FP Growth and Apriori.

Xiang and Gong (2005) show that unsupervised techniques may achieve better performance in comparison with supervised techniques, since human labelers tend to fill gap in unseen, supposed events and so mislabeling/ errors present in training data. Unsupervised learning techniques generally take a clustering approach for anomaly detection, in which training data points are clustered together based on some pre-defined rules e.g. distance between different class of data. In anomaly detection scenario, events in the training dataset are grouped then abnormalities are detected based on the distance of a test data point from the nearest cluster center. This method assumes that normal and abnormal events are fairly disconnected in the feature space, and that the training data contain mostly normal activities/points. Several propose techniques categorize an event as anomalous if the distance between the test data point and the nearest cluster center is greater than the predefined threshold, threshold may be adaptive, or user specified. The Battacharyya distance is a commonly used for distance calculation (Sharif, Uyaver, & Djeraba, 2010). In many techniques, threshold is not directly a distance measure. Goshorn et al. (2008) used the number of fluctuations essential to cause the test data point to fit into a trained cluster. Each fluctuation is given a cost and if the total cost surpasses a predefined threshold, then the action/event is labelled as abnormal. Other techniques (Li & Han, 2010; Xiang & Gong, 2005) use clustering methods based on the computation of probability of the test data point. If the probability is lower than threshold defined previously, then the point is labelled as abnormal such as the log likelihood (Xiang & gong, 2005) and the likelihood ratio test. Unsupervised learning techniques have application in many areas such as, social network analysis, customer segmentation in marketing, climatology, image segmentation and many others.

Semi-Supervised Learning

In the previous two types of methods, either labels are present for entire dataset or there are no labels for all the data points in the dataset. Semi-supervised learning lies in between these two methods. In many practical scenarios, the cost of labeling of data is significantly high, because it involves skilled expert persons to do labeling of data. So, in the unavailability of labels in most of the data points but available in some data points, semi-supervised algorithms are the best suitable choice for building the model. These approaches exploit the knowledge that although the group memberships of the unlabeled data are not known, this data carries substantial information about the group parameters.

When building semi-supervised learning models (Chui et al., 2017) one of the following assumptions should be followed:

- **Smoothness Assumption**: It is an idea that feature vector or data points, which are nearby to each other have a higher chance to have the similar output label as they share comparable feature in feature space (Shi, Zurada, & Guan, 2015).
- The cluster assumption is based on the fact that the datapoints have a propensity to form clusters where datapoints in the same clusters have higher probability to be the similar output label (Reitmaier & Sick, 2015).

Figure 8. Data required for semi supervised learning

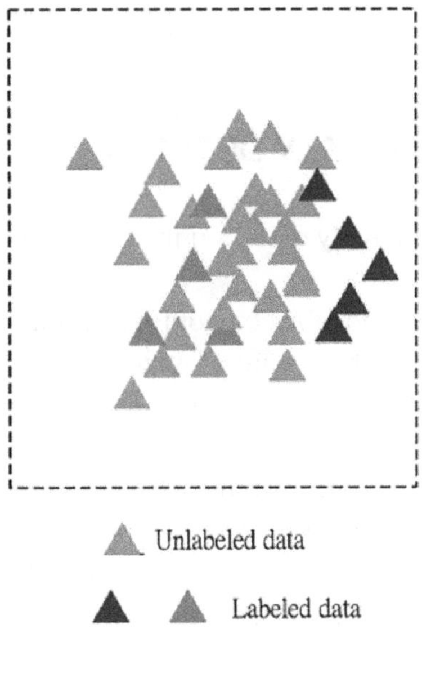

- This assumption is like the smoothness assumption except stating to clustering, manifold assumption means that data lie on a low-dimensional manifold implanted in a higher-dimensional space (Wang et al., 2015). It is noticed that manifold is a topological space that locally resembles Euclidean space near each data point.

Graph-based model (Nie et al., 2015), co-training (Jin et al., 2016), self-training (Cvetković et al., 2015), generative model (Ashfaq et al., 2017), heuristic model (Yan, Chen, & Tjhi, 2013) and low-density separation (Zhang et al., 2015).

Reinforcement Learning

These methods objective is to gathered observations from the interface with the environment to take activities that would minimize the risk or maximize the reward (Ivan, n.d.). Reinforcement learning methods endlessly updates iteratively from the environment. In this type of techniques, the agent update itself from its understandings of the environment until it discovers the full range of probable solutions.

To produce intelligent programs (also called agents), reinforcement learning goes through the following steps:

- First agents observe input state.
- A decision-making function is used to instruct the agent, to perform a set of actions.
- After the actions are completed, the agent gets prize or reinforcement from the environment.
- The state-action pair data about the prize is stored.

CHALLENGES IN ANOMALY DETECTION

The major challenges in anomaly detection in video surveillance systems are:

- Extraction of foreground objects from background objects (salient region detection): foreground objects are of main use in any video surveillance but it's a challenging task due to occlusion or may be foreground object is similar looking to background.
- Extraction of optimal video features that can describe the scene objects or different type of objects appropriately, Appropriate features should be fitted/ chosen to the type of application such as event detection, motion tracking, pedestrian detection, and utilize multimodal aspects (audio, text, and video).

- Selection of data mining technique: Selection of data modelling algorithm can greatly affect the performance of anomaly detection task.
- Computational efficiency: There is always trade-off between accuracy and computational power. Mostly complex algorithms give better performance but require high computational power and these type algorithms are not able to perform in real time.

CONCLUSION

Video surveillance systems are being used around us for a couple of decades but gain more interest from last decade. Currently available automated video surveillance systems in market, can process video and equipped to do nearly all crucial low-level functions/task, such as object tracking, segmentation, motion detection, and object classification with decent precision, but it fails in some scenarios. Recently, researcher's curiosity in video surveillance systems, has changed from low-level activity detection to more complicated activity analysis for the detection of object and/ or human behaviors, i.e., behavior of activities/ events (emotion, fighting etc.), for immediate risk finding and prevention.

Current event / behavior analysis systems emphasis on the predefined behaviors/ events, e.g., it combines the output of an automated video surveillance system along with spatiotemporal analysis about each object with respect to other objects and background areas in the scene. As smart cities are developing intelligent behavior analysis/ event analysis systems are needed, to automatically detect and define (learn) new activities/ behaviors by pattern detection, and further pass this information to authorities' persons to take appropriate action and suggest some possible action to take. The growing demand for advanced video surveillance systems and the transformation from analog to digital video surveillance structure, has changed automated video surveillance into a large-scale data analysis and management challenge.

This chapter reviews and exploits developments and general framework involved in video surveillance systems for anomaly detection based on behavior analysis. Human behavior analysis in visual surveillance systems, needs the cutting-edge and complex researches in computer vision and image processing with the use of artificial intelligence.

REFERENCES

Albusac, J., Castro-Schez, J., Lopez-Lopez, L., Vallejo, D., & Jimenez-Linares, L. (2009). A supervised learning approach to automate the acquisition of knowledge in surveillance systems. *Signal Processing*, *89*(12), 2400–2414. doi:10.1016/j.sigpro.2009.04.008

Albusac, J., Vallejo, D., Jimenez-Linares, L., Castro-Schez, J., & Rodriguez-Benitez, L. (2009). Intelligent surveillance based on normality analysis to detect abnormal behaviours. *International Journal of Pattern Recognition and Artificial Intelligence*, *23*(7), 1223–1244. doi:10.1142/S0218001409007612

Andersson, M., Rydell, J., & Ahlberg, J. (2009). Estimation of crowd behavior using sensor networks and sensor fusion. *Proc. 12th Int. Conf. Inf. Fusion*, 396–403.

Andrade, E. L., Blunsden, S., & Fisher, R. B. (2006). Modelling crowd scenes for event detection. *Pattern Recognition*, *1*, 175–178.

Ashfaq, R. A. R., Wang, X. Z., Huang, J. Z., Abbas, H., & He, Y. L. (2017). Fuzziness based semi-supervised learning approach for intrusion detection system. *Inf. Sci.*, *378*, 484–497. doi:10.1016/j.ins.2016.04.019

Calderara, S., Cucchiara, R., & Prati, A. (2007). Detection of abnormal behaviors using a mixture of von mises distributions. *Proc. IEEE Conf. Adv. Video Signal Based Surveillance*, 141–146. 10.1109/AVSS.2007.4425300

Chen, H., Tsai, M. J., & Chan, C. C. (2010). A hidden Markov model-based approach for recognizing swimmer's behaviors in swimming pool. *Proceedings - International Conference on Machine Learning and Cybernetics*, *5*, 2459–2465.

Chen, Y., Liang, G., Ka, K. L., & Xu, Y. (2007). Abnormal behavior detection by multi-SVM-based Bayesian network. *Proc. Int. Conf. Inf. Acquis.*, 298–303. 10.1109/ICIA.2007.4295746

Chui, K. T., Alhalabi, W., Pang, S. S. H., Pablos, P. O. D., Liu, R. W., & Zhao, M. (2017). Disease Diagnosis in Smart Healthcare: Innovation, Technologies and Applications. *Sustainability*, *9*(12), 2309. doi:10.3390u9122309

Cvetković, B., Kaluža, B., Gams, M., & Luštrek, M. (2015). Adapting activity recognition to a person with Multi-Classifier Adaptive Training. *Journal of Ambient Intelligence and Smart Environments*, *7*, 171–185.

Diamantopoulos, G., & Spann, M. (2005). Event detection for intelligent car park video surveillance. *Real-Time Imaging*, *11*(3), 233–243. doi:10.1016/j.rti.2005.02.002

Dong, N., Jia, Z., Shao, J., Xiong, Z., Li, Z., Liu, F., . . . Peng, P. (2010). Traffic abnormality detection through directional motion behavior map. *Proc. 7th IEEE Int. Conf. Adv. Video Signal Based Surveillance*, 80–84. 10.1109/AVSS.2010.61

Dong, Q., Wu, Y., & Hu, Z. (2009). Pointwise motion image (PMI): A novel motion representation and its applications to abnormality detection and behavior recognition. *IEEE Transactions on Circuits and Systems for Video Technology*, *19*(3), 407–416. doi:10.1109/TCSVT.2009.2013503

Duque, D., Santos, H., & Cortez, P. (2007). Prediction of abnormal behaviors for intelligent video surveillance systems. *Proc. IEEE Symp. Comput. Intell. Data Mining*, 362–367. 10.1109/CIDM.2007.368897

Duque, D., Santos, H., & Cortez, P. (2006). The OBSERVER: An intelligent and automated video surveillance system. *Proc. 3rd Int. Conf. Image Anal. Recognition*, 898–909. 10.1007/11867586_81

Ermis, E. B., Saligrama, V., Jodoin, P., & Konrad, J. (2008). Abnormal behavior detection and behavior matching for networked cameras. *Proc. 2nd ACM/IEEE Int. Conf. Distrib. Smart Cameras*, 1–10. 10.1109/ICDSC.2008.4635728

Goshorn, R., Goshorn, D., Goshorn, J., & Goshorn, L. (2008). Abnormal behavior-detection using sequential syntactical classification in a network of clustered cameras. *Proc. 2nd ACM/IEEE Int. Conf. Distrib. Smart Cameras*, 1–10. 10.1109/ICDSC.2008.4635732

Guo, P., & Miao, Z. (2007). A home environment posture and behavior recognition system. *Proc. Int. Conf. Convergence Inf. Technol.*, 175–180. 10.1109/ICCIT.2007.35

Hsieh, C., & Hsu, S. (2007). A simple and fast surveillance system for human tracking and behavior analysis. *Proc. 3rd IEEE Int. Conf. SignalImage Technol. Internet-Based Syst.*, 812–818. 10.1109/SITIS.2007.128

Hu, W., Tan, T., Wang, L., & Maybank, S. (2004). A survey on visual surveillance of object motion and behaviors. *IEEE Transactions on Systems, Man and Cybernetics. Part C, Applications and Reviews*, *34*(3), 334–352. doi:10.1109/TSMCC.2004.829274

Ivan, M. (n.d.). *Types of machine learning algorithms*. Retrieved from https://en.proft.me/2015/12/24/types-machine-learning-algorithms/

Jan, T., Piccardi, M., & Gunes, H. (2003). Suspicious behavior assessment for visual surveillance using neural network classifiers. *Proc. Int. Conf. Imaging Sci., Syst. Technol.*, *2*, 65–71.

Jin, L., Xue, Y., Li, Q., & Feng, L. (2016). Integrating human mobility and social media for adolescent psychological stress detection. In *Proceedings of the International Conference on Database Systems for Advanced Applications*. Dallas, TX: Springer. 10.1007/978-3-319-32049-6_23

Lao, W., Han, J., & de With, P. H. N. (2009). Automatic video-based human motion analyzer for consumer surveillance system. *IEEE Transactions on Consumer Electronics*, *55*(2), 591–598. doi:10.1109/TCE.2009.5174427

Lee, C., Lim, K., & Woon, W. (2010). Statistical and entropy based multipurpose human motion analysis. *Proc. 2nd Int. Conf. Signal Process. Syst.*, 734–738.

Li, H., & Han, Z. (2010). Catching attacker(s) for collaborative spectrum sensing in cognitive radio systems: An abnormality detection approach. *Proc. IEEE Symp. New Frontiers Dyn. Spectr.*, 1–1. 10.1109/DYSPAN.2010.5457898

Li, J., Gong, S., & Xiang, T. (2009). On-the-fly global activity prediction and anomaly detection. *Proc. IEEE 12th Int. Conf. Comput Vision Workshops*, 1330–1337. 10.1109/ICCVW.2009.5457455

Mehran, R., Oyama, A., & Shah, M. (2009). Abnormal crowd behavior detection using social force model. *Proc. IEEE Conf. Comput. Vision Pattern Recognit.*, 935–942. 10.1109/CVPR.2009.5206641

Nie, L., Zhao, Y. L., Akbari, M., Shen, J., & Chua, T. S. (2015). Bridging the vocabulary gap between health seekers and healthcare knowledge. *IEEE Transactions on Knowledge and Data Engineering*, *27*(2), 396–409. doi:10.1109/TKDE.2014.2330813

Pruteanu-Malinici, I., & Carin, L. (2008). Infinite hidden Markov models for unusual-event detection in video. *IEEE Transactions on Image Processing*, *17*(5), 811–822. doi:10.1109/TIP.2008.919359 PMID:18390385

Reitmaier, T., & Sick, B. (2015). The responsibility weighted Mahalanobis kernel for semi-supervised training of support vector machines for classification. *Inform. Sci.*, *323*, 179–198. doi:10.1016/j.ins.2015.06.027

Safara, F., & Eftekhari-Moghadam, A. (2006). Knowledge discovery of traffic/people behaviors based on image mining approach. *Proc. Geom. Model. Imaging-New Trends*, 255–258. 10.1109/GMAI.2006.31

Saykol, E., Bastan, M., Gudukbay, U., & Ulusoy, O. (2010). Keyframe labeling technique for surveillance event classification. *Optical Engineering (Redondo Beach, Calif.)*, *49*(11), 1–12. doi:10.1117/1.3509270

Shah, M., Javed, O., & Shafique, K. (2007). Automated visual surveillance in realistic scenarios. *IEEE MultiMedia*, *14*(1), 30–39. doi:10.1109/MMUL.2007.3

Sharif, M. H., Uyaver, S., & Djeraba, C. (2010). Crowd behavior surveillance using bhattacharyya distance metric. *Proc. CompIMAGE*, 311–323. 10.1007/978-3-642-12712-0_28

Shi, D., Zurada, J., & Guan, J. (2015). A Neuro-fuzzy System with Semi-supervised Learning for Bad Debt Recovery in the Healthcare Industry. In *Proceedings of the 2015 48th Hawaii International Conference on System Sciences (HICSS)*. IEEE Press. 10.1109/HICSS.2015.376

Tang, Y., Wang, X., & Lu, H. (2009). Intelligent video analysis technology for elevator cage abnormality detection in computer vision. *Proc. 4th Int. Conf. Comput. Sci. Convergence Inf. Technol.*, 1252–1258. 10.1109/ICCIT.2009.206

Vallejo, D., Albusac, J., Jimenez, L., Gonzalez, C., & Moreno, J. (2009). A cognitive surveillance system for detecting incorrect traffic behaviors. *Expert Systems with Applications*, *36*(7), 10503–10511. doi:10.1016/j.eswa.2009.01.034

Varadarajan, J., & Odobez, J. (2009). Topic models for scene analysis and abnormality detection. *Proc. IEEE 12th Int. Conf. Comput. Vision Workshops*, 1338–1345. 10.1109/ICCVW.2009.5457456

Wang, Y., Chen, S., Xue, H., & Fu, Z. (2015). Semi-supervised classification learning by discrimination-aware manifold regularization. *Neurocomputing*, *147*, 299–306. doi:10.1016/j.neucom.2014.06.059

Wu, X., Gong, H., Chen, P., Zhong, Z., & Xu, Y. (2009). Surveillance robot utilizing video and audio information. *J. Intell. Robot. Syst.: Theory Appl.*, *55*(4–5), 403–421. doi:10.100710846-008-9297-3

Xiang, T., & Gong, S. (2005). Video behaviour profiling and abnormality detection without manual labelling. *Proc. 10th IEEE Int. Conf. Comput. Vision*, *2*, 1238–1245. 10.1109/ICCV.2005.248

Xiang, T., & Gong, S. (2005). Video behaviour profiling and abnormality detection without manual labelling. *Proc. 10th IEEE Int. Conf. Comput. Vision*, *2*, 1238–1245. 10.1109/ICCV.2005.248

Yan, Y., Chen, L., & Tjhi, W. C. (2013). Fuzzy semi-supervised co-clustering for text documents. *Fuzzy Sets and Systems*, *215*, 74–89. doi:10.1016/j.fss.2012.10.016

Zhang, C., Chen, W., Chen, X., Yang, L., & Johnstone, J. (2010). A multiple instance learning and relevance feedback framework for retrieving abnormal incidents in surveillance videos. *Journal of Multimedia*, *5*(4), 310–321. doi:10.4304/jmm.5.4.310-321

Zhang, J., & Liu, Z. (2010). Abnormal behavior of pedestrian detection based on fuzzy theory. *Moshi Shibie Yu Rengong Zhineng/Pattern Recognit. Artificial Intelligence*, *23*(3), 421–427.

Zhang, J., & Liu, Z. (2008). Detecting abnormal motion of pedestrian in video. *Proc. Int. Conf. Inf. Autom.*, 81–85.

Zhang, X., Guan, N., Jia, Z., Qiu, X., & Luo, Z. (2015). Semi-supervised projective non-negative matrix factorization for cancer classification. *PLoS One*, *10*(9), e0138814. doi:10.1371/journal.pone.0138814 PMID:26394323

Zhicheng, W., & Jun, Z. (2008). Detecting pedestrian abnormal behavior based on fuzzy associative memory. *Proc. 4th Int. Conf. Nat. Comput.*, 143–147.

Zhou, G., & Wu, Y. (2009). Anomalous event detection based on self-organizing map for supermarket monitoring. *Proc. Int. Conf. Inf. Eng. Comput. Sci.*, 1–4. 10.1109/ICIECS.2009.5364586

Zweng, A., & Kampel, M. (2010). Unexpected human behavior recognition in image sequences using multiple features. *Proc. 20th Int. Conf. Pattern Recognit.*, 368–371. 10.1109/ICPR.2010.98

Chapter 8
Ontological Design for Cognitive Cities:
The New Principle for Future Urban Management

Sara D'Onofrio
University of Fribourg, Switzerland

Astrid Habenstein
University of Bern, Switzerland

Edy Portmann
University of Fribourg, Switzerland

ABSTRACT

Based on the advancements of a smart city, the cognitive city focuses on the communication between the city's stakeholders and cognitive systems to create a human-machine symbiosis in which human and machine can interact directly with each other. To develop the cognitive city means to design this sociotechnical relationship. This requires a variety of approaches, methods, and tools. To this purpose, the authors of this chapter suggest the concept of ontological design. This principle helps to develop a research methodology toolbox that can be applied to create research and development strategies suitable for shaping the relationship between human and environment. This chapter is an essay with the aim to encourage the reader to reflect. Ontological design means that it is necessary to become aware of the influences of today's actions on the future. With the help of an illustrative use case, this chapter wants to demonstrate why and to what extent the concept of ontological design can support urban development.

DOI: 10.4018/978-1-5225-8085-0.ch008

INTRODUCTION

Due to growing urbanization, today's cities face major challenges, such as the increasing scarcity of resources and the preservation of human well-being. The vision of a cognitive city is an important approach to meeting these challenges in the interests of the city and its stakeholders (e.g., inhabitants, companies and nonprofit-organizations) (Finger & Portmann, 2016). The cognitive city is based on the concept of the smart city and aims to compensate for its deficits. The focus often lies on technological possibilities, which represent a way to solve the problems of urban efficiency and sustainability; however, this strategy tends to be accompanied by a lack of stakeholder involvement to improve cities' processes in line with stakeholders' needs. However, as a sociotechnical system, the good functioning of a city depends on the well-balanced coexistence and cooperation of technology and human minds. It is therefore important to create an environment for collaboration to foster urban development, among other aspects, to create a cognitive city.

In terms of challenges to and problems of cities, there exist several scientific problem-solving approaches. If the problem is to be answered by identifying behavioral patterns or explaining a cause, empirical methods such as interviews or surveys can be applied. However, it is important to be aware that these approaches are useful for a specific class of problems but cannot entirely address the complex problems confronting a city. For instance, Krasnova, Abramova, Notter and Baumann (2016) carried out a survey about "*why phubbing is toxic for your relationship*" to understand the role of smartphone jealousy among "*generation Y*" users. This is an important topic in our society and we must address it, but it is very specific to an object (i.e., smartphone) and to a stakeholder group (i.e., generation Y). The findings of this study can be applied to urban development only to a certain extent. It helps draw the attention of the digital generation to certain points but addresses only one of many problems. It does not offer solution approaches to overcoming this problem. Another example is "The Swiss Software Industry Study 2017". This study helps to obtain "*in-depth information about the current state, emerging trends, and long-term developments of the Swiss software industry*" (Hurni, Huber & Dibbern, 2017, p. 3). However, similar to the previous study, it is very limited (geographically, thematically and methodically) and can therefore hardly, or only to a certain extent, be applied to urban development.

Innovative approaches such as action design research (Sein, Henfridsson, Purao, Rossi & Lindgren, 2011) can be utilized to address a specific problem. However, even if they are future-oriented, these research venues also have limitations. For instance, Petersson and Lundberg (2016) applied action design research to develop working methods to "*promote the technical development of turnouts*" in the railway sector. Starting from the assumption that "*failure of turnouts can lead to significant delays and societal costs*", solution approaches are developed by applying action design research. In comparison to empirical research, solution approaches are created to help to solve specific problems (including the consideration of stakeholder needs). However, the probability that the same approach can be transferred to other areas is rather low. The same applies to the prototype "*personal digital assistant 2.0*" developed by Kaltenrieder, Altun, D'Onofrio, Portmann and Myrach (2016), which was specifically designed for the mobility, as well as for the improved organization of busy people (i.e., improvement of calendar management). To a certain degree, their findings can be adapted to other areas, but adjustments will be necessary, depending on the target group, the application area and other aspects. Therefore, collaborative methods are vital to tackling multifaceted urban problems, and a well-considered mix of methods proves most promising in shaping the city of the future. It is vital to think outside the box and find new ways while avoiding conventional ones. This is where the fundamental idea of ontological design comes into play.

If stakeholders are willing to actively participate in the exchange of knowledge (e.g., through advanced information and communication technologies) and to be open to new ideas and ways (of research), a new era can be created for them based on the principle of ontological design. Ontological design means that everything humans have designed influences their behavior and life, too (Willis, 2006). Given this mindset, it is thus crucial to consider how a city should be developed; for instance, determining which scientific research methods are simultaneously compatible for practice and which resources are needed. The authors are engaged with this topic because they consciously try to address the gap of unconscious behaviors respecting research habits (e.g., often the research mode of the mentor is adopted). To the best of the authors' knowledge, so far, no research has questioned the entire concept of conducting research and analyzed individual methods over different research directions in detail. Only a few studies (e.g., Müller, D'Onofrio, & Portmann, 2018; D'Onofrio & Tschudi 2018) have recently begun to combine various methods, but no one has yet developed a toolbox that helps to think about how a problem or a challenge should be approached and to what extent the approach taken will influence society and the environment. Thus, the authors try to adapt the principles of ontological design in the research environment and to develop a research methodology toolbox that helps to select appropriate approaches, methods and tools during the preparation phase.

In the following essay, the authors wish to demonstrate why and to what extent ontological design research can support urban development. First, the authors explain what constitutes a smart city and where the limitations of this concept lie to illustrate why it is important to think about cognitive cities. Second, a selection of methods is presented, accompanied by a use case to illustrate the advantages and disadvantages of the method mentioned. Thereby the authors gradually lead the reader to the definition of ontological design research, which they propose as a toolbox of methods and concepts that could help in realizing the cognitive city. Lastly, this article concludes with a discussion.

FROM SMART TO COGNITIVE CITIES

Features of Smart Cities

There is no generally accepted definition among researchers and practitioners of what a smart city is (Kummitha & Crutzen, 2017); however, many smart city concepts describe the application of advanced information and communication technologies (e.g., sensor-based technologies) for a more sustainable, social and ecological design of urban areas as one of its essential characteristics (Albino, Berardi, & Dangelico, 2015; Tabacchi, Portmann, Seising, & Habenstein, 2018). By means of technology, existing processes within a city can be designed more efficiently so that input can be transferred more quickly to output and, when possible, costs can be reduced through automation (D'Onofrio & Portmann, 2017). For example, processes such as applying for a visa and registering in a new city could be carried out much faster and easier.

Another essential characteristic of smart cities is the collection, analysis and processing of data, the basis for possible automation through technologies: Cities can become smarter by collecting urban data and making it available to the relevant stakeholders, such as nonprofit organizations or research institutions (Hurwitz, Kaufman, & Bowles, 2015). The availability of data enables one to identify patterns and to improve or develop new processes based on these data (e.g., the behavior and needs of stakeholders). This data-driven approach, applied to topics such as smart environment, smart living and smart governance, is strongly focused on increasing efficiency and sustainable development using smart solutions. Smart city projects are thus responding to the central challenges facing the modern city: Cities' resources in space, money, skills and time are limited yet must meet exponentially growing demands (e.g., energy, transport, water). Cities must become more economically, socially and ecologically sustainable, a goal to which a more efficient use of existing resources makes a crucial contribution. However, this approach is a highly technical, if not technocratic, approach to urban management and has limitations.

Limitations of Smart Cities

The first limitation concerns the creation and preservation of urban knowledge. In the future, the survival of cities will depend on the ability to continually build, maintain and access a city's knowledge reservoirs. Increasing efficiency and sustainability, which a smart city strives for, requires more than the use of smart systems to conserve the city's resources. Stakeholders' behavior must be changed significantly: They must consciously contribute to urban development. One way to do this is through the exchange of information; for instance, by using advanced information and communication technologies or participating in public events and discussions. By revealing one's own perceptions, knowledge and experiences, information can be generated, which in turn helps one extract knowledge and learn from it (Müller et al. 2018). The arsenal of data analysis tools can help to process the vast amount of data (e.g., stakeholders' information) and can thus contribute to individual and collective learning.

The aim is to build up and especially to use the ever-evolving "urban intelligence", the collective intelligence of the city (Moyser & Uffer, 2016). This refers to a network of individual intelligences (i.e., people and/or objects) that together are more effective than by themselves (supported by the theory of transactive memory (Hollingshead, 2001)). Together they form a network that is more than the sum of its elements and that better contributes to solving problems than individually (Malone & Bernstein, 2015). This combination and interplay of individual intelligences is best described as "connectivism" (Siemens, 2005). Thus, learning is a process in which the learning object forms networks by linking to nodes (i.e., people and/or objects). Each node has its own networks that the learning object can access through interaction and communication by connecting to the corresponding node. As illustrated in Figure 1, the support of collective learning between human and machine by means of connectivism, so-called connected learning (Ito et al., 2013), could help smart cities to evolve into cognitive cities and simultaneously create a human-machine symbiosis leading to a superorganism (D'Onofrio, Müller, Papageorgiou, & Portmann, 2018; Malone, 2018).

The second limitation concerns an aspect already mentioned in the first limitation: The involvement of stakeholders. Urban challenges cannot be reduced to questions of efficiency and sustainability. Smart city models, which focus strongly on technology as a driver of development, are increasingly falling behind due to the underlying technocratic understanding of smart city – in favor of attempts to take the role of stakeholders into greater consideration (Fredericks, Hespanhol, Parker, Zhou, & Tomitsch, 2018; Kummitha & Crutzen, 2018). In addition to projects geared primarily to increasing sustainability and efficiency, it is also necessary to focus more strongly on people and their needs.

This can take various forms: On the one hand, it should involve personalizable applications, as people want to be perceived as individuals and to shape their lives as they see fit. On the other hand, it is about more civic participation in decision-making processes (e.g., regarding governmental actions) but also in co-creation with the city (e.g., launching projects). Co-creation is the more active involvement in urban development, while (smart) participation can take passive (e.g., information

Figure 1. Connectivism
(Portmann & Metzger, 2017)

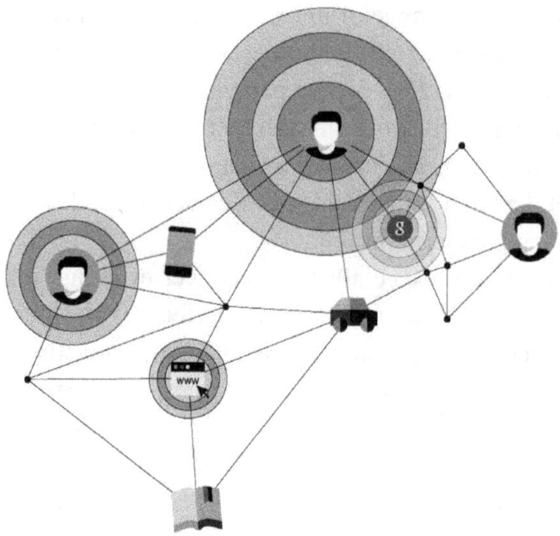

sheets for voting) as well as active (e.g., public panels) forms to different degrees. By foregrounding city stakeholders and giving them the possibility of taking an increasingly active role in urban development (e.g., civic tech movements) and becoming so-called “drivers of change”, new forms of participatory governance can arise (Müller et al., 2018; Oliveira & Campolargo, 2015). A smart city that follows these principles can be called a human smart city (Metzger et al., 2018): A city that is not only more efficient and sustainable but also more human-centered.

For both aspects – the increasing importance of urban knowledge ecosystems and the requirements of a human smart city – the collaboration between human and machine, machine and machine as well as human and human is central. This is where the concept of the cognitive city comes into play, understood as the design of communication, interaction and interplay of the city (as a system of systems (Hurwitz et al., 2015)) by means of cognitive systems.

Cognitive Cities and Cognitive Computing

Information and communication technologies tend to have only the ability to collect, aggregate and analyze (real-time) data and to use it again in a targeted manner (e.g., solving traffic jams through improved traffic light guidance based on sensory data). For the most part, statistical methods are used to optimize the efficiency of urban processes based on data. The human ability to process information (i.e., human cognition) is thereby widely ignored. However, to realize the vision of a cognitive city, existing information and communication technologies must be complemented by cognition-related methods.

The emerging technology of cognitive computing turns out to be useful here. Focusing on fuzzy logic (Zadeh, 1988) and the already mentioned connectivism-model by Siemens (2005), the authors understand cognitive computing as comprising soft computing methods as well as cognition and learning theories. Soft computing techniques allow the inclusion of imprecise, uncertain and vague data, creating a tolerance for the changing complexity (in terms of perceptions) within a city. Cognition and learning theories additionally allow human skills (i.e., information processing; e.g., the decomposition of complex facts into several simpler components, or learning from experience) to be integrated into the development of smart systems (D'Onofrio & Portmann, 2017; Müller et al., 2018). This results in cognitive systems that not only try to cope with the complexity of urban challenges but also do justice to stakeholders in a city. Thus, cities can become more sensitive, understanding and responsive to changes in their environment by improving the capabilities of urban systems (Hurwitz et al., 2015).

Cognitive cities enable one to address the first limitation of a smart city by supporting collective learning between humans and machines and creating an urban intelligence (Moyser & Uffer, 2016). Every stakeholder represents a valuable and important node of knowledge in the urban network (D'Onofrio et al., 2018). Cognitive systems also address the second limitation by facilitating civic participation and optimizing cooperation between those responsible in a city and their stakeholders. Therefore, neither the involvement of stakeholders nor the investment of cognitive systems in urban development should be underestimated. Only this combination and the right balance between technological and social aspects, as well as viewing the city as a sociotechnical system, allows for the emergence of cognitive cities and the possibility of coping with complex challenges, such as urbanization and digitalization.

THE CITY AS A SOCIOTECHNICAL SYSTEM

The concept of the sociotechnical system has various roots (Pfäffli, Habenstein, Metzger, & Portmann, 2018): In German-speaking research, the term derives from the sociology of technology, understood as the combination of technical artefacts and social action that is fundamental to the formation of technical systems. In those systems, technical and social components are linked to form a network, and their interaction ensures that the system works (Hughes, 1987). For instance, Ropohl (1988) calls these networks sociotechnical systems, as all technical systems are based on the interaction of humans and technical artefacts. Moreover, a widespread concept of sociotechnical systems in Anglo-Saxon research stems from the sociology of work and organization, which in the 1950s and 1960s originally referred to work systems that included a complex interaction between human, machine and environmental factors. From the central idea that all these factors must be equally considered in the development of such systems for successful collaboration, two approaches were developed: An analytical, observational approach and approaches that take sociotechnical systems from a design perspective (Baxter & Sommerville, 2011; Eason, 2014).

Today, theories of sociotechnical systems are also discussed in areas that are not necessarily related to the sociology of work and organization (Baxter & Sommerville, 2011; Eason, 2014). In the context of a city, it makes sense to combine different research perspectives on social technical systems, as cities can per se be described as complex sociotechnical systems (Finger & Razaghi, 2017; Hillier, 2012; Patorniti, Stevens, & Salmon, 2017). Cities comprise two subsystems: The physical-material subsystem (e.g., buildings, roads, infrastructure) and the human-social subsystem (human movements, interactions and activities, etc.). These subsystems are interconnected by human cognition and influence each other (Finger & Razaghi, 2017; Hillier, 2012). From this perspective, the symbiosis of social action and space/matter as well as of humans and technology is fundamental for the city. This relationship must be shaped so that the city remains or becomes a habitat worth living in for people.

Smart and cognitive cities are special forms of urban sociotechnical systems in which technical and social components are connected to form a network. At the material level, smart cities are a combination of urban infrastructure systems and digitization. To become more economically, socially and environmentally sustainable, cities must use existing resources in a more efficient way. Therefore, the physical infrastructure layers of a city (energy, transport, water, etc.) are duplicated by means of data respectively information layers, whereby the infrastructure systems and consumption can be optimized based on the generated data and information (Finger & Razaghi, 2017). The cognitive city goes one step further insofar as not only is the infrastructure enriched with information and communication technologies or becomes part of the infrastructure itself but relies on the communication between a city's stakeholders and cognitive systems to create a human-machine symbiosis in which human and machine can interact directly with each other. Hence, the proper functioning of the sociotechnical system (in this case the city) depends on the interaction and balance of these components and must therefore be analyzed and designed to be adapted and improved to fulfill its purpose: Enhancing the quality of life in the city, among other things, with the help of digital technologies (Pfäffli et al., 2018).

SHAPING THE COGNITIVE CITY

Smart Participation as Illustrative Example

Based on the definition of a sociotechnical system, consisting of physical-material and human-social subsystems, it is now necessary to find problem-solving approaches that address both subsystems and simultaneously tackle urban challenges on a long-term basis. To understand the advantages and disadvantages of the selected research directions and approaches, one of the current research projects of the authors is used as an illustrative example for pedagogical reasons. This example, in which the concept of a smart city is central to a focus on smart participation, is a research project on which the universities of Bern and Fribourg (both in Switzerland) are currently working. Therefore, this section shows how all these research approaches can be applied to the research topic of smart participation and how important they are for urban development.

The term "participation" refers to an intensive direct involvement of other parties, such as inhabitants, in decision-making regarding governmental actions. Such involvements can include public meetings, surveys, initiatives, and many other models. With rapid technological development, interest increases to foster urban participation through advanced information and communication technologies, to enable an influencing and meaningful change in public policy and governance. With the use of advanced technologies, such an involvement can then be called "smart participation" and therefore make a city smarter (Viale Pereira, Cunha, Lampoltshammer, Parycek, & Testa, 2017).

A good example of possible cooperation between authorities and citizens (i.e., smart participation) is "Honolulu Answers,"[1] a website for the inhabitants of Honolulu, Hawaii. In a simple search mask, questions or keywords on topics from the field of municipal government can be entered. Users then receive understandable answers formulated in natural language. Both the questions and the answers are provided by individuals. The database of this website consists of a collection of prepared question-answer pairs written in so-called Civic Write-a-thons. These are regular events where interested citizens can meet the city's officials. They agree on the best possible answers to posed questions and formulate them in a way that everyone understands. Honolulu Answers illustrates a typical web-based question-answering system that remains online but is no longer maintained. Therefore, the search function is limited (Habenstein, D'Onofrio, Portmann, Stürmer, & Myrach, 2016). The discontinuation of online platforms can be explained by the fact that maintenance following the creation of a platform was probably underestimated, as was the time it would take customers to become accustomed to switching from analog to digital using a new communication channel.

However, the authors are convinced that such a concept has potential and therefore have applied it in their research project to advance this kind of cooperation (i.e., fostering smart participation in Switzerland). To illustrate the importance of using a combination of research methods to address a challenge, such as increasing participation rates in Switzerland, a selection of methods is introduced and explained using the example of Honolulu Answers. The individual sections are for understanding and illustrative reasons and have not yet, or only to a certain degree, been executed in this manner. The examples, which are only roughly discussed and do not go into too much detail, help to explain why a single method would be unable to propose a possible solution approach for smart participation, and show why ontological design research can be useful here.

Perspectives of Design Theory

Throughout the history of our planet, no other species has shaped the natural world as much as humans have. Marx (1867) spoke of the "toolmaking animal", a conception that Bergson (1907) and Scheler (1926) transformed into the famous phrase "homo faber". Hence, the notion does not only imply that humans are able to shape the world according to their needs, but that shaping the world is a human need itself. However, it is difficult to define "design". In the 19th century, it acquired today's basic semantic content: The preparatory, modeling process and its result – the designed artefact. The concept currently encompasses a general understanding of technical and organizational planning, conceptualizing and problem-solving, and therefore a very wide range of discourses, fields of action, methods and artefacts (Mareis, 2014; Tabacchi et al., 2018).

Debates in design theory about what design is, its purposes and principles, as well as its social, economic, scientific, philosophical or pragmatic implications, focus mainly on the following aspects (Tabacchi et al., 2018): (1) **Technology-driven design principles versus the ideal of human-centered design**, (2) **Design as a sociocultural process**, (3) **Actors**, and (4) **"Epistemology of design"**. More information on these aspects and their application to the presented example are shown in Table 1.

The different aspects of design theory are discussed and applied to the example at a rather superficial level; the authors deliberately avoid going into too much detail, as this is only an overview of the most important points. There is no claim to completeness. However, it is already evident that such an urban challenge (with the aim of increasing participation within a smart city) requires more than just a certain solution. Based on these aspects, the question now arises as to which method (or combination of methods) is the most suitable for improving communication between the Swiss state and the Swiss population.

Table 1. Design Theory according to Tabacchi et al. (2018)

Aspect	Description	Example
Technology-driven design principles vs. the ideal of human-centered design	The shift in focus from the categories of usefulness and functionality to the perception, reception and usage of design artefacts.	*Technology:* The IT artefact is a question-answering system with the aim of improving interaction between urban stakeholders and the city's responsible agents. *Human-Centered:* The so-called write-a-thons are a starting point to improve the interaction between urban stakeholders and the city's responsible agents.
Design as a sociocultural process	The impact of the social, cultural, economic or political environment on design processes.	*Facts:* Switzerland has a well-functioning infrastructure, is one of the richest countries in the world and aims to be innovative. *Goals:* Transparency (e.g., keeping the stakeholders informed) and participation opportunities. *Conditions:* Keeping control of governmental processes, and "testing before implementing" (i.e., rather cautious about new approaches because of citizens' divergent attitudes towards new technologies). *Challenges:* Multilingualism (i.e., four languages: German, French, Italian, Romansh) and different cultures (i.e., successful implementation of a project in one region is not necessarily scalable to the others).
Actors	Users as active participants in the design process; designing should not be conducted solely by experts; the impact of inanimate artefacts as nonhuman actors.	*Design process:* Not only experts (e.g., scientists, engineers, administrators) are involved in the project but also the users of the IT artefact, especially in the process of writing adequate answers to posed questions. *Inanimate artefact:* Question-answering system that will ultimately be used.
"Epistemology of design"	Designing produces specific forms of knowledge that can be distinguished from those of the natural sciences or the humanities.	"*Knowledge through making*": Insights are gained into the development, design and implementation of a question-answering system as well as into the existing needs, abilities and habits of the Swiss population.

In short, all aspects of design theory are important, and methods, theories and frameworks deriving from them could be part of ontological design research and thus contribute to the development of the cognitive city. Hence, an important connecting point is design science research.

Design Science Research

Design science research does not only comprise computer science but ranges across many fields such as economics, mathematics and psychology. It should be necessarily cross-disciplinary and if possible transdisciplinary, and as such rather a research paradigm (Baskerville, 2008). Cross-interdisciplinarity involves at least two different disciplines relevant to the problem and integrates their insights into a more comprehensive understanding (Newell, 2001), while transdisciplinarity supplements scientific disciplines with expertise from industry (Hirsch Hadorn et al., 2008). Thus, as a science, design science research is about systematically creating new knowledge (i.e., "*knowledge through making*") (Vaishnavi & Kuechler, 2015; Baskerville, 2008). It aims to understand and improve "*the search among potential components in order to construct an artefact that is intended to solve a problem*" (Baskerville, 2008, p. 441).

Meanwhile, design science research has a long tradition in information systems research, at least dating back to the 1960s, when Fuller (1963) called for a *Design Science Decade* and Simon (1969) published his famous book *The Sciences of the Artificial* (cited from Peffers, Tuure, & Niehaves, 2018, p. 129). Today, design science research is an important research approach and an umbrella term encompassing numerous approaches (cf. Baskerville, 2008; Winter, 2008). For instance, Peffers et al. (2018) identify five genres of design science research:

1. **Information Systems Design Theory:** Elaboration of generic design constructs and their interrelationships (cf. Walls, Widmeyer & El Sawy, 1992; 2004).
2. **Design Science Research Methodology:** Design and construction of applicable artefacts, such as systems, applications and methods (cf. Peffers, Tuunanen, Rothenberger & Chatterjee, 2007).
3. **Design-Oriented Information Systems Research:** Development and instruction of the design and operation of information systems and innovative concepts, with a specific set of evaluation methods (cf. Österle, Becker, Frank, Hess, Karagiannis, Krcmar, Loos, Mertens, Oberweis & Sinz, 2011).
4. **Explanatory Design Theory:** Examination of design features and their effects on the environment, for example, on users (cf. Baskersville & Pries-Heje, 2010).
5. **Action Design Research:** Focus on the organizational context of design science research and how this affects the development and use of an artefact (cf. Sein et al., 2011).

All of these contain aspects that could be useful for ontological design research and thus for designing a cognitive city. For reasons of clarity, the authors introduce only one genre of design-scientific research – *action design research*. From their point of view, this is sufficient to understand in which direction design science is heading in general and which problems can be approached with it. As already mentioned, the aim of this article is to offer an overview of different research approaches and there is no claim to completeness.

Action Design Research

Action design research is a research method that combines design science with action research and aims at merging research and practice in information science (Sein et al., 2011). According to Sein et al. (2011), the core of information science is the development and use of IT artefacts (i.e., "ensemble artefacts" resulting from the interplay of design and context (Gregor & Jones, 2007)) in organizational contexts. Therefore, a research method is requested that explicitly recognizes artefacts as ensembles "*emerging from design, use, and ongoing refinement in an organizational context*", shaped by the "*interests, values and assumptions of a wide variety of communities of developers, investors and users*" (Orlikowski & Iacono, 2001, p. 131; cited from Sein et al., 2011, p. 38). Through building and evaluating ensemble IT artefacts, researchers of action design research generate prescriptive design knowledge, reflecting "*not only the theoretical precursors and intent of the researchers but also the influence of users and ongoing use in context*" (Sein et al., 2011, p. 40).

According to Sein et al. (2011) and as illustrated in Figure 2, the following four phases are required to carry out a project based on action design research: (1) Problem formulation, (2) Building, intervention, evaluation, (3) Reflection and learning, and (4) Formalization of learning.

In Table 2, the four phases are described and applied to the example of smart participation to show how this approach would work, since for building the IT artefact in the form of a question-answering system, design science research methods are

Figure 2. Phases of action design research according to (Tabacchi et al., 2018)

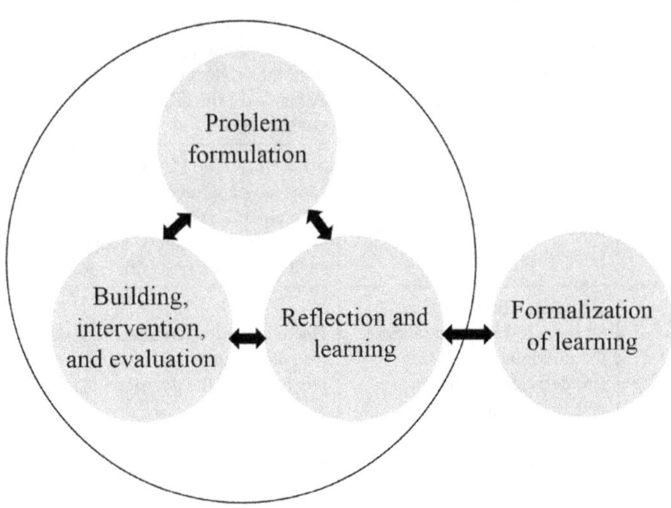

suitable. It is described from the viewpoint of a research group and strives for a collaboration with a partner from the administration.

In contrast to other approaches of design research, action design research is characterized by the fact that working with experts from practice makes finding solutions more of a real possibility. However, action design research focuses on the artefact and its context. Its primary goal is not to integrate the user into the design process in the sense of human-centered design, but rather to improve cooperation between research and practice by applying design knowledge methods. This does not

Table 2. Action design research according to Sein et al. (2011)

Phase	Description	Example
Problem formulation	Development of a research question; decision-making of how to proceed and elaboration of a research design.	*Facts:* In Switzerland, information about urban processes is distributed at three levels (i.e., municipality, cantons, and federal government); thus information retrieval is complicated. *Goals:* Designing an IT-based tool that helps users to find the information easily and intuitively (regardless of where and how it was stored). *(Possible) Solution approach:* Building one question-answering system based on open source. *Implications:* Evaluating which skills and which actors and functions (from research and from practice) are required for this project and decision-making in which environment this will be developed and tested before it is implemented and becomes available to everyone.
Building, intervention, evaluation	Designing, shaping and reshaping the ensemble IT artefact and intervening in organizational work practices are interwoven with ongoing evaluation to intensify the mutual learning of the partners of action design research.	*Design cycles:* The question-answering system is designed, developed and made available for testing in the form of prototypes. The test subjects provide feedback (regarding to usability and content of the product), which, in turn, is incorporated into the existing prototypes. These activities are repeated until the product is performing satisfactorily. Afterwards, the IT artefact is implemented in a real environment.
Reflection and learning	From building a solution for a specific problem to applying that knowledge to a broader class of problems.	*Scalability:* The IT artefact can be scaled in the same language area at the same level (e.g., only in municipalities). *Reuse:* The idea can be adapted to other domains, such as tourism or health care.
Formalization of learning	Documentation of the results on three levels: Generalizing the problem/the solution and deriving the design principles that connect the generalized outcomes to a class of solutions and a class of problems.	*Documentation:* To ensure that knowledge is not lost, all the steps that have been taken, the feedback received, and the decisions made are documented and retained. *Knowledge transfer:* Transferring knowledge is a necessity.

mean that the (individual) stakeholders are unimportant for action design research; on the contrary. However, the human factor is more one aspect among others belonging to the context of an IT artefact. However, design approaches are only one part of ontological design research. Methods of social research should also be considered.

Social Science Research

Social science research can be divided into quantitative and qualitative research, whereby a combination of both approaches is possible as well. Quantitative and qualitative research can be distinguished as follows: In quantitative research the aim is to discover facts and obtain quantifiable evidence of a social phenomenon. The (mostly numerical) data is collected by measurements and its analysis is done by means of numerical comparisons and statistical inferences. Lastly, the results are reported through statistical analysis (Balarabe Kura, 2012). In contrast, qualitative research aims to understand human behavior in natural settings. Therefore, data are collected from informants (i.e., targeted groups) by "*watching, asking or examining*" them (Punch, 2013, p. 86). Data, however, are rarely numerical but more often in other forms, such as text, body movements and feelings. Depending on the objective and the data, the analysis is done based on selected topics (mostly coming from the informants) and results are reported in the words of the informants (Minichiello, Aroni, & Hays, 2008; Punch, 2013).

There are numerous methods available to collect data, especially in qualitative research. The most common are observation, survey, interview and focus group, briefly described below. This list is not exhaustive.

- **Observation:** Acquiring data by observing something or someone in natural or artificial settings. It is about the discovery and use of ethnographic questions to guide what there is to see and hear (Spradley, 2016).
- **Survey:** Acquiring data using a list of questions that a target group can complete through different means, depending on the communication channel (e.g., Internet, telephone, e-mail) (Sue & Ritter, 2012).
- **Interview:** Acquiring data using a list of questions that a target person or group can answer in a face-to-face conversation (Minichiello, 2008).
- **Focus group:** Acquiring data in a planned discussion, whereby participants are deliberately selected according to demographic differences, among other factors (Breen, 2007).

These approaches can be useful for ontological design research as well and therefore are helpful in building a cognitive city. To gain an understanding of the range of methods, the authors introduce *survey research* as one of the qualitative analysis methods. Again, the aim of this article is not to obtain a detailed analysis of every method, but a sense of the available different research approaches.

Survey Research

Survey research is "*the collection of information from a sample of individuals through their responses to questions*" (Check & Schutt, 2012, p. 160, cited from Ponto, 2015). It is primarily a matter of finding people who are willing to take part in a survey, whereby researchers can obtain opinions (data) from them to answer their research questions. This type of method is often used to study human behavior and thus can often be found in social and psychological scientific fields (Ponto, 2015).

According to Sue and Ritter (2012), the following sequential steps are necessary to conduct a survey in general: (1) Definition of survey objectives, (2) Evaluation of survey methods and sampling options, (3) Creation of questionnaire, (4) Pretesting (and adjustment), (5) Launch of survey, and (6) Data collection, analysis and reporting. These steps are explained in Table 3. Furthermore, with today's habits of doing a lot (if not everything) using the Internet, the authors use an online survey as an example and show to what extent this can be useful for the situation presented here.

In this approach, the focus is on humans and their needs and perceptions, which are collected with the help of questions. In contrast to design science research, the focus is on perceptions (e.g., to determine to what extent humans would accept a new IT artefact) rather than on the development of the IT artefact itself. Compared to design science research, social science research is past-oriented and tries to collect existing information, but this is just as important for building a cognitive city. However, not only should individual research directions be examined but also their combinations.

Combining Social Science Research With Design Science Research

Combining social science methods with design science research has already been suggested (cf. Bider, Johannesson, & Perjons, 2013; Gregor & Baskerville, 2012). Researchers following this research strategy want to bridge the gap between the social sciences and the design sciences, as they are convinced that each of these fields of research is often incomplete without the other.

For instance, Bider et al. (2013, p. 464) identified three general possibilities to combine design science research with empirical research:

Table 3. Survey research according to Sue and Ritter (2012)

Phase	Description	Example
Definition of survey objectives	Development of distinct research objectives, by acquiring the findings of other researchers, learning from their experiences and identifying research gaps. It is also possible to contact experts directly to learn more about their research and to ask for support.	*Hypotheses:* Urban stakeholders want a more transparent and simpler exchange of information. *Status quo:* There are already numerous publications. They must be examined. Studies from Switzerland are to a certain degree scalable; studies from countries in (South) America or Asia are not scalable, but they can provide information on how a survey might look. *Expert meetings:* It makes sense to meet with experts from research and practice to discuss the topic based on their experiences.
Evaluation of survey methods and sampling options	Decision through which communication channel (e.g., Internet, telephone, e-mail) the study should be conducted, and whether a random group of respondents or a selected target group should complete the survey.	*Communication channel:* The survey is made available via the Internet. *Target group:* No demographic data criteria are defined to address a specific target group. *Differentiation:* To distinguish between experts and nonexperts (as it is possible that experts who are familiar with smart participation also participate), a filter is used at the beginning, which then separates the two groups. This helps to better classify the results.
Creation of questionnaire	Decision about closed or open questions; writing valid and reliable questions using simple and direct language; choice of tool to conduct the survey.	*Question type:* Closed questions are formulated with a selection of answers, whereby at the end the interviewee can provide additional information in the form of text. *Language:* The questions are formulated in a manner that is appropriate for the target group (i.e., for experts in more detail, for nonexperts as simply as possible), but they must nevertheless be comparable in terms of content. *Tool*: A Web-based survey tool supported by the University of Fribourg is used.
Pretesting (and adjustment)	Trial of survey with a selected target group; comparison of the data received with the defined research questions (i.e., is it possible to answer the research questions); based on generated data adjustments in questions.	*Trial:* To ensure that the questions retrieve the information that the research group needs (e.g., acceptance or rejection of new technologies such as question-answering systems), (usually two) pretests are carried out. *Adjustments:* Based on received feedbacks, adjustments (e.g., in structure and content) are made.
Launch of survey	Activation of the survey; sending of survey invitations and subsequent reminders; monitoring of the survey.	*Activation:* When the answers of the last pretest appear useful, the survey is put online (on a web page). *Invitation:* The survey link is sent to potential participants via e-mails, messages and other means. *Reminders:* Sometimes reminders are needed to encourage people to participate. *Monitoring:* Continuous monitoring of whether and how the survey form is completed is a main task.
Data collection, analysis and reporting	Data collection during the survey depending on the tool and deciding which instruments are to be used for analysis. Documentation of the results and answering the research questions.	*Data collection:* The data is collected automatically through the online survey tool. *Data analytics:* Simple analysis tools are already available in the tool. Further calculations can be made, for instance, using the SPSS program. *Documentation:* The results obtained confirm or reject the defined hypotheses and are summarized in a research report.

1. **Empirical Research as Basis:** Empirical research and/or results of previous empirical research studies are used when designing and testing a new solution/ artefact (cf. Gnewuch, Morana, Adam & Maedche, 2018).
2. **Empirical Research as Evaluation Step:** Design science research creates a new reality that can be investigated using empirical research methods. This happens after a solution/artefact built in a design science research project has been adopted in practice (cf. Müller et al., 2018).
3. **Empirical Research automated by Design Science:** Design science research is used for creating tools and conducting experiments within empirical research projects (cf. Doyle, Sammon & Neville, 2015).

Gregor and Baskerville (2012, p. 7/8) proposed a fusion model that connects social science and design science research. In this model, there are five linked and partly overlapping research phases, illustrated in Figure 3:

However, the question is to what extent the combination of design science and social science research phases could lead to a better result. According to Gregor and Baskerville (2012, p. 7/8) and the two other described approaches (in this context, action design research and survey research), the steps in a fusion of research directions would be as follows:

In this example and according to Bider et al. (2013), survey research serves as an evaluation in the development of an IT artefact. Something new is created and must be evaluated in the real environment to observe its influences on society. Based on this example, it is obvious that the two research directions do not compete but complement each other.

Figure 3. The fusion model for design and social science research according to Gregor and Baskerville (2012)

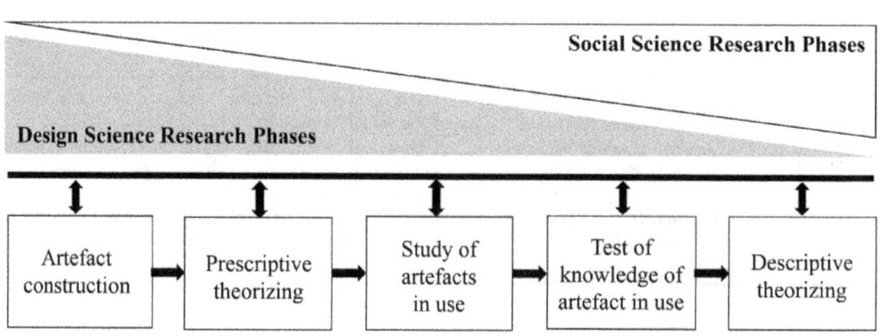

Table 4. The fusion model for design and social science research according to Gregor and Baskerville (2012)

Phase	Description	Example
Artefact construction	Development, implementation and testing of the artefact (i.e., the typical research activities of design science research).	*Construction:* The question-answering system is based on open source; designed, developed and made available for testing; continuously improved until its performance is satisfactory.
Prescriptive theorizing	Formalization of generalizable design knowledge.	*Design knowledge:* Findings and possible scalability to other areas are documented.
Study of the artefact in use	Utilization of the artefact in a real environment. Hypotheses may derive from the study of the artefact's interaction with its environment.	*Real environment:* The question-answering system is implemented in a real environment and is now available to users. *Hypotheses:* For example, urban stakeholders want a more transparent and simpler exchange of information.
Tests of knowledge of artefacts in use	Testing and validation of findings observed in phase 3.	*Validation:* An online survey can be launched in which participants can express their perceptions of the IT artefact. Questions about usability, the potential of the tool and personal opinions can be asked.
Descriptive theorizing	Formulation of the knowledge gained as descriptive knowledge and its integration into theory.	*Documentation:* The results obtained confirm or reject the defined hypothesis. *Knowledge transfer:* It helps to advance the theory about user acceptance in the area of question-answering systems.

Ontological Design Research

Design science research represents future-oriented research (e.g., construction of new IT artefacts), while social science research represents past-oriented research (e.g., collection of existing impressions when using an IT artefact). Even if these two research directions differ and, as the examples above illustrate, do not interfere but complement each other, many researchers continue to use only one approach of one research direction and do not consider looking at all research methods and then choosing the appropriate method or mixing different methods depending on the challenge (and not depending on skills!). Therefore, the authors of this article proposed this short description of various methods, one that might prove useful to researchers.

The vision of ontological design research or the idea of the research methodology toolbox is not to provide a predefined combination of research methods enabling the construction of a cognitive city. Rather, the aim is to demonstrate the diversity of the spectrum of approaches, methods and tools that might be applied to achieve a cognitive city. In a cognitive city, humans and cognitive systems learn from and with each other, and they mutually influence and change each other (Finger & Portmann, 2016). Formulated as a principle, this can be described as "ontological design": "*We shape our world as this world affects and shapes us*" (Willis, 2006, p. 70). The challenge is now to systematically use this principle to design and build a cognitive city.

The change of perspective respecting the awareness of reflecting one's actions is vital because everyone contributes to urban development. Therefore, it is essential that everyone consider what kind of future we would like and how we can achieve this vision. Such considerations must be incorporated into everyday actions. In terms of the research environment, this would imply that when a challenge arises, researchers must first consider which outcome is desirable (among others for society) and whether and to what extent the individual available methods create added value for the elaboration of this problem. This also means that habits must be reflected. Even if previous approaches and methods have worked, this does not mean that there are no better ways of solving the problem (Tschudi & D'Onofrio, 2018). Reflection is crucial to advancing urban development in a way that is profitable for society and the environment.

As explained above, our vision of a cognitive city involves the collaboration between human and cognitive systems to achieve, among other things, a human-machine symbiosis. It is about working together, reflecting together and collectively developing solution approaches to urban problems. Ontological design research allows one to create new thought patterns, take new paths and support urban development in a more holistic way.

CONCLUSION

Cities are complex sociotechnical systems that only function if the technical and social components, for instance information and communication technologies and humans, are interacting with each other. Design science methods are suitable for the development of an IT artefact that should help to solve a specific problem. In contrast, empirical methods aim to describe phenomena or to point out possible reasons for a development. Empirical approaches, thus, pursue quite different objectives from those of design science methods. Neither one research method nor the other is the optimal solution to more complex problems within a smart city in terms of a cognitive city. Depending on the problem, the objective to be achieved, the available resources (skills, money, data, etc.), as well as the urgency, some methods are applicable while others are not. The most important thing, however, is to become aware that the different methods can be intertwined, as Bider et al.

(2013) and Gregor and Baskerville (2012) have shown. Based on the principle of ontological design, it is essential for stakeholders to be aware of how the city they live in is moving forward. Every decision that is made has an influence on one's own self-development. This also applies to researchers: With their work, they influence the continued existence of a (research) environment. Therefore, they must be aware in advance which approaches to consider in addressing a particular problem, and if they want to tackle the challenge alone or in collaboration with others. Ontological design research is a research methodology toolbox in which exactly this aspect is considered: To respond to urban challenges in the best possible way, regardless of whether new paths are needed or whether the familiar ones can be taken. In the end, the goal is to build an efficient, sustainable and resilient city: The cognitive city (Finger & Portmann, 2016). Thus, it makes sense to develop a research methodology toolbox that could be called "ontological design research". This can be applied to develop research and development strategies that are suitable for shaping the relationship between human and environment; in our case, the sociotechnical human-machine symbiosis of the cognitive city by adopting ontological design as a design principle.

FUTURE RESEARCH DIRECTIONS

This article discussed the limitations of smart cities and questioned the clear distinction of research directions with regard to creating a cognitive city. On this basis, the following future directions are planned:

1. **Building an Understanding of the Term "City":** Before proposing solution approaches for cities, it is necessary to understand the concept of a sociotechnical system, as well as the city (i.e., "smart city," "human smart city," and "cognitive city") and to acquire a common understanding of that concept.
2. **Development of Ontological Design Research:** Ontological design research is a research methodology toolbox to be developed. The aim is to create a toolbox in which various approaches, methods and tools are available to develop suitable strategies for problem-solving in the cognitive city.

REFERENCES

Albino, V., Berardi, U., & Dangelico, R. M. (2015). Smart Cities. Definitions, Dimensions, Performance, and Initiatives. *Journal of Urban Technology*, *22*(1), 3–21. doi:10.1080/10630732.2014.942092

Balarabe Kura, S. Y. (2012). Qualitative and Quantitative Approaches to the Study of Poverty: Taming the Tensions and Appreciating the Complementarities. *Qualitative Report, 17*(20), 1–19.

Baskerville, R. (2008). What Design Science is Not. *European Journal of Information Systems, 17*(5), 441–443. doi:10.1057/ejis.2008.45

Baskerville, R., & Pries-Heje, J. (2010). Explanatory Design Theory. *Business & Information Systems Engineering, 2*(5), 271–282. doi:10.100712599-010-0118-4

Baxter, G., & Sommerville, I. (2011). Socio-Technical Systems: From Design Methods to Systems Engineering. *Interacting with Computers, 23*(1), 4–17. doi:10.1016/j.intcom.2010.07.003

Bergson, H. (Ed.). (1907). *L'Evolution créative*. Paris: Aclan.

Bider, I., Johannesson, P., & Perjons, E. (2013). Using Empirical Knowledge and Studies in the Frame of Design Science Research. Design Science at the Intersection of Physical and Virtual Design, 1939, 463-470. doi:10.1007/978-3-642-38827-9_38

Breen, R. L. (2007). A Practical Guide to Focus-Group Research. *Journal of Geography in Higher Education, 30*(3), 463–475. doi:10.1080/03098260600927575

Check, J., & Schutt, R. K. (2012). Survey research. In J. Check & R. K. Schutt (Eds.), *Research methods in education* (pp. 159–185). Thousand Oaks, CA: SAGE Publications. doi:10.4135/9781544307725.n8

D'Onofrio, S., Müller, S. M., Papageorgiou, E. I., & Portmann, E. (2018). Fuzzy Reasoning in Cognitive Cities – An Exploratory Work on Fuzzy Analogical Reasoning Using Fuzzy Cognitive Maps. In *International Conference on Fuzzy Systems (FUZZ-IEEE 2018)*. Rio de Janeiro, Brazil: Academic Press. 10.1109/FUZZ-IEEE.2018.8491474

D'Onofrio, S., & Portmann, E. (2017). Cognitive Computing in Smart Cities. *Informatik Spektrum, 40*(1), 46–57. doi:10.100700287-016-1006-1

D'Onofrio, S., & Tschudi, F. (2018). Design Thinking: Ein Videoprojekt zu Computing with Words schlägt neue Brücken zwischen Forschung und Praxis. In A. Meier & R. Seising (Eds.), Vague Information Processing (pp. 510-527). Springer Fachmedien Wiesbaden.

Doyle, C., Sammon, D., & Neville, K. (2015). Building an evaluation framework for social media-enabled collaborative learning environments (SMECLEs). *Journal of Decision Systems, 24*(3), 298–317. doi:10.1080/12460125.2015.1054694

Eason, K. (2014). Afterword: The Past, Present and Future of Sociotechnical Systems Theory. *Applied Ergonomics*, *45*(2), 213–220. doi:10.1016/j.apergo.2013.09.017 PMID:24183568

Finger, M., & Portmann, E. (2016). What Are Cognitive Cities? In E. Portman & M. Finger (Eds.), *Towards Cognitive Cities. Advances in Cognitive Computing and Its Application to the Governance of Large Urban Systems* (pp. 1–11). Heidelberg, Germany: Springer International Publishing.

Finger, M., & Razaghi, M. (2017). Conceptualizing "Smart Cities". *Informatik Spektrum*, *40*(1), 6–13. doi:10.100700287-016-1002-5

Fredericks, J., Hespanhol, L., Parker, C., Zhou, D., & Tomitsch, M. (2018). Blending Pop-up Urbanism and Participatory Technologies: Challenges and Opportunities for Inclusive City Making. *City. Cultura e Scuola*, *12*, 44–53.

Gnewuch, U., Morana, S., Adam, M. T. P., & Maedche, A. (2018). Faster Is Not Always Better: Understanding the Effect of Dynamic Response Delays in Human-Chatbot Interaction. *European Conference on Information Systems (ECIS 2018).*

Gregor, S., & Baskerville, R. (2012). The Fusion of Design Science and Social Science Research. *Information Systems Foundation Workshop.*

Gregor, S., & Jones, D. (2007). The Anatomy of a Design Theory. *Journal of the Association for Information Systems*, *8*(5), 312–335. doi:10.17705/1jais.00129

Habenstein, A., D'Onofrio, S., Portmann, E., Stürmer, M., & Myrach, T. (2016). Open Smart City: Good Governance für smarte Städte. In A. Meier & E. Portmann (Eds.), *Smart City: Strategie, Governance und Projekte* (pp. 47–71). Wiesbaden: Springer Vieweg. doi:10.1007/978-3-658-15617-6_3

Hillier, B. (2012). The City as a Socio-technical System: A Spatial Reformulation in the Light of the Levels Problem and the Parallel Problem. *Digital Urban Modeling and Simulation*, *242*, 24–48. doi:10.1007/978-3-642-29758-8_3

Hirsch Hadorn, G., Hoffmann-Riem, H., Biber-Klemm, S., Grossenbacher-Mansuy, W., Joye, D., Pohl, C., ... Zemp, E. (2008). *Handbook of Transdisciplinary Research.* Springer. doi:10.1007/978-1-4020-6699-3

Hollingshead, A. B. (2001). Cognitive Interdependence and Convergent Expectations in Transactive Memory. *Journal of Personality and Social Psychology*, *81*(6), 1080–1089. doi:10.1037/0022-3514.81.6.1080 PMID:11761309

Hughes, T. P. (1987). The Evolution of Large Technological Systems. In W. E. Bijker, T. P. Hughes, & T. J. Pinch (Eds.), *The Social Construction of Technological Systems. New Directions on the Sociology and History of Technology* (pp. 51–82). Cambridge, MA: MIT Press.

Hurni, T., Huber, T. L., & Dibbern, J. (2017). Swiss Software Industry Survey 2017. Current State, Emerging Trends & Long-term Developments in the Swiss Software Industry. *ICT Switzerland*. Retrieved from: https://ictswitzerland.ch/media/dateien/Studien/Swiss_Software_Industry_Survey_2017-Report.pdf

Hurwitz, J., Kaufman, M., & Bowles, A. (Eds.). (2015). *Cognitive Computing and Big Data Analytics*. Indianapolis, IN: John Wiley and Sons.

Ito, M., Gutiérrez, K., Livingstone, S., Penual, B., Rhodes, J., Salen, K., ... Watkins, S. C. (Eds.). (2013). *Connected Learning: An Agenda for Research and Design*. Irvine, CA: Digital Media and Learning Research Hub.

Kaltenrieder, P., Altun, T., D'Onofrio, S., Portmann, E., & Myrach, E. (2016). Personal digital assistant 2.0 – A software prototype for cognitive cities. *IEEE International Conference on Fuzzy Systems (FUZZ-IEEE 2016)*. 10.1109/FUZZ-IEEE.2016.7737872

Krasnova, H., Abramova, O., Notter, I., & Baumann, A. (2016). Why Phubbing is Toxic for Your Relationship: Understanding the Role of Smartphone Jealousy among "Generation Y" Users. *European Conference in Information Systems (ECIS 2016)*.

Kummitha, R. K. R., & Crutzen, N. (2017). How Do We Understand Smart Cities? An Evolutionary Perspective. *Cities (London, England)*, *67*, 43–52. doi:10.1016/j.cities.2017.04.010

Malone, T. W. (Ed.). (2018). *Superminds: The Surprising Power of People and Computers Thinking Together*. Kindle Edition.

Malone, T. W., & Bernstein, M. S. (Eds.). (2015). *Handbook of Collective Intelligence*. Cambridge: MIT Press.

Mareis, C. (Ed.). (2014). *Theorien des Design*. Hamburg, Germany: Junius-Verlag.

Marx, K. (Ed.). (1959). Das Kapital: Vol. I. Marx/Engels-Gesamtausgabe. Berlin: Dietz Verlag. (Originally Published 1867)

Metzger, S., Portmann, E., Finger, M., Habenstein, A., Riedle, A., & Witschi, R. (2018). Human Smart City – der Mensch im Zentrum. *Transforming Cities*, *1*, 62–67.

Minichiello, V., Aroni, R., & Hays, T. N. (Eds.). (2008). *In-Depth Interviewing: Principles, Techniques, Analysis*. Pearson Education Australia.

Moyser, R., & Uffer, S. (2016). From Smart to Cognitive: a Roadmap for the Adoption of Technology in Cities. In E. Portmann & M. Finger (Eds.), *Towards Cognitive Cities. Advances in Cognitive Computing and Its Application to the Governance of Large Urban Systems* (pp. 13–35). Heidelberg, Germany: Springer International Publishing. doi:10.1007/978-3-319-33798-2_2

Müller, S. M., D'Onofrio, S., & Portmann, E. (2018). Fuzzy Analogical Reasoning in Cognitive Cities – A Conceptual Framework for Urban Dialogue Systems. *20th International Conference on Enterprise Information Systems (ICEIS 2018)*. 10.5220/0006816103530360

Newell, W. H. (2001). A Theory of Interdisciplinary Studies. *Issues in Integrative Studies*, *19*, 1–25.

Oliveira, A., & Campolargo, M. (2015). From Smart Cities to Human Smart Cities. *48th International Conference on System Sciences*, 2336-2344. 10.1109/HICSS.2015.281

Orlikowski, W., & Iacono, C. (2001). Desperately Seeking the "IT" in IT Research – A Call to Theorizing the IT Artifact. *Information Systems Research*, *12*(2), 121–134. doi:10.1287/isre.12.2.121.9700

Österle, H., Becker, J., Frank, U., Hess, T., Karagiannis, D., Krcmar, H., ... Sinz, E. J. (2011). Memorandum on design-oriented information systems research. *European Journal of Information Systems*, *20*(1), 11–15. doi:10.1057/ejis.2010.55

Österle, H., & Otto, B. (2010). Consortium Research: A Method for Researcher-Practitioner Collaboration in Design-Oriented IS Research. *Business & Information Systems Engineering*, *2*(5), 283–293. doi:10.100712599-010-0119-3

Patorniti, N. P., Stevens, N. J., & Salmon, P. M. (2017). A Systems Approach to City Design: Exploring the Compatibility of Sociotechnical Systems. *Habitat International*, *66*, 42–48. doi:10.1016/j.habitatint.2017.05.008

Peffers, K., Tuunanen, T., & Niehaves, B. (2018). Design Science Research Genres: Introduction to the Special Issue on Exemplars and Criteria for Applicable Design Science Research. *European Journal of Information Systems*, *27*(2), 129–139. doi:10.1080/0960085X.2018.1458066

Peffers, K., Tuunanen, T., Rothenberger, M. A., & Chatterjee, S. (2007). A Design Science Research Methodology for Information Systems Research. *Journal of Management Information Systems, 24*(3), 45–78. doi:10.2753/MIS0742-1222240302

Petersson, A. M., & Lundberg, J. (2016). Applying action design research (ADR) to develop concept generation and selection methods. *Procedia CIRP, 50*, 222–227. doi:10.1016/j.procir.2016.05.024

Pfäffli, M., Habenstein, A., Metzger, S., & Portmann, E. (2018). Eine Architektur zur Transformation von Städten in Human Smart Cities. In M. Knoll (Ed.), *IT-Architektur*. HMD der Wirtschaftsinformatik. doi:10.136540702-018-00451-z

Ponto, J. (2015). Understanding and Evaluating Survey Research. *Journal of the Advanced Practitioner in Oncology, 6*(2), 168–171. PMID:26649250

Portmann, E., & Metzger, S. (2017). PostGrid: Das smarte Netzwerk der Schweizerischen Post für eine intelligentere Stadt. *Informatik Spektrum, 40*(1), 88–99. doi:10.100700287-016-1012-3

Portmann, E., & Pletscher, C. (2016). *Transdisziplinarität als Antwort auf die digitale Transformation. In BeWL Departement Betriebswirtschaftslehre* (pp. 33–35). Universität Bern.

Punch, K. F. (Ed.). (2013). *Introduction to Social Research: Quantitative and Qualitative Approaches*. SAGE.

Rophol, G. (1988). Allgemeine Technologie der Netzwerke. *Technikgeschichte, 56*, 153–162.

Scheler, M. (Ed.). (1960). Die Wissensformen und die Gesellschaft. Bern: Francke. (Originally published 1926)

Sein, M. K., Henfridsson, O., Purao, S., Rossi, M., & Lindgren, R. (2011). Action Design Research. *Management Information Systems Quarterly, 35*(1), 37–56. doi:10.2307/23043488

Siemens, G. (2005). Connectivism: A learning theory for the digital age. *International Journal of Instructional Technology and Distance Learning, 2*(1), 3–10.

Simon, H. A. (Ed.). (1969). *The Sciences of the Artificial.* Cambridge, MA: The MIT Press.

Smith, R. C., Bossen, C., & Kanstrup, A. M. (2017). Participatory design in an era of participation. *CoDesign, 13*(2), 65–69. doi:10.1080/15710882.2017.1310466

Spradley, J. P. (Ed.). (2016). *Participant Observation*. Waveland Press.

Sue, V. M., & Ritter, L. A. (Eds.). (2012). *Conducting Online Surveys*. SAGE. doi:10.4135/9781506335186

Tabacchi, M. E., Portmann, E., Seising, R., & Habenstein, A. (2018) Introduction. In E. Portmann, M.E. Tabacchi, R. Seising, & A. Habenstein (Eds.), Designing Cognitive Cities. Heidelberg, Germany: Springer International Publishing.

Tschudi, F., & D'Onofrio, S. (2018). Wie wir durch unsere Denk- und Handlungsmuster beeinflusst werden. In A. Meier & R. Seising (Eds.), Vague Information Processing (pp. 567-571). Springer Fachmedien Wiesbaden.

Vaishnavi, V., & Kuechler, W. (2015). *Design Science Research in Information Systems*. Retrieved from http://desrist.org/desrist/article.aspx

Viale Pereira, G., Cunha, M. A., Lampoltshammer, T. J., Parycek, P., & Testa, M. G. (2017). Increasing collaboration and participation in smart city governance: A cross-case analysis of smart city initiatives. *Information Technology for Development*, *23*(3), 526–553. doi:10.1080/02681102.2017.1353946

Walls, J. G., Widmeyer, G. R., & El Sawy, O. A. (1992). Building an Information Systems Design Theory for Vigilant EIS. *Information Systems Research*, *3*(1), 36–59. doi:10.1287/isre.3.1.36

Walls, J. G., Widmeyer, G. R., & El Sawy, O. A. (2004). Assessing Information System Design Theory in Perspective: How useful was our 1992 Initial Rendition? *Journal of Information Technology Theory and Application*, *6*(2), 43–58.

Willis, A. M. (2006). Ontological Designing. *Design Philosophy Papers*, *4*(2), 69–92. doi:10.2752/144871306X13966268131514

Winter, R. (2008). Design Science Research in Europe. *European Journal of Information Systems*, *17*(5), 470–475. doi:10.1057/ejis.2008.44

Zadeh, L. A. (1988). Fuzzy Logic. *IEEE Computer*, *21*(4), 83–93. doi:10.1109/2.53

ADDITIONAL READING

D'Onofrio, S., Müller, S. M., Papageorgiou, E. I., & Portmann, E. (2018). Fuzzy Reasoning in Cognitive Cities – An Exploratory Work on Fuzzy Analogical Reasoning Using Fuzzy Cognitive Maps. In *International Conference on Fuzzy Systems (FUZZ-IEEE 2018)*. Rio de Janeiro, Brazil. 10.1109/FUZZ-IEEE.2018.8491474

Hurwitz, J., Kaufman, M., & Bowles, A. (Eds.). (2015). *Cognitive Computing and Big Data Analytics*. Indianapolis: John Wiley and Sons.

Malone, T. W. (Ed.). (2018). *Superminds: The Surprising Power of People and Computers Thinking Together*. Kindle Edition.

Müller, S. M., D'Onofrio, S., & Portmann, E. (2018) Fuzzy Analogical Reasoning in Cognitive Cities – A Conceptual Framework for Urban Dialogue Systems. In *20th International Conference on Enterprise Information Systems (ICEIS 2018)*. Funchal, Portugal. 10.5220/0006816103530360

Portman, E., & Finger, M. (Eds.). (2016). *Towards Cognitive Cities – Advances in Cognitive Computing and its Application to the Governance of Large Urban Systems*. Heidelberg: Springer International Publishing.

Portmann, E., Tabacchi, M. E., Seising, R., & Habenstein, A. (Eds.). (2018). *Designing Cognitive Cities*. Heidelberg: Springer International Publishing.

Siemens, G. (2005). Connectivism: A learning theory for the digital age. *International Journal of Instructional Technology and Distance Learning*, *2*(1), 3–10.

Townsend, A. M. (Ed.). (2013). *Smart Cities. Big Data, Civic Hackers, and the Quest for a New Utopia*. New York: W.W. Norton Inc.

KEY TERMS AND DEFINITIONS

Cognitive Systems: Cognitive systems can not only recognize the meaning of processed information but also create cognitive linkages between information and learn from it. They are able to process fuzzy data and to include the abilities of human cognition to a certain extent in information processing.

Design: Design is about creating an artefact. The aim is to prepare, model, and then execute the development of an artefact. The whole process is understood as design, and the goal is ultimately a designed artefact.

Ontological Design (Research): Based on the idea that humans shape their world in such a way that this world affects and shapes them, ontological design research refers to a research methodology toolbox that can be used to develop strategies to shape the relationship between humans and the environment.

Question-Answering System: A question-answering system, regarded as an IT artefact, aims to automatically deliver relevant and concise answers to questions. It represents a dialogue system that is accessible through an interface, such as a web application, to conduct a conversation with a human.

Smart City: The term smart city is often used to describe cities that enrich functions relevant to the city using advanced information and communication technologies that can contribute to an efficient and sustainable development of the social-ecological design of urban space.

Smart Participation: Smart participation is about giving urban stakeholders the opportunity to be involved in the development of the city. Various possibilities for participation, such as discussion rounds (from passive listening to active involvement), are conceivable.

Sociotechnical System: A sociotechnical system refers to a network of technical and social components of which proper functioning depends on the interaction and balance of these components. A city that is a system of systems is such a sociotechnical system.

Urban Intelligence: Urban intelligence is the cognitive intelligence of the city. A network based on individual intelligences (humans, electronic devices, etc.) creates unique and valuable insights from the city to foster its development.

ENDNOTE

1 cf. https://hnlgovanswers.herokuapp.com

Chapter 9
Role of Electronics Devices for E-Health in Smart Cities

Davinder Singh Rathee
Adama Science and Technology University, Ethiopia

Kiran Ahuja
DAV Institute of Engineering and Technology, India

Tadesse Hailu
Adama Science and Technology University, Ethiopia

ABSTRACT

As the world's population ages, those suffering from diseases will increase. Researchers in electronics, computer, networking, and medical fields need to work more seriously to make the broad vision of smart healthcare/e-health system. To achieve the objectives of e-healthcare in smart cities, there is a need to create new system that allows the acquisition of health information smartly, automatically, and transparently in order to take efficient decisions provided by the supporting system. Such systems may be designed technically by embedded together communication, smart signals, internet of things, network of sensors.

INTRODUCTION

In present scenario developments of smart cities are in early stages (Maldague et al, 2001) mainly in terms of traffic management and pollution control (Modest et al, 2013; Vollmer et al, 2011), water management system (Zissis et al, 1978), public Wi-Fi, historical heritage perseveration (Gaussorgues et al, 1994). Also, IoT entering into every aspect of life i.e. health care, industrial control, security,

DOI: 10.4018/978-1-5225-8085-0.ch009

building health and in emergency services (Holler et al, 2014; Kortuem et al, 2010; Romer et al, 2010; Guinard et al, 2010). Due to the impact of ICT (Information and Communications Technologies), smart cities can be imagined. Where most of the services are based on monitoring of different variables via sensors in order to extract and analyze useful information ("Orchestrating infrastructure", n.d.) for applications like monitoring of pollutions, traffic control, water supply, public Wi-Fi, library database, power management ("Smart cities", n.d; Nasirudin et al, 2011; Leccese et al, 2014). In this chapter, our main concentration is on new health care system which may not only promise for attractive applications for health monitoring like elderly care, the fitness of individuals, remote health monitoring i.e. B.P sugar etc. seamlessly and also suggest treatment and medication at the user end. To build such system undoubtedly researcher in the field of electronics, ICT and medical field need to work out more seriously in order to make the broad vision of smart e-health system. So, there is a need of diagnostic and imaging devices, sensors having the capability to connect with the cloud for providing health care services at reduced cost to increase the quality of life (Vasanth et al, n.d.). Additionally, here main concentration is on health monitoring for human, animals, and machines in smart cities using Infrared thermography (IRT) ("ISTAG", 2013). The basic principle to collect the information is capturing of radiations emitting from the object under investigation (Meola et al, 2012). IRT has many applications in the field of medicine, maintenance and process monitoring (Gorostiza et al, 2011), electrical as well as mechanical maintenance in order to avoid costly breakdowns (Lahiri et al, 2012). Fault detection is confirmed due to abnormal patterns of temperature measurement especially in electrical equipment (Purohit et al, 2008). Also, excessive fraction may be noticeable in case of mechanical equipment.

IRT is a science which acquires thermal patterns from the object with the help of pyrometer, process it. The beauty of the data acquisition method is to contact less measurement (Bouge et al, 2013). The fundamental unit in this system is pyrometer which acquires the information through infra radiations emitted from objects and process in order to transform into an electrical signal.

IRT is a non-contact technology. The devices used are not in contact with the source of heat, so it is safe to measure extremely hot as well as a dangerous product like acids. Main Advantages of IRT methods are given below:

- Real-time 2-D thermal information can be acquired by this method.
- IRT does not radiate harmful radiations like X-Ray imaging, so suitable for repeated use.
- IRT is painless and remotely accessible, so can be used for measuring the health of not only animals but other variables as well.
- IRT can be a diagnostic tool for various medical applications.

- IRT can be an early indicator of health as it can measure small changes in body temperature. So, increase chances of survival by early detection of body tumors, breast cancer etc
- Also, this method is helpful in diagnoses of vascular disorders, fever screening, dermatological infections and for heart operations (Al-Kassir et al, 2005).

HOW INFRARED THERMOGRAPHY WORKS

The basic principle is the collect the radiated heat from the samples. In normal conditions, the heat radiated by objects is different than the faulty conditions. Variation in such energy is directly proportional to the temperature of the material. To collect exact radiations, there is a need to concentrate on the dissipated energy from any of the method liked absorption, transmission, and reflection (Ibarra-Castanedo, Tarpani, & Maldague, 2013). So, all these three parameters are wavelength dependent and known as the absorptivity (αA), transmissivity αT and reflectivity (αR) of the body (Mollmann, Karstadt, Pinno, & Vollmer, 2005). The relation between these parameters is shown in equation 1 and Figure 1

$$\alpha A + \alpha T + \alpha R = 1 \tag{1}$$

Further, the Wien's Law presents the relation between temperature and Wavelength

$$\lambda peak = 0.0029/ T \tag{2}$$

Figure 2 describes the relation between peak wavelength and temperature graphically.

The total hemispherical radiation intensity of a blackbody can be measured using Stefan–Boltzmann formula, which states about the power radiated from black body in terms of its temperature and described in energy radiated per unit surface area across all wavelengths per unit time as shown below in Figure 3.

Further, Figure 3 gives the concept of acquiring temperature from the body of interest. The basic concept is that every material shows different thermal behavior. The object will dissipate different radiations with respect to heating and cooling of the source at various rates. Then, these variations result in temperature variations on the surface of the body under inspection, which can be observed by using the infrared camera. The IRT method can be implemented in two ways i.e. active approach, which is more common in thermography. Firstly, stimulation i.e. optical, mechanical or inductive will be applied on samples due to that heat will generate and propagates as a thermal wave from the surface through the specimen, which

Figure 1. Planck's law: EM radiation Vs Wavelength at a definite temperature. (a) Middle IR wave radiated at high temperature (b) Long wave radiation corresponding low temperature

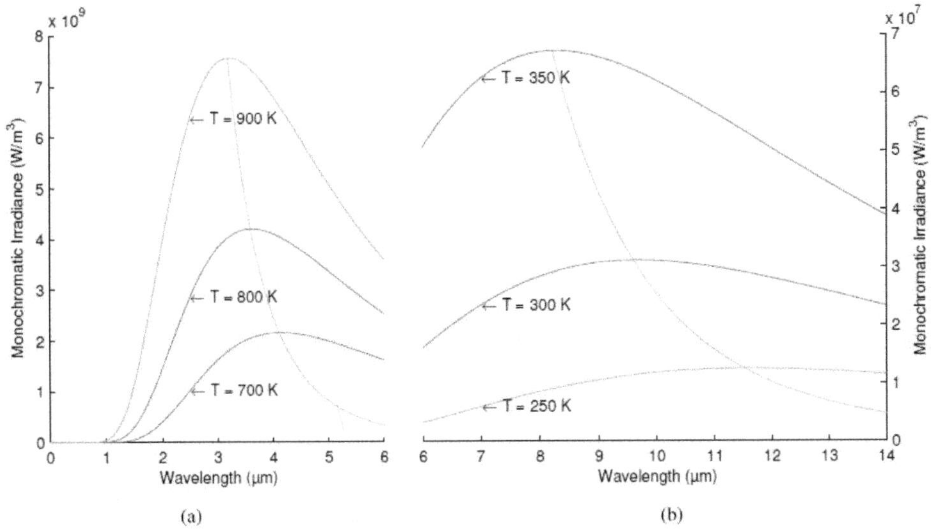

Figure 2. Wien's displacement law

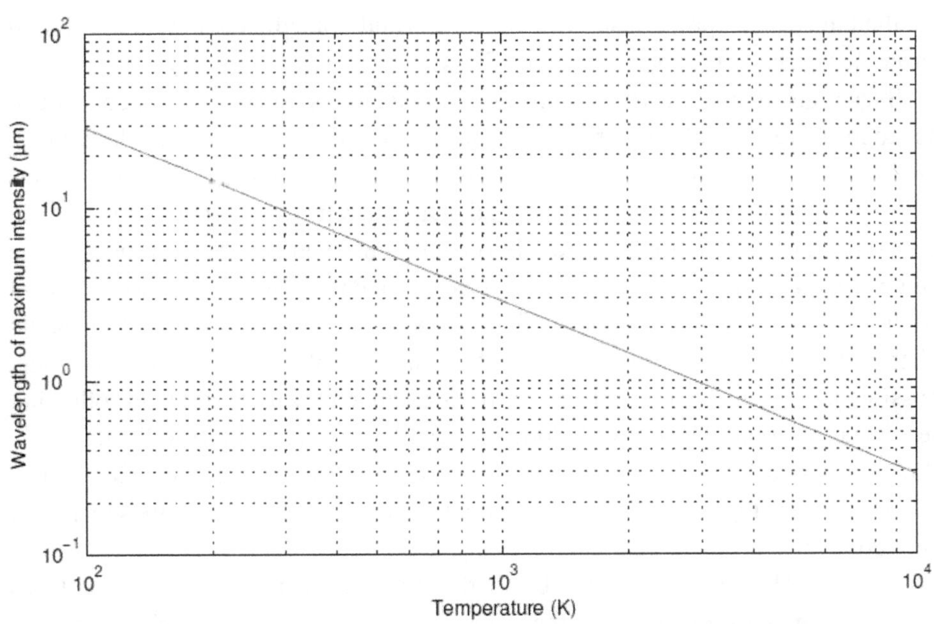

Figure 3. Stefan–Boltzmann law: energy radiated per unit surface area across different wavelengths per unit time

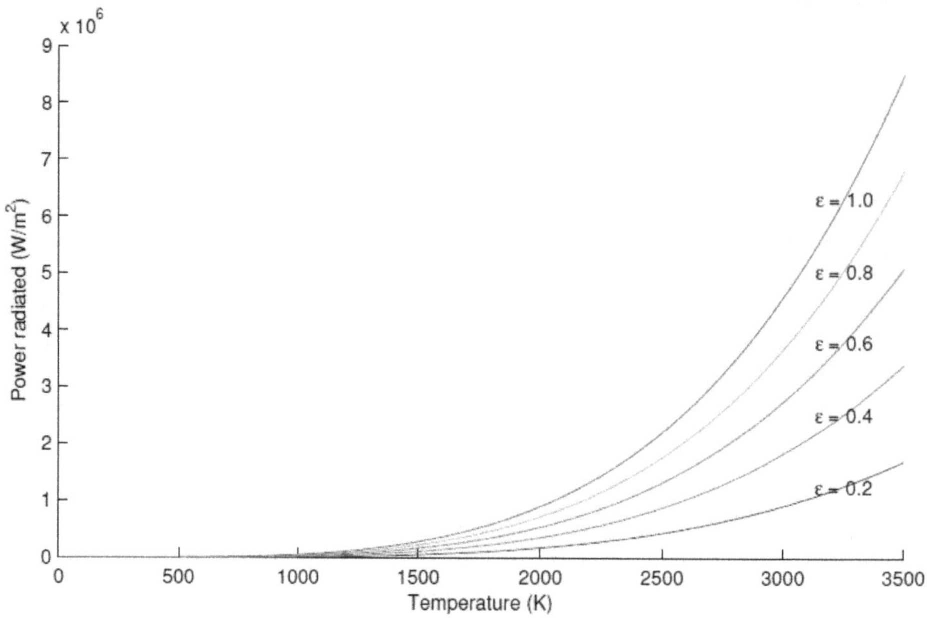

resulted in changed propagation rate on defects and immediately produces thermal variations surface (Usamentiaga et al., 2014).

Such tools can be effective for industrial applications. Recently solar cells are in use to measure the dark forward current density in order to detect possible failures. Such tools may be suitable for various industrial sector applications. The automotive industry, where there is an ever-increasing demand for reliable, non-destructive testing methods. Mass production requires short cycle times and automated processes, so defect selective NDT-techniques are very appreciated for such applications. Recent studies (Meola & Carlomagno, 2010) allow a quantitative analysis of the spatial distribution of the dark forward current density of solar cells in order to detect possible failures. Also, the IRT applications extended for inspections in following sectors i.e. solar cells (Meola & Carlomagno, 2010), aeronautical sector (Breitenstein, Rakotoniaina, & Rifai, 2003) and concrete inspection (Fan, Sun, & Yang, 2005). The continuous development by researchers made it clear that the defects are superficial enough to enable IRT to detect them easily.

Further, IRT techniques are suitable for the inspection of metallic parts of airplanes (Sham, Chen, & Long, 2008; Luong, 1992; Plekhov et al., 2005; Noethen, Wolter, & Meyendorf, 2010). Though IRT has experienced a great evolution even then the exact analysis of data is in need of required the operator's skill. Also, after capturing

Figure 4. Basic setup for thermographic inspection
(Howell, Siegel, & Menguc, 2010)

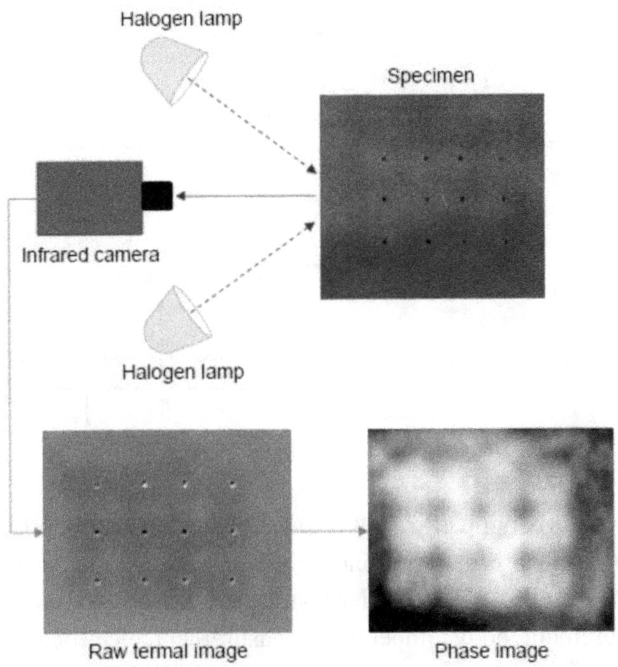

information further, we need to process and share with a central office or global. So, we have to look forward to the latest digital components like microcontrollers, converters, MPUs, drivers and sensors etc.

ROLE OF ELECTRONICS DEVICES

The remarkable growth of digital devices, such as sensors, actuators, smartphones, and smart appliances give freedom to interconnect and communicate with each other by IoT services through internet (Noethen, Wolter, & Meyendorf, 2010). Due to excel in technologies, it is possible to design some public health information centers, e.g., real-time location and utilization, the condition of subscriber (human/animal/machines) health, occupancy at health centers, and other data like weather conditions, air, and noise pollution status. To design such an intelligent and automated platform, there is a need of different technologies to be clubbed to address the specific features of each application. Figure 5 and Figure 6 show a conceptual futuristic e-health care system.

Figure 5. Conceptual block diagram for modern e-health care system
(Islam, Kwak, Kabir, Hossain, & Kwak, 2015)

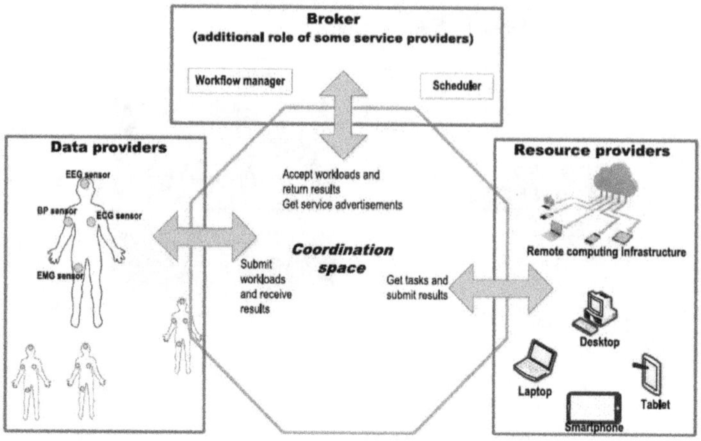

Two methods can be applied to analyze data from patient one via contact and another contactless. Figure 3 describes the concept of via contact method, where monitoring devices need to install in a body under investigation. Numerous sensors like blood pressure measurement, body temperature measurement, electrocardiograms (ECG) etc. can monitoring the health status capable to transfer information electronically using Wi-Fi, GSM either to centralized system/broker or even to an individual on their smart hand handled devices such as phones, iPods and laptops (Islam et al., 2015). In the contactless method, infrared thermography can be the best candidate in future because it is a fast, clean and safe technology. Conventional clinical method for monitoring body temperature is the measurement by the thermometer, while IRT provides fast, a non-contact alternative method to monitor body surface temperature remotely. Due to rapid growth in thermal imaging camera, now days successfully used in the detection of breast cancer, gynecology, heart, fever screening, neuropathy, vascular disorders and for brain imaging (Lahiri, Bagavathiappan, Jayakumar, & Philip, 2012).

So, these advanced cameras can also be implemented in a field with the network wireless sensors (WSNs) employed RFIDs in order to obtained movement and temperature related to human/machine/animals. So, the increasing demand for heterogeneous platforms for smart cities is also the good field of research for the future. These platforms may help to cover the communication between machines, human, animals, sensors, and the application suppliers.

The infrared sensors are not so widespread, because of high cost and lack of training. However, fast developing technologies opening scope of IRT in the near future, integrating with IoT architecture as shown in the above figure may be useful

Figure 6. Block diagram of a wireless sensor node

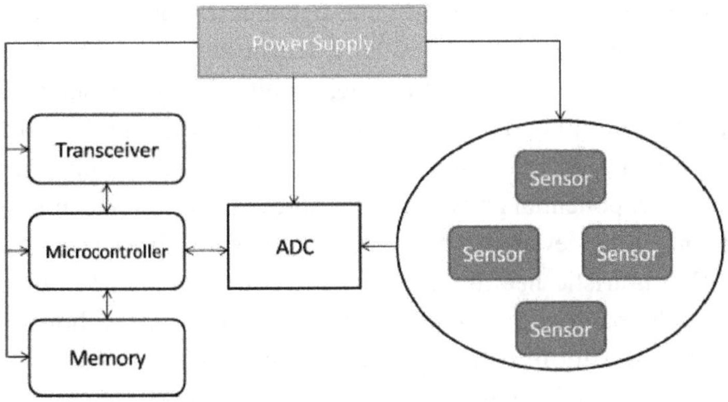

to monitor health for the human/animal/machine. So the next step is to integrate the above concept with IoT. Such sensor networks can be utilized a huge number of e-checkup of human and machines. Indeed, through transforming temperature and other data into electronics devices/sensors such devices can be linked up via the internet.

As a typical component of e-health monitoring system, smart health monitoring stations can be deployed at various public places like parks, gyms, market in our daily living environments. IR cameras and other sensors will detect the temperature above a threshold value and immediately will inform the concerned person as well as e-health center on priority bases. Also, a citizen of smart cities can access health care facilities by using their hand handled devices from their home viz. (1) Free time slot for consultation (2) Primary and initial recommendations regarding treatment (3) Can reserve an appointment by online payment method after deployment in a daily surrounding. So, each citizen must have a specific RFID tag to build an account, to keep checking availability and to reserve by the internet. Health monitoring centers may be addressed by using two models: public place entry gate and specific health monitoring centers. Each entry gate and e-health center will be assembled with an IR camera or advanced sensor to detect the health of human/animal/vehicles. To implement such a huge system, there is a need to adopt some effective approach. Here, for the proposed model of the e-health monitoring system, we agreed on the KATHODIGOS approach [40], currently, this approach is implemented for smart parking system. When this approach is assembled with wireless sensors, then it gives information about free space status in the parking area. Here, in e-health monitoring system sensors utilize IRT and motioning will be the detection of the temperature of various objects under investigation. This complete system provides a contactless health monitoring system called e-health system, which will be capable to provide prior information about critical health issues in advance. Because of the low cost of hardware and emerging technology of radio communication, numerous topologies adopted to deploy the sensors. Additionally, sensor nodes are capable to perform data storage, smart sensing, target tracking, monitoring and efficient routing for efficient data transmission. The era of wireless networks provides popular services like monitoring environmental conditions such as temperature, humidity, vibrations, oxygen etc due to exponential growth in electronics equipment (like microcontroller and various computing devices). The exponential growth of new generation sensors may define the futuristic health care system for a better life. Presently, various sensors helping humanity for various healthcare applications as shown in Table 1

Hence smart e-health monitoring system may be built with sensors networks integrated with communication system and internet. The broad view is shown in Figure 7.

As the E-health system is one of the most extensively growing fields of this era where people looking for smart health solutions. To achieve the same we are proposing a system which mainly can be built by few IOT components which are shown in Figure 7 in form of block diagram. The real implementation can be achieved by integrating all components shown in Figure 8 according to block diagram in Figure 7. This system can be designed with the help of electronics engineer can be called the doctor for components. This person must be fully experts in the mechanism, operation, and maintenance of this equipment. Electronics dedicatedly serving in various medical procedures like magnetic resonance imaging (MRI), automating insulin injections, troubleshooting skills in analog circuits, digital circuits, and processors for medical equipment. Future e-health system is feasible with the collaboration of life scientists, chemists, and medical scientists and electronic engineers. The simplest idea is proposed in Figure 7 for the future E-health expert system.

Arduino Uno

The Arduino UNO is an open-source microcontroller board equipped with sets of digital and analog input/output (I/O) pins. The board contains 14 Digital Pin, 6 analog pins and easily can be programmed with the Arduino IDE (Integrated Development Environment) via a type B USB cable by the single click.

- **Microcontroller:** Microchip ATmega328P
- **Operating Voltage:** 5 Volt
- **Input Voltage:** 7 to 20 Volts
- **Digital I/O Pins:** 14 (of which 6 provide PWM output)
- **Analog Input Pins:** 6
- **DC Current per I/O Pin:** 20 mA

Table 1. Role of electronics in the present era for e-health

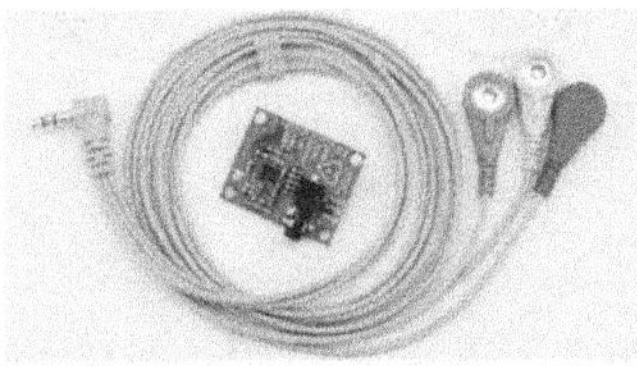 Electro Cardiogram (ECG) Measure hear rate , diagnosis multifaceted arrhythmias, myocardial ischemia, AD8232 module	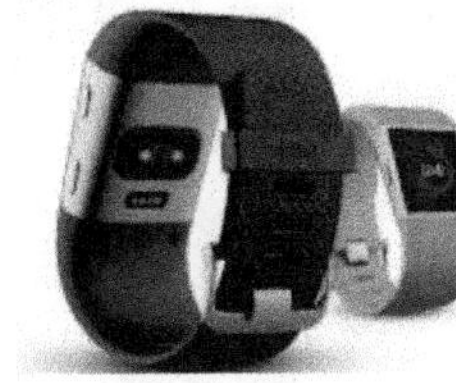 Fitness Tracker HR, calories burned, sleep quality, number of steps, climbing, running, swimming
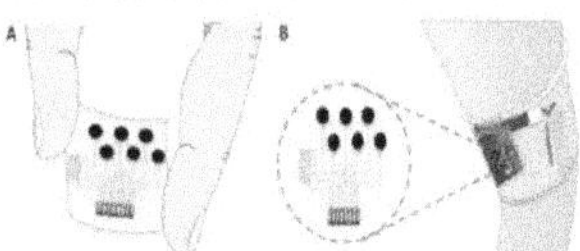 Glucometer Blood glucose monitoring for sugar patients Main components are blood sugar collector and any connectivity hand handled device	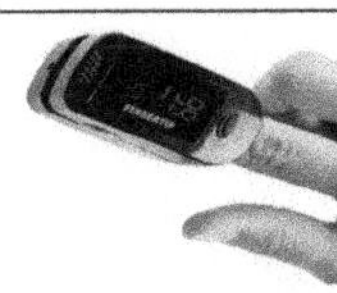 Blood Pressure (BP) meter, Composed of a BP apparatus and wireless, the communication module
 Body temperature sensor non-contact detect body tempe infrared detection and transmit with RFID module, MLX9 AAA Non-Contact human body infrared IR temperature sens	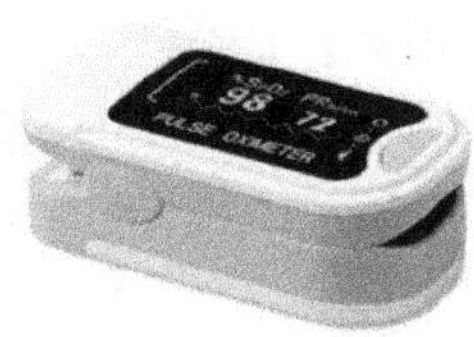 The oximeter can be used to measure huma hemoglobin saturation and heart rate through finge Also, can continuously measure and transmit data t WSN with Bluetooth module.

Nowadays disabled person can use smart wheelchairs with full automation equipped with PIR sensor, proximity sensors, ultrasonic sensors, gyroscope, motors, and actuators etc.

Figure 7. E-health monitor set up hardware

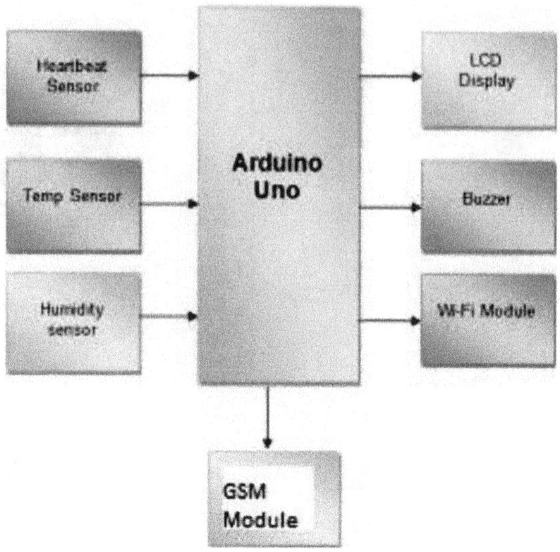

- **DC Current for 3.3V Pin:** 50 mA
- **Flash Memory:** 32 KB of which 0.5 KB used by boot loader
- **SRAM:** 2 KB
- **EEPROM:** 1 KB
- **Clock Speed:** 16 MHz

The following technical feature indicates that it's the strongest candidate for future automation and controlling services

Figure 8. LCD Display

LCD

LCDs (Liquid Crystal Display) used to display information like sensors data and any other information. These are very popular because of its cheap cost, handy size and smartest way of presenting information.

LCD is 16 pin devices where the leftmost pin is GND, the second Pin is VCC connected with 5 V on the Arduino board. For sending data or command RS pin is enabled i.e. RS=0, used for sending command while RS=1 used for sending data to the LCD. Next pin R/W used to select the read or write on LCD. Pin E is used to write on the register. D0 to D7 are known as data pins. The last two pins A and K are for the backlight.

GSM

Various GSM modules are available quad-band GSM 850/EGSM 900/DCS 1800/ PCS 1900 module which easily can be integrated with Aurdino. GSM provides us almost all services like SMS text message, TCP/IP etc. The SM5100B available in the market with the following features two UARTS, Li-ion battery charging, an SPI interface, and two 10-bit ADCs, Input/output as speaker and LCD interface.

Wi-Fi Module

ESP8266 module is capable to establish Wi-Fi connection and easy to integrate with Aurdino deice. In addition to that, it can be integrated with sensors as well Its available in market at extremely low cost, wide applications in IOT industry. This module can process on board and can connect with sensors or other modules thorough GPIOs. It also supports VOIP applications and Bluetooth connectivity. Its really having limitless functionality and connectivity for any expert system, especially for IOT.

Figure 9. GSM Module and Wi-Fi Module

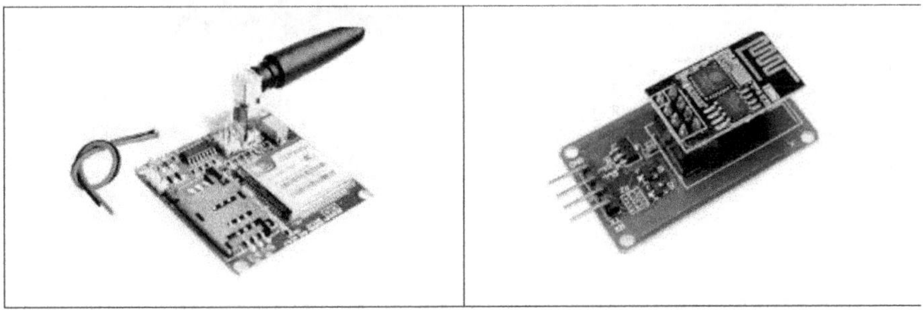

Temperature/ Humidity Sensors: This sensor used to measure the amount of heat from surrounding and convert the received data understandable for other devices in the network either by the principle of thermal coupling or Resistor temperature detectors (RTD). Various kinds of temperature sensors is shown in Figure 10.

- **Thermocouples:** Change its voltage with respect to (w.r.t) change in temperature.
- **RTD:** Change in Resistance is directly proportional to temperature change.
- **Thermistors:** Changes its physical resistance w.r.t temperature. It is a temperature sensitive resistor that changes its physical resistance with the change in temperature.
- **IC (Semiconductor):** This wonderful sensor provides a temperature reading directly on digital displays and works on the principle of conductivity changes with temperature variations.

Humidity sensors: Humidity monitoring is one of the important parameters of the industry manufacturing process. Humidity can be measured in terms of relative humidity (RH), dew/frost point (D/F PT) and parts per million (PPM).

The concept of e-health system can be designed with the help of the above basic sensors and equipment. We broadly categorized mainly in three different layers sensing, networking and user according to their functions.

Figure 10. Temperature and humidity sensors

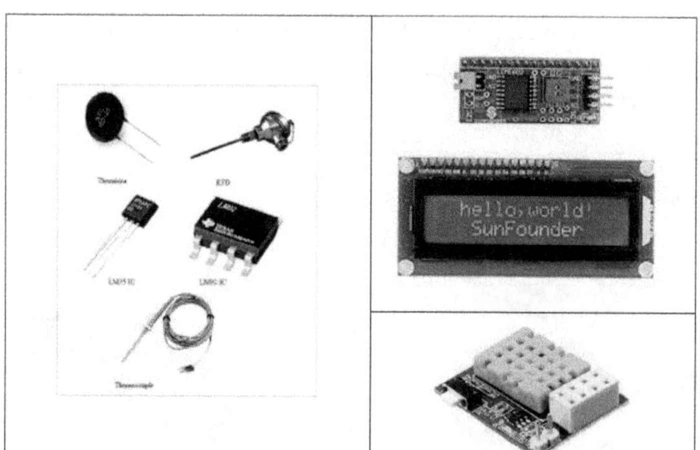

Humidity sensors

The Sensing Layer

These layers consist of various sensors shown in figure 5 and table 1 in order to investigate the patients/samples. The main role of sensors will be to collect information (slave node) form sample and to transmit (master node) to the expert system and to advise the patient on behalf of the expert system.

The Networking Layer

The main function of this layer will be established a reliable communication (anchor node) between all sensors, the expert system as well as the end user. In addition to that all networks must be capable to communicate anytime from anywhere.

End Users

Clients' devices may be connected with networks via TCP/IP protocol to access different services based on the health customer database. In order to build such an advanced system, the network of sensors must be deployed in public places and health centers either by WSN technology or by radio frequency identification (RFID) networks. The optimal placement of nodes in networks are possible by various topologies like dynamic programming (Talari et al., 2017) genetic algorithm (Rose, 2001; Han, Park, Choi, & Park, 2001), and Tabu search algorithm (Meunier, Talbi, & Reininger, 2000).

Sensor Placement Model

The master sensors can be deployed at the location of the expert system, salves sensors under investigation and anchors may perform efficient communication in the heterogeneous environment via Ethernet protocols or wireless links as shown in Figure 11.

Further, we would like to discussion on communication between different networks, the basic unit is obviously signal, which is used to transmit particular information in a faster way. But for an advanced e-health monitoring system, these signals must be smart enough and capable to integrate with IoT devices and networks. Signals is the most important part of the e-health system, it must be intelligent enough to warn about the critical stage of a patient to avoid risks.

So, undoubtedly the smart signals will be one of the important parts of IoT and its applications. According to Talari et al. (2017) traditional signal can be transformed into smart by following up a few important features as below:

Figure 11. Proposed model for e-health monitoring

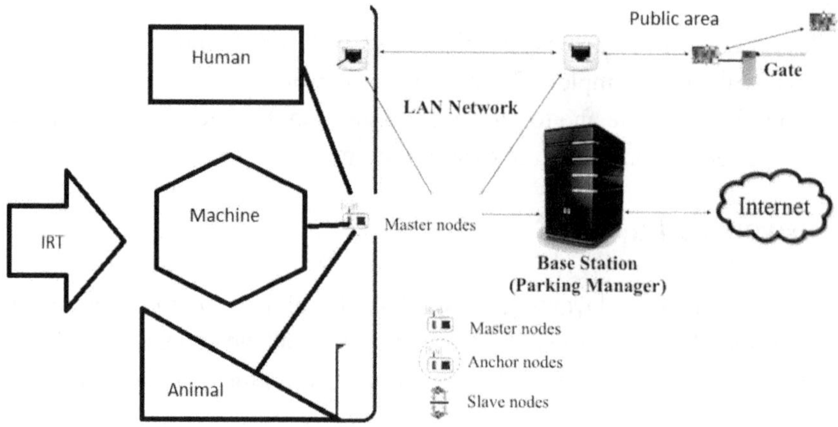

- **Message:** Messages must be visible by indicating blinking or fading etc on the device according to a different level of circumstances.
- **Control:** After sensing the behaviors or signal, it must have communication capabilities.
- **Communications:** Signal must have capabilities to communicate with surrounding i.e. for extremely critical situation signal must inform to the patient, health center as well as concerned bodies.
- **Context:** The signal should have full knowledge about its hierarchy for communication.

The basic feature of the smart signal includes conditions and inform or to take relevant action optimally all the time. It must have the skill to make communication between other signals and capable to share information automatically, which actually will be part of future intelligent networks. Presently, this technology is available for monitoring of urban traffic flow management (The University of Minnesota, 2019). Similarly, it can be applied for the e-health monitoring system.

The ultimate aim of the smart signal is a measurement of signal priority wise and tuned to best and controllable action automatically. So smart signal may be composed of a different building block as shown in Figure 12.

Interaction with object/human under investigation then smart signal must communicate with all registered entities, provide signaling effectively and immediately on priority bases like an emergency on demand. Also, should displaying systems of mandatory messages with some beep sound or different color lights as in smartphones. So, these main components can display different kinds of messages for some other interfacing device for human/machines.

Figure 12. Smart signal concept
(Dhillon & Chakrabarty, 2003)

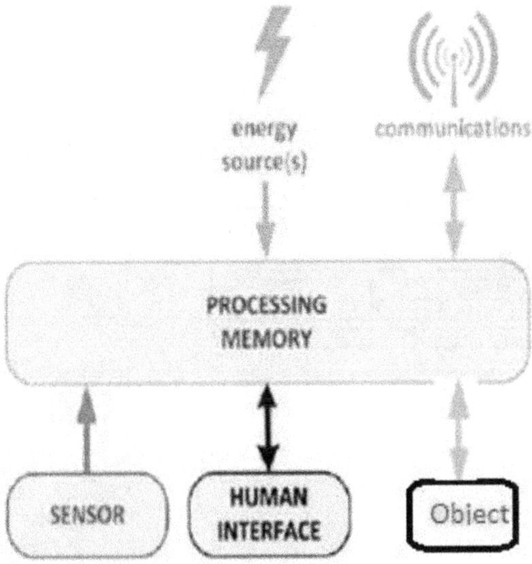

Communication between a different kinds of user like the man to machine, machine to machine must be granted. All types of users' communication human and machines as well as sensors should be connected and can exchange data mutually in codified form. So, we can use some already existed protocols i.e. wired and wireless. Some commonly used protocols are Ethernet, CAM, Bluetooth, ZigBee, Wi-Fi, GSM etc. Currently to meet the demand of new system IEEE802.15.4 protocol for providing facility to communicate with different kind of networks.

Obviously, the component needs the energy sources to operate. So, energy sources are also a key design aspect, it may be a long life battery, energy harvesting system, or solar energy source. The energy will be consumed for losses in components, maintaining communication between all network components and also with their signaling. So, to build such a system there is a need to combine traditional signal with smart decision-making devices as shown in Figure 13.

The smart signals communicate among them wirelessly through a mesh ZigBee network, backed by some dedicated, mains-powered devices, which route the messages exchanged (ZigBee Routers).

Figure 13. Block diagram of the electronic device
(Wikipedia, n.d.)

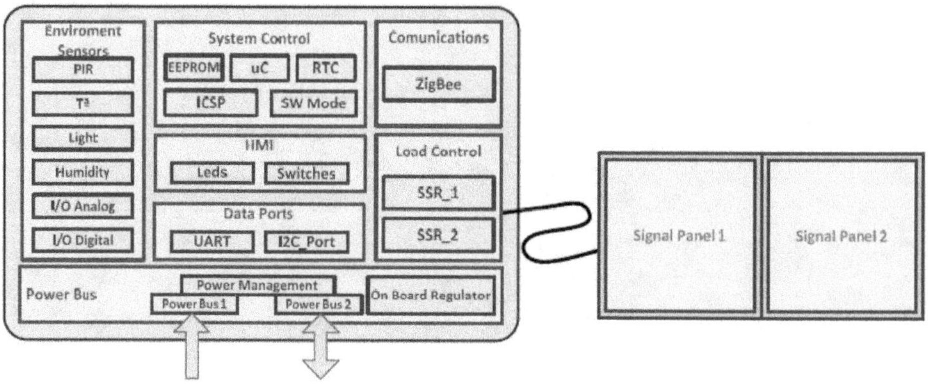

CONCLUSION

Actually, e-health is a broad term which may be intergration of electronics, information and communication technologies in the field of healthcare. To develop such systems are not only develop the human life, but also develop the way of thinking & attitude. The success of this system depends on commitment of doctors, electronic engineers, ICT developers; cost of electronics equipments, sensors etc in order to improve health care locally, regionally and worldwide. In this chapter we are proposing such expert e health care system at initial level, which mainly consist of few electronics components. Firstly we discussed about IRT techniques its advantages and requirements. IRT is a fast, clean and safest technology, which can be applied for inspections purpose and diagnose or early detection of various diseases. This technology is limited, highly dependent on the selection of sensors and adequate training for the user. Also, improved sensors are expected for thermal resistance measurement in order to create a visible thermal contract, which is a limitation of a human eye nowadays. Also, improvement in human interference, signal and image processing tools may result in better e-health system. Collaborative training schedules for communication and embedded system engineers, doctors, ICT experts, and IoT experts are required to develop such future e-health systems.

REFERENCES

Al-Ali, A. R., Zualkernan, I., & Aloul, F. (2010). A Mobile GPRS-Sensors Array for Air Pollution Monitoring. *IEEE Sensors Journal, 10*(10), 1666–1671. doi:10.1109/JSEN.2010.2045890

Al-Kassir, A. R., Fernandez, J., Tinaut, F., & Castro, F. (2005). Thermographic study of energetic installations. *Applied Thermal Engineering, 25*(2-3), 183–190. doi:10.1016/j.applthermaleng.2004.06.013

Alamri, A., Ansari, W. S., Hassan, M. M., Hossain, M. S., Alelaiwi, A., & Hossain, M. A. (2013). A Survey on Sensor-Cloud: Architecture, Applications, and Approaches. *International Journal of Distributed Sensor Networks, 9*(2), 917923. doi:10.1155/2013/917923

Bogue, R. (2013). Sensors for condition monitoring: A review of technologies and applications. *Sensor Review, 33*(4), 295–299. doi:10.1108/SR-05-2013-675

Breitenstein, O., Rakotoniaina, J., & Rifai, M. A. (2003). Quantitative evaluation of shunts in solar cells by lock-in thermography. *Progress in Photovoltaics: Research and Applications, 11*(8), 515–526. doi:10.1002/pip.520

Chang, K.-H., & Mason, B. (2012). The IEEE 802.15.4g Standard for Smart Metering Utility Networks. *Proceedings of the 2012 IEEE Third International Conference on Smart Grid Communications (SmartGridComm).* 10.1109/SmartGridComm.2012.6486030

Dhillon, S. S., & Chakrabarty, K. (2003). Sensor placement for e_ective coverage and surveillance in distributed sensor networks. *Proceedings of the 2003 IEEE Wireless Communications and Networking Conference*, 3, 1609–1614.

Fan, C., Sun, F., & Yang, L. (2005). Investigation on nondestructive evaluation of pipelines using infrared thermography. *Proceedings of the Joint 30th International Conference on IEEE, Infrared and Millimeter Waves and 13th International Conference on Terahertz Electronics, 2*, 339–340.

Gaussorgues, G. (1994). *Infrared Thermography.* Berlin: Springer. doi:10.1007/978-94-011-0711-2

Gimeno, A., Marco, A.M., Marín, R.B., & Nebra, R.C. (2014). Protocol and Architecture to Bring Things into Internet of Things. *Int. J. Distrib. Sens. Netw., 2014*, 158252:1–158252:18.

Gorostiza, E. M., L'azaro Galilea, J. L., Meca Meca, F. J., Salido Monzú, D., Espinosa Zapata, F., & Pallarés Puerto, L. (2011). Infrared sensor system for mobile-robot positioning in intelligent spaces. *Sensors (Basel), 11*(5), 5416–5438. doi:10.3390110505416 PMID:22163907

Guinard, Trifa, & Wilde. (2010). A resource oriented architecture for the Web of Things. *Proc. Internet Things (IOT),* 1-8.

Han, J. K., Park, B. S., Choi, Y. S., & Park, H. K. (2001). Genetic approach with a new representation for base station placement in mobile communications. *Proceedings of the 2001 IEEE Vehicular Technology Conference*, 4, 2703–2707.

Höller, J., Tsiatsis, V., Mulligan, C., Karnouskos, S., Avesand, S., & Boyle, D. (2014). *From Machine-to-Machine to the Internet of Things: Introduction to a New Age of Intelligence*. Amsterdam: Elsevier.

Howell, J. R., Siegel, R., & Menguc, M. P. (2010). *Thermal Radiation Heat Transfer* (5th ed.). New York: Taylor and Francis. doi:10.1201/9781439894552

Ibarra-Castanedo, C., Tarpani, J. R., & Maldague, X. P. (2013). Nondestructive testing with thermography. *European Journal of Physics, 34*(6), S91–S109. doi:10.1088/0143-0807/34/6/S91

Information Society Technologies Advisory Group (ISTAG). (2011). *Orientations for EU ICT R&D & Innovation beyond 2013*. Available online http://cordis.europa.eu/fp7/ict/istag/documents/ stag_key_recommendations_beyond_2013_full.pdf

Islam, S. M. R., Kwak, D., Kabir, H., Hossain, M., & Kwak, K. S. (2015). *Internet of Things for Health Care: A Comprehensive Survey.* doi:10.1109/ACCESS.2015.2437951

ISO/IEC. (2014). *Smart Cities: Preliminay Report.* Available online: http://www.iso.org/ iso/smart_cities_reportjtc1.pdf

Kortuem, Kawsar, Fitton, & Sundramoorthy. (2010). Smart objects as building blocks for the Internet of Things. IEEE Internet Comput., 14(1), 44-51.

Lahiri, B., Bagavathiappan, S., Jayakumar, T., & Philip, J. (2012). Medical applications of infrared thermography: A review. *Infrared Physics & Technology, 55*(4), 221–235. doi:10.1016/j.infrared.2012.03.007

Lahiri, B. B., Bagavathiappan, S., Jayakumar, T., & Philip, J. (2012). "Medical applications of infrared thermography: A review" July 2012. *Infrared Physics & Technology, 55*(4), 221–235. doi:10.1016/j.infrared.2012.03.007

Lasobras, J., Alonso, R., Carretero, C., Carretero, E., & Imaz, E. (2014). Infrared Sensor-Based Temperature Control for Domestic Induction Cooktops. *Sensors (Basel), 14*(3), 5278–5295. doi:10.3390140305278 PMID:24638125

Leccese, F., Cagnetti, M., Calogero, A., Trinca, D., di Pasquale, S., Giarnetti, S., & Cozzella, L. (2014). A New Acquisition and Imaging System for Environmental Measurements: An Experience on the Italian Cultural Heritage. *Sensors (Basel), 14*(5), 9290–9312. doi:10.3390140509290 PMID:24859030

Leccese, F., Cagnetti, M., & Trinca, D. (2014). A Smart City Application: A Fully Controlled Street Lighting Isle Based on Raspberry-Pi Card, a ZigBee Sensor Network and WiMAX. *Sensors (Basel), 14*(12), 24408–24424. doi:10.3390141224408 PMID:25529206

Luong, M. P. (1992). Infrared thermography of fatigue in metals. *Proceedings of the SPIE*, 1682, 222–233. 10.1117/12.58539

Maldague, X. (2001). *Theory and Practice of Infrared Technology for Nondestructive Testing*. New York: Wiley.

Meola, C. (2012). *Infrared Thermography: Recent Advances and Future Trends*. New York: Bentham Science. doi:10.2174/97816080514341120101

Meola, C., & Carlomagno, G. M. (2004). Recent advances in the use of infrared thermography. *Measurement Science & Technology, 15*(9), R27–R58. doi:10.1088/0957-0233/15/9/R01

Meola, C., & Carlomagno, G. M. (2010). *Recent Advances in Non-Destructive Inspection*. New York: Nova Science Publisher Inc.

Meunier, H., Talbi, E., & Reininger, P. (2000). A multiobjective genetic algorithm for radio network optimization. *Proceedings of the 2000 Congress on Evolutionary Computation*, 1, 317–324. 10.1109/CEC.2000.870312

Modest, M. F. (2013). *Radiative Heat Transfer*. Waltham, MA: Academic Press. doi:10.1016/B978-0-12-386944-9.50023-6

Mollmann, K., Karstadt, D., Pinno, F., & Vollmer, M. (2005). Selected critical applications for thermography: Convections in fluids, selective emitters and highly reflecting materials. *Proceedings of the Infrared Camera Calibration Conference*, 6, 161–173.

Nasirudin, M. A., Za'bah, U. N., & Sidek, O. (2011). Fresh water real-time monitoring system based on Wireless Sensor Network and GSM. *Proceedings of the IEEE Conference on Open Systems (ICOS2011)*, 354–357. 10.1109/ICOS.2011.6079290

Noethen, M., Wolter, K. J., & Meyendorf, N. (2010). Surface crack detection in ferritic and austenitic steel components using inductive heated thermography. *Proceedings of the 2010 33rd International Spring Seminar on IEEE Electronics Technology (ISSE)*, 249–254. 10.1109/ISSE.2010.5547297

Plekhov, O., Palin-Luc, T., Saintier, N., Uvarov, S., & Naimark, O. (2005). Fatigue crack initiation and growth in a 35CrMo4 steel investigated by infrared thermography. *Fatigue & Fracture of Engineering Materials & Structures*, *28*(1-2), 169–178. doi:10.1111/j.1460-2695.2005.00856.x

Purohit, R., Turner, T. A., & Pascoe, D. D. (2008). Use of infrared imaging in veterinary medicine. In N. A. Diakides & J. D. Bronzino (Eds.), *Medical Infrared Imaging*. Boca Raton, FL: CRC Press.

Romer, K., Ostermaier, B., Mattern, F., Fahrmair, M., & Kellerer, W. (2010). Real-time search for real-world entities: A survey. *Proc. IEEE*, 98(11), 1887-1902. 10.1109/JPROC.2010.2062470

Rose, R. (2001). A smart technique for determining base-station locations in an urban environment. *IEEE Transactions on Vehicular Technology*, *50*(1), 43–47. doi:10.1109/25.917869

Sham, F., Chen, N., & Long, L. (2008). Surface crack detection by flash thermography on concrete surface. Insight-Non-Destructive Test. *Cond. Monit.*, *50*, 240–243.

Sudevalayam, S., & Kulkarn, P. (2011). Energy Harvesting Sensor Nodes: Survey and Implications. *IEEE Communications Surveys and Tutorials*, *13*(3), 443–461. doi:10.1109/SURV.2011.060710.00094

Talari, S., Shafie-Khah, M., Siano, P., Loia, V., Tommasetti, A., & Catalão, J.P.S. (2017). A Review of Smart Cities Based on the Internet of *Things. Concept Energies*, *10*, 421. doi:10.3390/en10040421

Tan, L., & Wang, N. (2010). Future Internet: The Internet of Things. *Proc.3rd Int. Conf. Adv. Comput. Theory Eng. (ICACTE)*, 5, V5-376-V5-380.

The University of Minnesota. (2019). *SMART Signal, Traffic Data Collection for Urban Traffic Flow*. Retrieved from http://license.umn.edu/technologies/z08214_smart-signal-traffic-data-collection-for-urban-traffic-flow

Usamentiaga, R., Venegas, P., Guerediaga, J., Vega, L., Molleda, J., & Bulnes, F. G. (2014). *Infrared Thermography for Temperature Measurement and Non-Destructive Testing Sensors*. Academic Press. doi:10.3390140712305 PMID:25014096

Vasanth, K., & Sbert, J. (n.d.). Creating solutions for health through technology innovation. *Texas Instruments*. Available: http://www.ti.com/lit/wp/sszy006/sszy006.pdf

Vollmer, M., & Mollmann, K. P. (2011). *Infrared Thermal Imaging: Fundamentals, Research and Applications*. Weinheim, Germany: Wiley.

Wikipedia. (n.d.). *Block Diagram*. Retrieved from https://en.wikipedia.org/wiki/Block_diagram

Zissis, G.J., & Wolfe, W.L. (1978). *The Infrared Handbook*. Technical report, DTIC document.

Chapter 10 Technological and Gamified Solutions for Pollution Control in Cognitive Cities

Pavneet Bhatia
Dr. B. R. Ambedkar National Institute of Technology, India

Parulpreet Singh
Lovely Professional University, India

ABSTRACT

This chapter presents an idea in the contribution to the concept of smart cities is to build a place which provides clean air to breathe. Advancements in the technology is acting as a bane to environmental balance leading to global warming, irregular rain cycles, increasing pollution, disturbed rainfalls. The need of the hour is to find a solution to the issue of increasing pollution that may lead to disastrous consequences. This chapter builds a framework of a cognitive city with attention to the impact and burden over environmental resources. This chapter is a perfect integration of traditional approaches, technological methods, and gamified solution as a remedy to growing pollution that can be and should be deployed in the project of smart city.

INTRODUCTION

The idea of cognitive cities seems to be fascinating to technological geeks as well as to environmental lovers. In the current scenario it is an area of keen interest and scope. The aim is to develop smart, cognitive and urban residential cities that are equipped with automated facilities to provide a comfortable and better living conditions to the

DOI: 10.4018/978-1-5225-8085-0.ch010

residents. Our idea in the contribution to the concept of smart cities is to build a place which provides clean air to breath in. Advancements in the technology is acting as a bane to environmental balance leading to global warming, irregular rain cycles, increasing pollution, disturbed rainfalls. The need of the hour is to find a solution to the issue of increasing pollution that may lead to disastrous consequences. We target to design a plan that measures pollution of a specified area and if the statistics fall below the standards, a notification is sent to registered residents of the area intimidating them about the danger and advising them to plant a tree or a sapling and even control the activities which are adding up to the issue. The awareness and regular updates about their space will motivate them to be take responsibility of the air quality. The setup will involve pollution sensors, microcontrollers, air quality indicators, GSM (Global System for Mobile Communication) modules, data analysis and other technical inputs. Also, the government and private organizations are made part of this revolution by imposing a rule that bound them to plant trees in proportion to the area they own. Even vertical planting will add well to the efforts promoting a green and clean smart city. All genuine efforts to the idea will be appreciated and promoted well by gamified rewards. This kind of design will be fruitful in meeting the basic requirement of a healthy living space and should be considered as a high priority objective.

CHALLENGES IN CITIES

Increasing Pollutants in the air are significantly contributing to the growing cases of ill health in India. According to Greenpeace India's report, 1.2 million deaths take place every year due to air pollution in India (Pti, 2017). The reports also claimed that none of 168 cities across the country where air quality test was performed, could live up to the standards prescribed by World Health Organization (WHO). Rural and urban India are both affected by poor air quality. Rapid urbanization and industrial development have adversely affected quality of air in urban areas (WHO, 2016). Industrial and vehicular emissions have made the case worse, however, the deterioration in air quality at rural places occurred due to use of cow dung, biomass fuels or coal as major energy sources for cooking and heating (Gargava et al, 2016; Dey et al, 2012). A study published by the World Bank in 2016 revealed that air pollution cost India approximately 8% of its GDP or $560 billion in 2013, as a result of lost productivity due to premature mortality and morbidity ("World Bank", 2016). Under National Air Monitoring Program (NAMP), PM10 (Particulate Matter having an aerodynamic diameter less than or equal to 10 μm), Sulphur dioxide (SO_2) and Nitrogen dioxide (NO_2) are considered as criteria pollutants to measure air quality. Figure 1 depicts the PM10 levels in top 10 polluted cities.

Figure 1. RSPM/PM10 in top 10 polluted cities in 2013

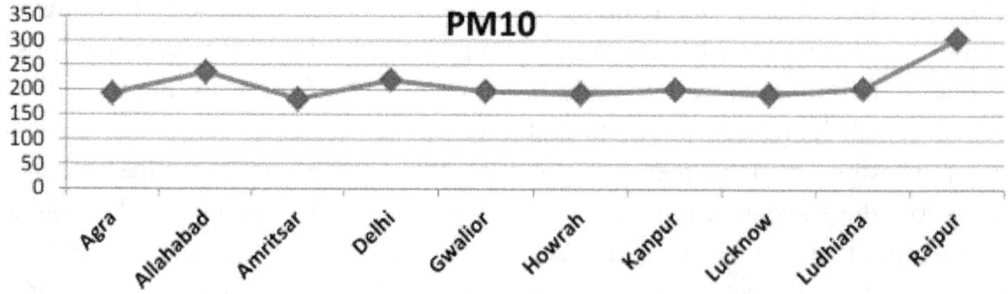

Raipur emerged as the most polluted city with RSPM (Respirable Suspended Particulate Matter) levels of 305 μg/m³ in 2013 followed by Allahabad, Gwalior, Agra and Delhi. In a WHO report published in 2016, it was reported that India's towns and cities were grappling with toxic air, due to increasing vehicular and industrial emissions as well as limited government interventions in solving and reducing the issue (Shukla, n.d.).

Figure 2 shows the mean SO_2 concentration over the years in different cities in India. There is definitely some reduction in SO_2 levels over the years due to reduction in Sulphur level in diesel, use of cleaner fuel such as CNG (Compressed Natural Gas) in metro cities as well as use of LPG (Liquified Petroleum Gas) as domestic fuel.

Figure 3 represents mean NO_2 concentration in different Indian cities, which was found to be almost stable over the year's despite of the increase in usage of vehicles. This have occurred might be due to improvement in vehicular technology and emerging alternate fuels.

Figure 2. Mean SO_2 concentration in different cities of India (2009 – 2013)

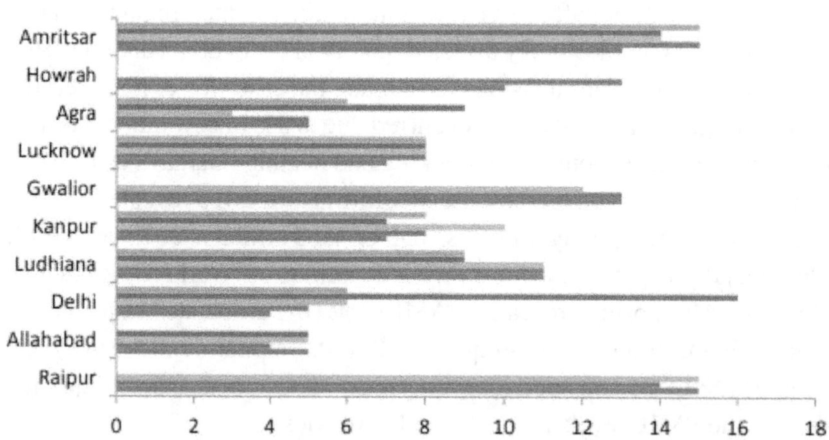

**For a more accurate representation see the electronic version.*

Figure 3. Mean SO_2 concentration in different cities of India (2009 – 2013)

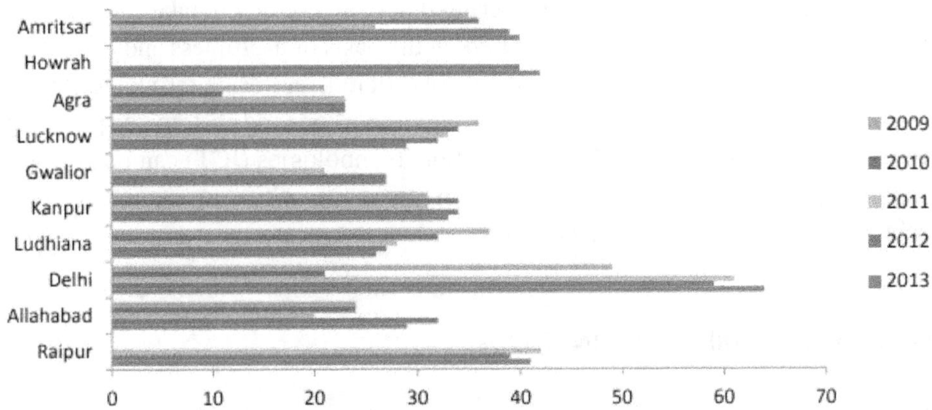

**For a more accurate representation see the electronic version.*

1.4 million lives are lost in 2013 due to air pollution in India as per a report by World Bank and the Institute for Health Metrics and Evaluation. The increasing pollutants in the air have serious health impacts like reduced lung function, asthma, chronic obstructive pulmonary disorder (COPD) and many respiratory conditions. Many urban cities like Delhi have been declared unfit as there are hazardous living conditions in such metropolitan polluted cities.

ATTRIBUTES OF A SMART CITY

Our current cities are forecasted to be more urbanized in the coming future. Population density, industrial expansion, manmade landscapes, buildings, automated systems, new-age facilities, educational institutions, medical centers, interconnectivity etc. are considered as parameters to consider a space as a city, however a city is not just a space but is a system of networks and flow (Batty, 2013). The definition of a city is definitely being altered with time. Smart city is a complex hierarchical system connecting networks, system flows, people, things and places. Smartness of a city should be quantized in terms of economy, society, mobility, maintenance, people, technology, governance, sustainability and environment. It promotes a planned space where functions and activities are visioned to be organized. The idea of Urbanization requires to focus on energy and resource management first and then should optimize the infrastructure through advanced technology. The set and maintain a technological connectivity within a city, requires a well-planned setup of connections, networks, data analyzers, sensors, applications, user interface, regulating rules and governance (Figure 4). A proper sustainable management of the resources as well as the environment and emission reduction should never be ignored in the blueprint of a cognitive city. Any automation through mobile applications should be evaluated before implication considering that there is a city to take care of. Any technological system should be quantified on the basis of usefulness and impact on the cognitive city. Cities around the world are different, suffering through different limitations and issues and hence require different solutions and customized planning accordingly. Information and Communication Technologies (ICT) can play a vital role in improving the efficiency and effectiveness of a city's management using technology as a means and not as a target.

SMART CITY MISSION IN INDIA

Smart City Mission is an urban renewal, retrofitting and redevelopment program launched by Government of India in 2015. A total budget of ₹98,000 crore was declared by Indian Cabinet for the development of 100 smart cities and reviving 500 others. The Union Ministry of Urban Development is responsible for implementing the mission in collaboration with the state governments of the respective cities. It is a five-year program in which every state and union territory except West Bengal has given name of at least one city for the challenge. 99 cities have been selected as of now to obtain the mission benefits. Financial aid has been proposed to be given by central and state governments between 2017-2022 and the expected results are supposed to be visible from 2022 (Smart Cities Mission, n.d.).

Mission is aimed to utilize data and technology to enhance sustainability, economic growth and improve quality of life for the people living and working in cities. Figure 5 glorifies the key objectives of the project.

Figure 4. Graphical elaboration of a smart city

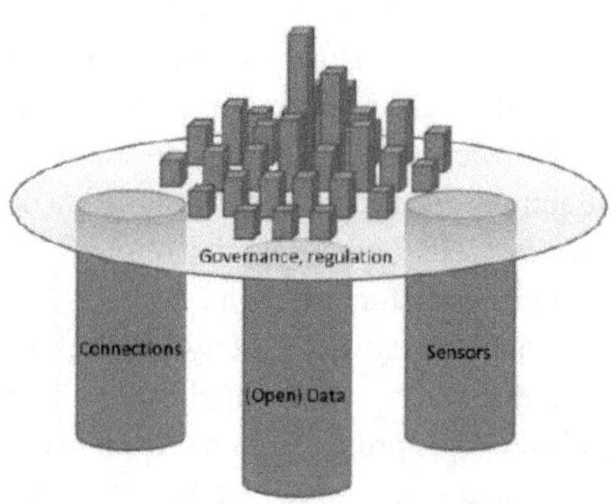

Figure 5. Smart city objectives

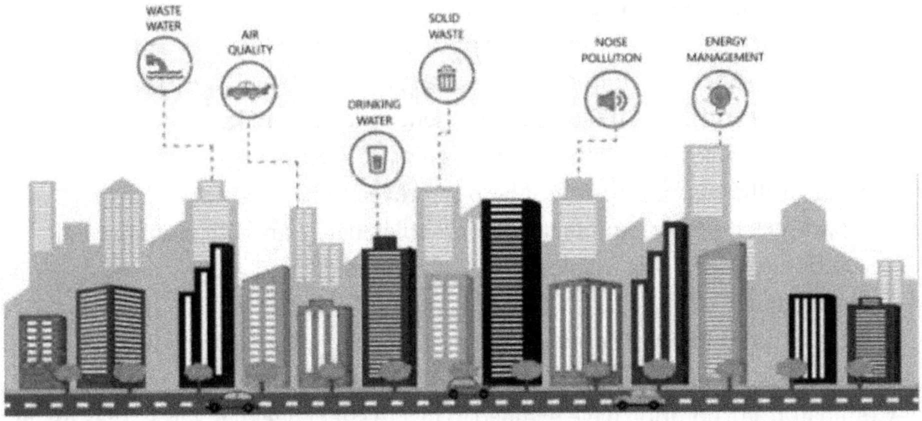

Pan city solution is aimed to collect data from different domains in the space like traffic, health, utilities or waste management using sensors, networks and IoT based applications. The collected data is stored and processed in command control systems and then the information is integrated to monitor things and operate accordingly. Pan city solutions are being designed for smart waste management, smart parking, smart pollution check, traffic management, mobility, healthcare and many more objectives.

Pan city solution has wide ranging applications and benefits in improving the living conditions like:

Figure 6. Smart city attributes

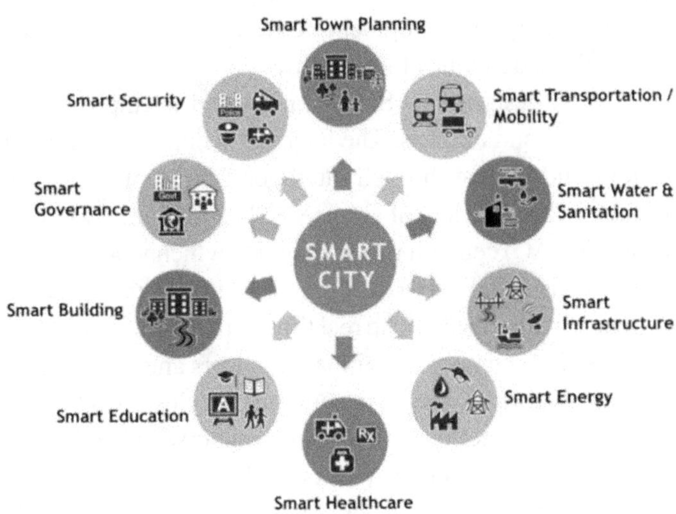

- Adequate water supply and management.
- Integrated municipal e-governance.
- Emergency response systems in healthcare centers improving the management systems and accessibility of doctors, hospitals and healthcare facilities.
- Enhancing educational facilities and hence increasing digital literacy.
- Smart surveillance system improves safety of citizens.
- Smart traffic management will help improve connectivity and transport.
- Smart meters, smart appliances, renewable energy and other energy efficient solutions reduce the exploitation of resources and helps in sustainable management.
- ICT solutions for waste control, water management and other pollution issues.

Coimbatore corporation have planned to install street benches across the city that provide a shady and cozy sitting space under the solar tree with free Wi-Fi facility (Figure 7). 'All Abilities' park has been proposed in Visakhapatnam enhancing the social infrastructure of the city according to the needs of specially -abled individuals (Figure 8). Public bicycle and electric vehicle sharing have been planned by Bhopal. 30 Docking smart stations have already been completed and 20 are in finishing stage (Figure 9). As an initiative in smart city concept Pune has designed intelligent transportation management system gathering traffic related data. Components like vehicle locators, GPS, passenger information systems etc. are being designed to formulate intelligent transportation network that can reduce journey times and even help in traffic management (Figure 10). These are some initiatives by various parts of the country to emerge as a smart, cognitive city. Various projects and proposals have been on cards for the local area development. The mission is drive economic growth as cities contribute 63% of GDP according Census 2011 and to improve the quality of life of people by redevelopment of system incorporating technology to develop smart, cognitive outcomes for citizens.

The pan city vision designed by Surat is purposed to improve the interaction between public services and citizens. The idea has been proposed highlighting the theme of Citizen Friendly Surat through Intelligent Transport and Connectivity. It is aimed to provide citizens variety of services in best possible way with minimum cost. A Smart City (SMAC) center has been planned which would serve as a control center for the city ensuring delivery of efficient services to the citizens. The center is designed to operate in four layers. The real time data of almost all the departments will be collected at SMAC center via automated sensors and systems. The collected information or data will be analyzed for meaningful conclusions. It is being assumed that such an initiative of connecting various departments with one control center will help maintaining a real time check over functionality of services hence promoting quick and better decisions in restoring services. Such a connected unit can play a

Figure 7. Smart street bench: Coimbatore

Figure 8. All Abilities park: Visakhapatnam

Figure 9. Public bicycle sharing: Bhopal

Figure 10. Smart transportation management system: Pune

vital role during emergency situations and managing rescue and relief operations. The city has planned an Integrated Traffic and Mobility Administration Center (IT-MAC) to regulate city Traffic and mobility. Application interface is being designed to help various agencies to coordinate well enabling a smooth traffic flow. A special transit system for BRTS is a part of the IT-MAC objectives aimed to extend the city bus operations providing real-time vehicle location and passenger information. It is being planned to put together all the stakeholder agencies like BRTS, City Bus, Traffic Police, Emergency services etc., on one integrated platform such that they can coordinate well through IT-MAC as well as citizens can connect with real time mobility scenario through messages, websites, mobile applications, alert messages etc. A common city smart card has been envisioned to offer wide range of services to citizens with the facility of easy payments and access. Services like transport, education, health, housing, recreational facilities etc. will be available to the people with tap and go features. The whole city will be connected with internet services enabling social and economic change as well reinventing the way with which government, business and citizens will interact. Free Wi-Fi will be provided covering all important areas of the city. This will definitely bridge the digital divide by enabling internet access to all sections of the society. It will also support the local economy by improving online shopping and bookings (Rasyid et al, 2016). ERP platforms will be built for health, education, woman and child care, municipal operations for integrated solutions on a common technology platform. It will address going on processes to citizens, will provide detailed reports of projects and will collect feedback and suggestions for performance analysis. For improving quality of services, ERP systems will be developed for all major departments. With the increasing urbanization the traffic congestion is expected to boost up. As a smart solution for traffic congestion, Intelligent Parking Management System has been proposed. The setup will involve parking sensors and data processors to manage the daily struggle of finding a parking space. It would help a smarter traffic, parking and transportation management system. Live updates of traffic flow, parking systems, congested areas over mobile applications will help people optimize their journey and pre-plan their activity. Moreover, intelligent waste disposal systems are also a part of the pan city solution for solid waste disposal. This will require proper coordination between vehicle tracking, bin tracking, solid waste collection, transportation and disposal system. SWM (Solid Waste Management) application is an integrated solution for waste monitoring and tracking. This project will be controlled from Smart city center (SMAC). An initiative named Smart Street Lighting Monitoring System has also been proposed with a vision to achieve better illumination using energy efficient technology. Considering energy saving as a prime factor all the conventional light fittings will be replaced by LED lights. Apart from the replacement, it will also monitor the proper usage, working and maintenance of Smart street light. LED lights are energy efficient and have longer life as compared to conventional lights. Automated monitoring system would ensure usage of lights only in required hours hence cutting off the wastage of electricity due to ignorance or manual errors. Also, usage of LED lights is an ecofriendly move for the city as these lights don't emit harmful ultra violet (UV) and infrared radiations. Automated monitoring system would result in Centralized control and maintenance of street lights. There would be an ease in gathering information related to power consumption, billing and monitoring.

Many more proposals have been invited by Surat government to emerge as one of the smart cities in the country. The vision of the pan city project in Surat is to enhance their economy and build as a major city preserving its culture and identity.

Figure 11. Site view after installation of LED lights

GAMIFIED SOLUTION FOR POLLUTION CONTROL IN A SMART CITY

Sustainability of cities is the main challenge in the concept of smart city. This work of ours present a gamified solution for pollution control and air quality monitoring in a smart city. We have proposed a framework that is a perfect integration of traditional approaches, technological methods and gamified solution as a remedy to growing pollution that can be and should be deployed in the project of smart city. The setup involves pollution sensors, microcontrollers, air quality indicators, GSM modules, data analysis, IoT technology and other technical inputs. The framework requires deployment of pollution sensors, pH sensors, smoke sensors that monitor air quality and report the type of molecules present in the air. Sensors used are capable of measuring ambient concentrations of air pollutants like NO_2, O_3, CO_2, VOC (volatile organic compounds). The whole city is required to be divided into different sectors and sensors are deployed in each area. Throughout provincial divisions, each region is installed with corresponding sensors. The information so acquired by the sensors is transferred to data centers for analysis and computation. The data is properly analyzed against presence of different pollutants and particulate matter like PM10, PM2.5 and the quality of air of the particular sector is categorized for being healthy/safe or unhealthy/unsafe. Classifiers are used in data processing unit for categorizing air quality index as unhealthy, safe or borderline according to the composition of air. The sensors send the real time collected information to the nearest data center through IoT technology. The air quality updates are monitored every day regularly both in day and night.

We even suggest to accompany sensors with camera and video recorders that are aimed to record the actions being carried out in the area. The information acquired from video processing will help identify the pollution causing activities adding pollutants to the air. After identifying the affected areas and the contributing actions, a notification is sent to the registered residents and the utilities of the place intimidating them about the danger. A full air quality report including the contributing factors will give people a real time check of the air they breathe in and will notify them their actions contributing to the issue. This effort will definitely motivate the utilities to take responsibility of the resources they have been utilizing and even exploiting from generations. Along with the danger alarming, an advisory message is sent to the responsible residents and utilities informing them about the recent activities adding up to the issue and making air unfit. The advisory message also includes immediate preventive steps like planting a sapling or controlling CO_2 emission from vehicles etc., required to control the amount of pollutants being added in the environment. This would require a proper GSM module setup connecting data centers and the users. The awareness and live updates about their space will motivate people to take responsibility of the air quality they breathe in. Figure 12 represents the control flow in between different sections of the proposed framework for ensuring better air quality.

Government have to be the controlling unit regulating the proper implementation of the system in all regional divisions. A government body/department or legal institution should be designed to regulate the proper implementation of the plan and to monitor the success of the project. All government as well as private organizations should be roped in the concept to contribute in the idea of clean and green city. If government, people and organizations are made to work together then a check can be imposed over the exploitation of the nature. The government and private workspaces should be made as a part of this revolution by imposing a rule that bound them to plant trees in proportion to the area they own. Even vertical planting will add well to the efforts. Vertical gardens should be installed over railway stations, bridges, flyovers and other man-made structures. Terrace farming at workspaces and offices should be appreciated and promoted well. This kind of design will be fruitful in meeting the basic requirement of a healthy living space and should be considered as a high priority objective. The utilities can be kept motivated by reinforcements. A gamified appreciating reinforcement strategy is also recommended to be designed for encouraging active participation of people and organization to make the air clean and healthy. All genuine efforts to the idea should be appreciated and promoted well by gamified rewards like if there are some notable pollution control efforts in a regional division then the area should be given some relaxation in taxes or appreciated well for the efforts. Best effort awards will work well for motivating organizations

Figure 12. Framework flowchart

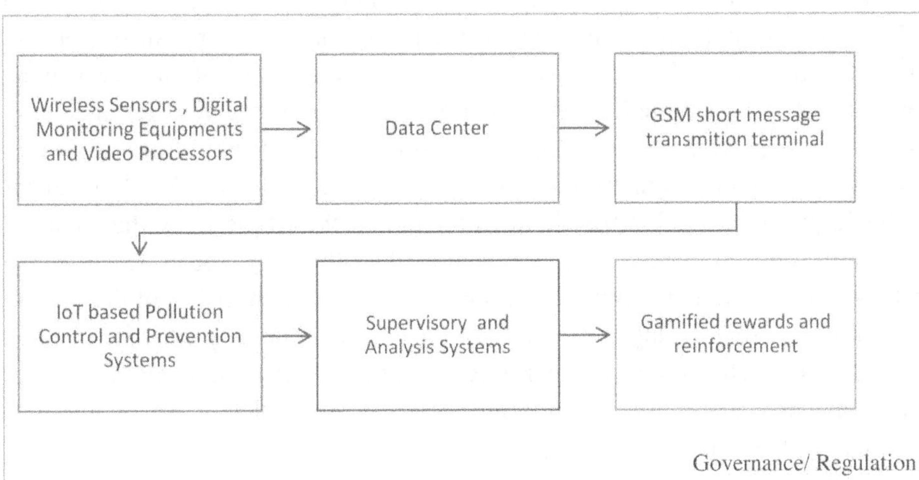

to participate in the drive. Inter-zonal competitions should be organized to create a challenging environment for sectors to contribute their best in the maintenance of the city. Mandatory rules should be imposed on utilities to overcome exploitation of the environment.

This whole concept can be made more interactive and accessible by taking it over an internet platform hence developing an android based application (which could be easily installed in mobile handsets) that includes all real time data, updated air quality index, live captured pollution causing activities, various solutions recommended etc. A complaint section is also suggested in the application so that citizens can add their issues and recommendations easily. Tickets should be allowed to be raised in no time on the application by the people asking the authorities to come in play. This will definitely help in eliminating the time delays encountered in the process and will make whole regulating system fast and efficient. A dashboard can also be designed in the application to keep a check on the performance of the sectors in the development of the area, awarding them according to their efforts. Funds can also be raised through it to ensure better results in the form of services and facilities to the citizens by the government and regulating bodies. Such an interactive user interface will definitely bridge the gap between the citizens and the government authorities opening a way for easy flow of information and services.

TECHNICAL ASPECTS

Sensors employed for the project need to be weather resistant i.e. it should be able to withstand ambient temperatures and humidity. Sensors should be highly sensitive to extreme levels of pollutant concentration without responding to non- polluting gases/ PM. It should be able to bypass those gases or pollutants which are not considered to be harmful or polluting while measuring the air quality. Mostly air sensors are battery operated so it should have more life and wide operational range. Since the collected data is transmitted to service stations so it should have Wi-Fi connectivity and data storage features according to the requirements of the framework. From past few years pollution monitoring systems have been based on Wireless Sensor Technology (Rasyid et al, 2016). In order to improve the efficiency of the pollution monitoring systems various data aggregation algorithms have been developed so that significant data is monitored and processed. The classification of data for its air quality and segregating it in the categories of being healthy or unsafe can be done using predictive models. Supervised or unsupervised learning algorithms are applied as classifiers. Images captured are also analyzed for identifying the activities adding up to the issue (Wang et al, 2017).

CHALLENGES IN THE IMPLEMENTATION PLAN

The quality of life in cities is one of the main challenges the current and future generations have to tackle. A cognitive approach to overcome the challenges require a blend of conventional approaches with best technologies available to achieve unprecedented success in the task. We have proposed an implementation plan that is aimed to provide a gamified solution for air pollution control and air quality monitoring in smart cities. There are definitely some challenges in the proper implementation of the framework. Deployment of the pollution and smoke sensors over the regional divisions is a challenging initiative involving high installation and maintenance cost. Moreover, a proper supervision is required over the setups for their proper functioning. Video capturing over public places to record the activities adding up pollutants in the air would require proper processing methods recovering appropriate data from fog, haze, smoke and other disturbances. This may require proper algorithms, microcontrollers, IoT systems and processing units to process data collected from sensors and video recorders. Also, to maintain the privacy of data, various encryption and decryption techniques are also required to be deployed over the information. Data centers have to be built in each sector incorporating the GSM modules and transmission towers as well for connecting people with their environment providing them live updates about the space they live in. This whole setup demands huge investments coming up from the government, public sectors and official authorities to build a well-connected system. It won't be an easy task to keep motivating people to add their bit in the initiative. Organizing promotional events, challenge weeks or even imposing mandatory rules would require funding. A proper supervisory or governance body requires to be formulated for correct implementation of the framework. This is definitely a challenging task to understand how the idea should be put correctly into the planning and implementing procedure. It should be well known that technology is a means and not the target. It is important to identify the beneficiaries and the objective should never be forgotten in a technology driven program during implementation.

CONCLUSION

Cities around the world are different and hence require different solutions for sustainable development. Technology can play a vital role in innovations like transport, energy, telecommunication, waste management, resource management, pollution control and other major departments. Many experiments in the concept of smart city have been proposed to improve quality of life of urban people. An appropriate blend of conventional techniques and new technological solutions can bring wonder's in city's success and proper functioning. The basic requirements of a healthy space needs to be fulfilled ensuring a sustainable growth before incorporating technological remedies.

REFERENCES

Al Rasyid, M. U., Nadhori, I. U., Sudarsono, A., & Alnovinda, Y. T. (2016, January 1). Pollution monitoring system using gas sensor based on wireless sensor network. *International Journal of Engineering and Technology Innovation.*, *6*(1), 79–91.

Batty, B. (2013). *The New Science of Cities*. Cambridge, MA: The MIT Press. doi:10.7551/mitpress/9399.001.0001

Batty, M. (2014). Can it happen again? Planning support, Lee's requiem and the rise of the smart cities movement. *Environment and Planning. B, Planning & Design*, *41*(3), 388–391. doi:10.1068/b4103c2

Dey, S., Di Girolamo, L., van Donkelaar, A., Tripathi, S. N., Gupta, T., & Mohan, M. (2012). Variability of outdoor fine particulate (PM 2.5) concentration in the Indian subcontinent: A remote sensing approach. *Remote Sensing of Environment*, *127*, 153–161. doi:10.1016/j.rse.2012.08.021

Gargava, P., & Rajagopalan, V. (2016). Source apportionment studies in six Indian cities—Drawing broad inferences for urban PM10 reductions. *Air Quality, Atmosphere & Health*, *9*(5), 471–481. doi:10.100711869-015-0353-4

Khedo, K. K., Perseedoss, R., & Mungur, A. (2010). *A wireless sensor network air pollution monitoring system.* arXiv preprint arXiv:1005.1737

Murgante, B., & Borruso, G. (2013). Cities and smartness: a critical analysis of opportunities and risks. *Lect. Notes Comput. Sci., 7973*, 630–642. doi:10.1007/978-3-642-39646-5_46

Murgante, B., & Borruso, G. (2013). Smart cities: un'analisi critica delle opportunità e dei rischi. *GEO Media, 17*(3).

Pti. (2017, January 11). *Air pollution causes 1.2 million deaths in India annually; Delhi most polluted: Greenpeace report - Times of India*. Retrieved from https://timesofindia.indiatimes.com/india/air-pollution-causes-12-lakh-deaths-in-india-annually-delhi-most-polluted-greenpeace-report/articleshow/56478622.cms

Shukla, P. (n.d.). *Air Pollution in Select Indian Cities - teriin.org*. Retrieved from http://www.teriin.org/library/files/Pollution-level-in-India.pdf

Smart Cities Mission. (n.d.). Retrieved from: http://smartcities.gov.in/content/

Wang, F., Jiang, M., Qian, C., Yang, S., Li, C., Zhang, H., . . . Tang, X. (2017). *Residual attention network for image classification*. arXiv preprint arXiv:1704.06904

World Bank and Institute for Health Metrics and Evaluation. (2016). *The Cost of Air Pollution: Strengthening the economic case for action*. Available from: http://documents.worldbank.org/curated/en/781521473177013155/pdf/108141-REVISEDCost-of-PollutionWebCORRECTEDfile.pdf

World Health Organization. (2016). *WHO Global Urban Ambient Air Pollution Database (Update 2016)*. WHO.

Compilation of References

Abdelhak, A., & Asmaa, E. H. (2016). *The Role of Communication Technologies in Building Future Smart Cities. In Smart Cities Technologies* (pp. 55–76). Intech.

Abderrezek, & Fathi, M. (2017). Experimental study of the dust effect on photovoltaic panels' energy yield. *Solar Energy*, *142*, 308–320. doi:10.1016/j.solener.2016.12.040

Access2All. (2010). *EU FP7 project, Mobility schemes ensuring accessibility of public transport for all users*. Retrieved from http://www.access-to-all.eu/

Adinoyi, M. J., & Said, S. A. M. (2013). Effect of dust accumulation on the power outputs of solar photovoltaic modules. *Renewable Energy*, *60*, 633–636. doi:10.1016/j.renene.2013.06.014

Agiwal, M., Roy, A., & Saxena, N. (2016). Next generation 5G wireless networks: A comprehensive survey. *IEEE Communications Surveys and Tutorials*, *18*(3), 1617–1655. doi:10.1109/COMST.2016.2532458

Agrawal & Das. (2011). *Internet of Things – A Paradigm Shift of Future Internet Applications*. Academic Press.

Ahn, J. H., Lee, S. H., & Lee, T. (2016). Anti-Collision Protocol for Coexistence of RFID and NFC P2P Communications. *IEEE Communications Letters*, *20*(11), 2185–2188. doi:10.1109/LCOMM.2016.2602865

Al Rasyid, M. U., Nadhori, I. U., Sudarsono, A., & Alnovinda, Y. T. (2016, January 1). Pollution monitoring system using gas sensor based on wireless sensor network. *International Journal of Engineering and Technology Innovation.*, *6*(1), 79–91.

Al-Ali, A. R., Zualkernan, I., & Aloul, F. (2010). A Mobile GPRS-Sensors Array for Air Pollution Monitoring. *IEEE Sensors Journal*, *10*(10), 1666–1671. doi:10.1109/JSEN.2010.2045890

AlAmmouri, A., Andrews, J. G., & Baccelli, F. (2018). SINR and Throughput of Dense Cellular Networks With Stretched Exponential Path Loss. *IEEE Transactions on Wireless Communications*, *17*(2), 1147–1160. doi:10.1109/TWC.2017.2776905

Alamri, A., Ansari, W. S., Hassan, M. M., Hossain, M. S., Alelaiwi, A., & Hossain, M. A. (2013). A Survey on Sensor-Cloud: Architecture, Applications, and Approaches. *International Journal of Distributed Sensor Networks*, *9*(2), 917923. doi:10.1155/2013/917923

Albino, V., Berardi, U., & Dangelico, R. M. (2015). Smart Cities. Definitions, Dimensions, Performance, and Initiatives. *Journal of Urban Technology*, *22*(1), 3–21. doi:10.1080/10630732.2014.942092

Albusac, J., Castro-Schez, J., Lopez-Lopez, L., Vallejo, D., & Jimenez-Linares, L. (2009). A supervised learning approach to automate the acquisition of knowledge in surveillance systems. *Signal Processing*, *89*(12), 2400–2414. doi:10.1016/j.sigpro.2009.04.008

Albusac, J., Vallejo, D., Jimenez-Linares, L., Castro-Schez, J., & Rodriguez-Benitez, L. (2009). Intelligent surveillance based on normality analysis to detect abnormal behaviours. *International Journal of Pattern Recognition and Artificial Intelligence*, *23*(7), 1223–1244. doi:10.1142/S0218001409007612

Al-Hasan, A. Y. (2016). A new correlation between photovoltaic panel's efficiency and amount of sand dust accumulated on their surface. Academic Press.

Al-Hasan, A. Y., & Ghoneim, A. A. (2005). A new correlation between photovoltaic panel's efficiency and amount of sand dust accumulated on their surface. *International Journal of Sustainable Energy*, *24*(4), 187–197. doi:10.1080/14786450500291834

Al-Kassir, A. R., Fernandez, J., Tinaut, F., & Castro, F. (2005). Thermographic study of energetic installations. *Applied Thermal Engineering*, *25*(2-3), 183–190. doi:10.1016/j.applthermaleng.2004.06.013

Al-Shabaan. (2016). *Effects of dust grain size and density on the monocrystalline PV output power* (Vol. 6). Academic Press.

Amin, M. (2015). *Smart Grid. Public Utilities Fortnightly.*

AMTS. (2006). Annual Data booklet. Ahmedabad Bus Operator Association.

An, F., & Sauer, A. (2004). *Comparison of passenger vehicle fuel economy and greenhouse gas emission standards around the world.* Pew Center on Global Climate Change.

Andersson, M., Rydell, J., & Ahlberg, J. (2009). Estimation of crowd behavior using sensor networks and sensor fusion. *Proc. 12th Int. Conf. Inf. Fusion*, 396–403.

Andrade, E. L., Blunsden, S., & Fisher, R. B. (2006). Modelling crowd scenes for event detection. *Pattern Recognition, 1*, 175–178.

Antonini, A., Barenghi, A., Pelosi, G., & Zonouz, S. (2014). Security challenges in building automation and SCADA. *2014 International Carnahan Conference on Security Technology (ICCST)*, 1-6. 10.1109/CCST.2014.6986996

Anwar, M. I., Khosla, A., & Singh, G. (2017). *Visibility enhancement with single image fog removal scheme using a post-processing technique.* In *Signal Processing and Integrated Networks (SPIN), 2017 4th International Conference on* (pp. 280-285). IEEE. 10.1109/SPIN.2017.8049960

Appels, R., Lefevre, B., Herteleer, B., Goverde, H., Beerten, A., Paesen, R., ... Poortmans, J. (2013). Effect of soiling on photovoltaic modules. *Solar Energy, 96*, 283–291. doi:10.1016/j.solener.2013.07.017

APTIE. (2011). *EU FP7 project.* Accessible Public Transport in Europe. Retrieved from http://www.aptie.eu/site

Araújo, D. C., Maksymyuk, T., Almeida, A. L. F., Maciel, T., Mota, J. C. M., & Jo, M. (2016). Massive MIMO: Survey and Future Research Topics. *IET Communications, 10*(15), 1938–1946. doi:10.1049/iet-com.2015.1091

Asadi, A., Wang, Q., & Mancuso, V. (2014). A Survey on Device-to-Device Communication in Cellular Networks. *IEEE Communications Surveys and Tutorials, 16*(4), 1801–1819. doi:10.1109/COMST.2014.2319555

Ashfaq, R. A. R., Wang, X. Z., Huang, J. Z., Abbas, H., & He, Y. L. (2017). Fuzziness based semi-supervised learning approach for intrusion detection system. *Inf. Sci., 378*, 484–497. doi:10.1016/j.ins.2016.04.019

Australia, I. (2010). *State of Australian cities 2010.* Canberra: Infrastructure Australia.

Baig, Z. A., Szewczyk, P., Valli, C., Rabadia, P., Hannay, P., Chernyshev, M., & Syed, N. (2017). Future challenges for smart cities: Cyber-security and digital forensics. *Digital Investigation, 22*, 3–13. doi:10.1016/j.diin.2017.06.015

Bala, I., Bhamrah. M.S., & Singh, G. (2015). Adaptive Power Control Scheme for Cognitive Radio Systems Based on Receiver Sensitivity. Computational Advancement in Communication Circuits and Systems, 69-79.

Bala, I., Bhamrah, M. S., & Singh, G. (2017). Capacity in Fading Environment Based on Soft Sensing Information Under Spectrum Sharing Constraints. *Wireless Networks*, *23*(2), 519–531. doi:10.100711276-015-1172-0

Balarabe Kura, S. Y. (2012). Qualitative and Quantitative Approaches to the Study of Poverty: Taming the Tensions and Appreciating the Complementarities. *Qualitative Report*, *17*(20), 1–19.

Barbu, O., & Fratu, O. (2011). *An enabler of interoperability in heterogeneous wireless networks. In Proceedings of Wireless Communication, Vehicular Technology, Information Theory and Aerospace & Electronic Systems Technology* (pp. 1–5). Chennai: Wireless VITAE.

Barth, M., & Boriboonsomsin, K. (2008). Real-world carbon dioxide impacts of traffic congestion. *Transportation Research Record: Journal of the Transportation Research Board*, *2058*(1), 163–171. doi:10.3141/2058-20

Bartlett, S., & Satterthwaite, D. (2016). *Cities on a finite planet: Towards transformative responses to climate change*. Routledge. doi:10.4324/9781315645421

Bartoli, A., Hernández-Serrano, J., Soriano, M., Dohler, M., Kountouris, A., & Barthel, D. (2011). Security and privacy in your smart city. *Proceedings of the Barcelona smart cities congress*, 292.

Baskerville, R. (2008). What Design Science is Not. *European Journal of Information Systems*, *17*(5), 441–443. doi:10.1057/ejis.2008.45

Baskerville, R., & Pries-Heje, J. (2010). Explanatory Design Theory. *Business & Information Systems Engineering*, *2*(5), 271–282. doi:10.100712599-010-0118-4

Batty, B. (2013). *The New Science of Cities*. Cambridge, MA: The MIT Press. doi:10.7551/mitpress/9399.001.0001

Batty, M. (2014). Can it happen again? Planning support, Lee's requiem and the rise of the smart cities movement. *Environment and Planning. B, Planning & Design*, *41*(3), 388–391. doi:10.1068/b4103c2

Bawany, N. Z., & Shamsi, J. A. (2015). Smart city architecture: Vision and challenges. *International Journal of Advanced Computer Science and Applications*, *6*(11), 246–255.

Baxter, G., & Sommerville, I. (2011). Socio-Technical Systems: From Design Methods to Systems Engineering. *Interacting with Computers*, *23*(1), 4–17. doi:10.1016/j.intcom.2010.07.003

Bergin, M. H. (2017). *Large Reductions in Solar Energy Production Due to Dust and Particulate Air Pollution*. Academic Press.

Bergson, H. (Ed.). (1907). *L'Evolution créative*. Paris: Aclan.

Bhattarai, S., Hong, L., & Shetty, S. (2013). Optimum transmit beamforming scheme for underlay cognitive radio networks. In *Proceedings of International Symposium on Wireless Personal Multimedia Communications (WPMC)* (pp. 1-6). Academic Press.

Bhushan. (2018). *India's very own Dust Bowl*. Retrieved from https://www.downtoearth.org.in/blog/india-s-very-own-dust-bowl-60913

Bider, I., Johannesson, P., & Perjons, E. (2013). Using Empirical Knowledge and Studies in the Frame of Design Science Research. Design Science at the Intersection of Physical and Virtual Design, 1939, 463-470. doi:10.1007/978-3-642-38827-9_38

Birkeland, J. (2014). Resilient and sustainable buildings. Resilient Sustainable Cities, 146(159).

Bogue, R. (2013). Sensors for condition monitoring: A review of technologies and applications. *Sensor Review*, *33*(4), 295–299. doi:10.1108/SR-05-2013-675

Botta, A., De Donato, W., Persico, V., & Pescapé, A. (2016). Integration of cloud computing and internet of things: A survey. *Future Generation Computer Systems*, *56*, 684–700. doi:10.1016/j.future.2015.09.021

Breen, R. L. (2007). A Practical Guide to Focus-Group Research. *Journal of Geography in Higher Education*, *30*(3), 463–475. doi:10.1080/03098260600927575

Breitenstein, O., Rakotoniaina, J., & Rifai, M. A. (2003). Quantitative evaluation of shunts in solar cells by lock-in thermography. *Progress in Photovoltaics: Research and Applications*, *11*(8), 515–526. doi:10.1002/pip.520

Brown. (2012). Soiling Test Methods and their Use in Predicting Performance of Photovoltaic Modules with and without 3M Anti-Soiling Coating. Renewable Energy Division, 3M Company.

Bulkeley, H., Broto, V., Hodson, M., & Marvin, S. (Eds.). (2010). Cities and Low carbon transitions. Routledge.

ByPAD. (2008). *EU FP6 project, Bicycle Policy Audit*. Retrieved from http://www.bypad.org

Cabanillas, R., & Munguía, H. (2011). Dust accumulation effect on efficiency of Si photovoltaic modules. *Journal of Renewable and Sustainable Energy*, *3*(4), 043114. doi:10.1063/1.3622609

Calderara, S., Cucchiara, R., & Prati, A. (2007). Detection of abnormal behaviors using a mixture of von mises distributions. *Proc. IEEE Conf. Adv. Video Signal Based Surveillance*, 141–146. 10.1109/AVSS.2007.4425300

Carli, R., Dotolia, M., & Pellegrino, R. (2018, November). Multi-criteria decision-making for sustainable metropolitan cities. *Journal of Environmental Management*, *226*(15), 46–61. doi:10.1016/j.jenvman.2018.07.075 PMID:30107310

Chandra. (2018, March 29). Solar Energy a Path to India's Prosperity. *Journal of Institution of Engineers India Ser. C.*

Chang, K.-H., & Mason, B. (2012). The IEEE 802.15.4g Standard for Smart Metering Utility Networks. *Proceedings of the 2012 IEEE Third International Conference on Smart Grid Communications (SmartGridComm)*. 10.1109/SmartGridComm.2012.6486030

Chao, H.-C. (2011). Internet of Things and Cloud computing for future Internet. In Ubiquitous Intelligence and Computing. Springer. doi:10.1007/978-3-642-23641-9_1

Check, J., & Schutt, R. K. (2012). Survey research. In J. Check & R. K. Schutt (Eds.), *Research methods in education* (pp. 159–185). Thousand Oaks, CA: SAGE Publications. doi:10.4135/9781544307725.n8

Chen, G., Zhou, H., & Yan, J. (2007). A novel method for moving object detection in foggy day. *8th ACIS international conference on software engineering. Artificial intelligence, networking, and parallel/distributed computing*, 53–58.

Chen, C., Do, M. N., & Wang, J. (2016). Robust image and video dehazing with visual artifact Suppression via gradient residual minimization. *Proc.Eur. Conf. Comput. Vis*, 576–591. 10.1007/978-3-319-46475-6_36

Chen, H., Tsai, M. J., & Chan, C. C. (2010). A hidden Markov model-based approach for recognizing swimmer's behaviors in swimming pool. *Proceedings - International Conference on Machine Learning and Cybernetics*, *5*, 2459–2465.

Chen, Y., Liang, G., Ka, K. L., & Xu, Y. (2007). Abnormal behavior detection by multi-SVM-based Bayesian network. *Proc. Int. Conf. Inf. Acquis.*, 298–303. 10.1109/ICIA.2007.4295746

Choo, S., & Mokhtarian, P. L. (2004). What type of vehicle do people drive? The role of attitude and lifestyle in influencing vehicle type choice. *Transportation Research Part A, Policy and Practice*, *38*(3), 201–222. doi:10.1016/j.tra.2003.10.005

Chuah, J. W. (2014). The Internet of Things: An Overview and New Perspectives in Systems Design. *International Symposium on Integrated Circuits*. 10.1109/ISICIR.2014.7029576

Chui, K. T., Alhalabi, W., Pang, S. S. H., Pablos, P. O. D., Liu, R. W., & Zhao, M. (2017). Disease Diagnosis in Smart Healthcare: Innovation, Technologies and Applications. *Sustainability*, *9*(12), 2309. doi:10.3390u9122309

Cisco. (2015). *Visual Networking Index*. White paper, Feb. 2015. Retrieved from www.Cisco.com

CITyLOG. (2012). *EU FP7 project, Sustainability and efficiency of city logistics*. Retrieved from http://www.city-log.eu

Citymobil Consortium. (2011). *CityMobil – Advanced Transport for the Urban Environment*. Final CityMobil Brochure.

Civitas Initiative. (2012). *Cleaner and Better Transport in Cities*. Retrieved from http://www.civitas.eu

Computer Vision System Toolbox. (n.d.). MATLAB. Retrieved from https://www.mathworks.com/products/computer-vision.html

Cvetković, B., Kaluža, B., Gams, M., & Luštrek, M. (2015). Adapting activity recognition to a person with Multi-Classifier Adaptive Training. *Journal of Ambient Intelligence and Smart Environments*, *7*, 171–185.

D. Associates. (2005). *A Study of innovations in the Funding and Delivery of Transportation Infrastructure Using Tolls*. Author.

D'Onofrio, S., & Tschudi, F. (2018). Design Thinking: Ein Videoprojekt zu Computing with Words schlägt neue Brücken zwischen Forschung und Praxis. In A. Meier & R. Seising (Eds.), Vague Information Processing (pp. 510-527). Springer Fachmedien Wiesbaden.

D'Onofrio, S., Müller, S. M., Papageorgiou, E. I., & Portmann, E. (2018). Fuzzy Reasoning in Cognitive Cities – An Exploratory Work on Fuzzy Analogical Reasoning Using Fuzzy Cognitive Maps. In *International Conference on Fuzzy Systems (FUZZ-IEEE 2018)*. Rio de Janeiro, Brazil: Academic Press. 10.1109/FUZZ-IEEE.2018.8491474

D'Onofrio, S., & Portmann, E. (2017). Cognitive Computing in Smart Cities. *Informatik Spektrum*, *40*(1), 46–57. doi:10.100700287-016-1006-1

Daly, H. (1996). *Beyond growth.* Boston: Beacon Press.

Darwish, Z. A., Kazem, H. A., Sopian, K., Al-Goul, M. A., & Alawadhi, H. (2015). Effect of dust pollutant type on photovoltaic performance. *Renewable & Sustainable Energy Reviews*, *41*, 735–744. doi:10.1016/j.rser.2014.08.068

Demarchou, E., Psomas, C., & Krikidis, I. (2018). Mobility Management in Ultra-Dense Networks: Handover Skipping Techniques. *IEEE Access: Practical Innovations, Open Solutions*, *6*, 11921–11930. doi:10.1109/ACCESS.2018.2810318

Department for Transport. (2004). *Bus Priority: The Way Ahead.* Department of Transport. Retrieved from http://www.dft.gov.uk/pgr/regional/buses/bpf/buspriority thewayahead12/)

Dey, S., Di Girolamo, L., van Donkelaar, A., Tripathi, S. N., Gupta, T., & Mohan, M. (2012). Variability of outdoor fine particulate (PM 2.5) concentration in the Indian subcontinent: A remote sensing approach. *Remote Sensing of Environment*, *127*, 153–161. doi:10.1016/j.rse.2012.08.021

Dhillon, S. S., & Chakrabarty, K. (2003). Sensor placement for e_ective coverage and surveillance in distributed sensor networks. *Proceedings of the 2003 IEEE Wireless Communications and Networking Conference*, 3, 1609–1614.

Diamantopoulos, G., & Spann, M. (2005). Event detection for intelligent car park video surveillance. *Real-Time Imaging*, *11*(3), 233–243. doi:10.1016/j.rti.2005.02.002

Dong, N., Jia, Z., Shao, J., Xiong, Z., Li, Z., Liu, F., . . . Peng, P. (2010). Traffic abnormality detection through directional motion behavior map. *Proc. 7th IEEE Int. Conf. Adv. Video Signal Based Surveillance*, 80–84. 10.1109/AVSS.2010.61

Dong, Q., Wu, Y., & Hu, Z. (2009). Pointwise motion image (PMI): A novel motion representation and its applications to abnormality detection and behavior recognition. *IEEE Transactions on Circuits and Systems for Video Technology*, *19*(3), 407–416. doi:10.1109/TCSVT.2009.2013503

Doyle, C., Sammon, D., & Neville, K. (2015). Building an evaluation framework for social media-enabled collaborative learning environments (SMECLEs). *Journal of Decision Systems*, *24*(3), 298–317. doi:10.1080/12460125.2015.1054694

Drexel, A.J. (2016). *Options for Achieving Deep Reductions in Carbon Emissions in Philadelphia by 2050*. Prepared by Drexel University for The Philadelphia Mayor's Office of Sustainability.

Duque, D., Santos, H., & Cortez, P. (2007). Prediction of abnormal behaviors for intelligent video surveillance systems. *Proc. IEEE Symp. Comput. Intell. Data Mining*, 362–367. 10.1109/CIDM.2007.368897

Duque, D., Santos, H., & Cortez, P. (2006). The OBSERVER: An intelligent and automated video surveillance system. *Proc. 3rd Int. Conf. Image Anal. Recognition*, 898–909. 10.1007/11867586_81

Eames, M., Dixon, T., May, T., & Hunt, M. (2013, October). City futures: Exploring urban retrofit and sustainable transitions. *Building Research and Information*, *41*(5), 504–516. doi:10.1080/09613218.2013.805063

Eason, K. (2014). Afterword: The Past, Present and Future of Sociotechnical Systems Theory. *Applied Ergonomics*, *45*(2), 213–220. doi:10.1016/j.apergo.2013.09.017 PMID:24183568

EBSF. (2012). *EU FP7 project*. European Bus System of the Future. Retrieved from http://ebsf.eu/

Ehlers, G. A., Howerton, R. D., & Speegle, G. E. (1996). *U.S. Patent No. 5,572,438*. Washington, DC: U.S. Patent and Trademark Office.

Elminir, H. K., Ghitas, A. E., Hamid, R. H., El-Hussainy, F., Beheary, M. M., & Abdel-Moneim, K. M. (2006). Effect of dust on the transparent cover of solar collectors. *Energy Conversion and Management*, *47*(18-19), 3192–3203. doi:10.1016/j.enconman.2006.02.014

El-Shobokshy, & Hussein, F. M. (1993). Degradation of Photovoltaic Cell Performance Due to Dust Deposition on to Its Surface. *Renewable Energy*, *3*(6-7), 585–590. doi:10.1016/0960-1481(93)90064-N

Energy, C. (2017) What is a carbon footprint? *Cloud Energy*. Retrieved from: https://cloudenergy.com.ng/blog/what-carbon-footprint

Ermis, E. B., Saligrama, V., Jodoin, P., & Konrad, J. (2008). Abnormal behavior detection and behavior matching for networked cameras. *Proc. 2nd ACM/IEEE Int. Conf. Distrib. Smart Cameras*, 1–10. 10.1109/ICDSC.2008.4635728

Ernston, H., van der Leeuw, S. E., Redman, C. L., Meffert, D. J., Davis, G., Alfsen, C., & Elmqvist, T. (2010). Urban transitions: On urban resilience and human-dominated ecosystems. *Ambio*, *39*(8), 531–545. doi:10.100713280-010-0081-9 PMID:21141773

ERTRAC Research and Innovation Roadmaps. (2011). *Implementation of the ERTRAC Strategic Research Agenda 2010*. Author.

European Commission. (1995). Green Paper, The Citizens' Network - Fulfilling the potential of public passenger transport in Europe. COM (95) 601 final, Brussels.

European Commission. (1998). Communication from the Commission Developing the Citizens' Network. COM (1998) 431 final, Brussels.

European Commission. (2006). Communication from the Commission on the Thematic Strategy on the Urban Environment. COM (2005) 718 final, Brussels.

European Commission. (2007a). Green Paper, Towards a New Culture for Urban Mobility. SEC (2007) 1209 final, Brussels.

European Commission. (2007b). Regulation of the European Parliament and of the Council on Public Passenger Transport Services by Rail and by Road. Regulation (EC) No 1370/2007, Brussels.

European Commission. (2009a). Communication from the Commission, Action Plan on Urban Mobility. SEC (2009) 1211, SEC (2009) 1212, final, Brussels.

European Commission. (2009b). *Directive of the European Parliament and the Council on the Promotion of Clean and Energy Efficient Road Transport Vehicles*. Directive 2009/33/EC, Brussels.

European Commission. (2010a). Communication from the Commission, Europe 2020 Flagship Initiative, Innovation Union. SEC (2010) 1161, Brussels.

European Commission. (2010b). Communication from the Commission, European Disability Strategy 2010-2020: A Renewed Commitment to a Barrier-Free Europe. SEC (2010) 1324, Brussels.

European Commission. (2010c). Communication from the Commission, Towards a European road safety area: policy orientations on road safety 2011-2020. COM (2010) 0389 final, Brussels.

European Commission. (2010d): Communication from the, The EU Internal Security Strategy in Action: Five steps towards a more secure Europe, COM (2010) 673 final, Brussels.

European Commission. (2011a). *White Paper, Roadmap to a Single European Transport Area, Towards a competitive and resource efficient transport system.* COM (2011) 144 final, Brussels.

European Commission. (2011d). Proposal for a Regulation of the European Parliament and of the Council, establishing Horizon 2020 – The Framework Programme for Research and Innovation. COM (2011) 809 final, Brussels.

European Commission. (2012a). *Road Transport - A change of gear.* Author.

European Commission. (2012b). CARE - EU road accidents database. Author.

European Commission. (2013a). *Communication Clean Power for Transport: European alternative fuels strategy* (COM(2013) 17).

European Commission. (2013b). *Proposal for a Directive on the deployment of alternative fuels infrastructure* (COM(2013) 18).

European Green Cars Initiative. (2012). *Public-Private Partnership European Green Cars Initiative.* Retrieved from http://www.green-cars-initiative.eu

Eu, Z. A., Winston, K., Seah, G., & Tan, H. P. (2008). A study of MAC schemes for wireless sensor networks powered by ambient energy harvesting. *Proceedings of 4th Annual International Conference on Wireless Internet (WICON '08)*, 9.

Evans, M. (2018). 7 ways to reduce CO2 emissions pollution in industry. *The Balance Small Business.* Retrieved from: https://www.thebalancesmb.com/reducing-co2-emissions-3157794

Faber, J., Sevenster, M., Markowska, A., Smit, M., Zimmermann, K., Soboh, R., & van 't Riet, J. (2012). *Behavioural Climate Change Mitigation Options: Domain Report Food.* Report Delft.

Faber, J., Schroten, A., Bles, M., Sevenster, M., Markowska, A., & Smit, M. (2012). *Behavioural climate change mitigation options and their appropriate inclusion in quantitative longer-term policy scenarios.* Delft: CE Delft.

Fan, C., Sun, F., & Yang, L. (2005). Investigation on nondestructive evaluation of pipelines using infrared thermography. *Proceedings of the Joint 30th International Conference on IEEE, Infrared and Millimeter Waves and 13th International Conference on Terahertz Electronics, 2*, 339–340.

Feng, D., Lu, L., Yuan-Wu, L., Li, G. Y., Feng, G., & Li, S. (2013). Device to-Device Communications Underlaying Cellular Networks. *IEEE Transactions on Communications, 61*(8), 3541–3551. doi:10.1109/TCOMM.2013.071013.120787

Fette, B. (2013). Fourteen Years of Cognitive Radio Development. In *Proceedings of International Military Communications Conference* (pp. 1166-1175). Academic Press.

Figgis, B. W. (2016). Solar test facility in Doha. Academic Press.

Figgis, B., Ennaoui, A., Ahzi, S., & Rémond, Y. (2017). Review of PV soiling particle mechanics in desert environments. *Renewable & Sustainable Energy Reviews, 76*, 872–881. doi:10.1016/j.rser.2017.03.100

Finger, M., & Portmann, E. (2016). What Are Cognitive Cities? In E. Portman & M. Finger (Eds.), *Towards Cognitive Cities. Advances in Cognitive Computing and Its Application to the Governance of Large Urban Systems* (pp. 1–11). Heidelberg, Germany: Springer International Publishing.

Finger, M., & Razaghi, M. (2017). Conceptualizing "Smart Cities". *Informatik Spektrum, 40*(1), 6–13. doi:10.100700287-016-1002-5

Fredericks, J., Hespanhol, L., Parker, C., Zhou, D., & Tomitsch, M. (2018). Blending Pop-up Urbanism and Participatory Technologies: Challenges and Opportunities for Inclusive City Making. *City. Cultura e Scuola, 12*, 44–53.

Furness, A. (2008). A Framework Model for The Internet of Things. *GRIFS/ CASAGRAS Workshop.*

Garcia, M., Marroyo, L., Lorenzo, D., & Perez, M. (2011). Soiling and other optical losses in solar-tracking PV plants innavarra. *Progress in Photovoltaics: Research and Applications, 19*(2), 211–217. doi:10.1002/pip.1004

Gargava, P., & Rajagopalan, V. (2016). Source apportionment studies in six Indian cities—Drawing broad inferences for urban PM10 reductions. *Air Quality, Atmosphere & Health, 9*(5), 471–481. doi:10.100711869-015-0353-4

Garima Yadav Saurabh Maheshwari Anjali Agarwal. (2014). *Contrast Limited Adaptive Histogram Equalization Based Enhancement For Real Time Video System.* Rajasthan: ICACCI.

Gaussorgues, G. (1994). *Infrared Thermography*. Berlin: Springer. doi:10.1007/978-94-011-0711-2

Geisler, K. (2013). *The relationship between smart grids and smart cities.* IEEE Smart Grid Newsletter.

Georgetown University. (n.d.). Simple tupes to reduce your carbon footpting. *Georgetown University.* Retrieved from: https://sustainability.georgetown.edu/getinvolved/thingsyoucando

Gershenfeld, N., Krikorian, R., & Cohen, D. (2004). The Internet of things. *Scientific American, 291*(4), 76–81. doi:10.1038cientificamerican1004-76 PMID:15487673

Ghazi, Sayigh, A., & Ip, K. (2014). Dust effect on flat surfaces-a review paper. *Renewable & Sustainable Energy Reviews, 33*, 742–751. doi:10.1016/j.rser.2014.02.016

Gholami, A., Saboonchi, A., & Alemrajabi, A. A. (2017). Experimental study of factors affecting dust accumulation and their effects on the transmission coefficient of glass for solar applications. *Renewable Energy, 112*, 466–473. doi:10.1016/j.renene.2017.05.050

Gimeno, A., Marco, A.M., Marín, R.B., & Nebra, R.C. (2014). Protocol and Architecture to Bring Things into Internet of Things. *Int. J. Distrib. Sens. Netw., 2014*, 158252:1–158252:18.

Girardet, H. (2014). *Creating regenerative cities*. Routledge. doi:10.4324/9781315764375

Glackin, S., & Newton, P. (2016). *Engaging the Greyfields: community engagement and co-design in residential redevelopment of public housing. Instruments of Planning*. Taylor & Francis.

Gnewuch, U., Morana, S., Adam, M. T. P., & Maedche, A. (2018). Faster Is Not Always Better: Understanding the Effect of Dynamic Response Delays in Human-Chatbot Interaction. *European Conference on Information Systems (ECIS 2018).*

Gorostiza, E. M., L'azaro Galilea, J. L., Meca Meca, F. J., Salido Monzú, D., Espinosa Zapata, F., & Pallarés Puerto, L. (2011). Infrared sensor system for mobile-robot positioning in intelligent spaces. *Sensors (Basel), 11*(5), 5416–5438. doi:10.3390110505416 PMID:22163907

Goshorn, R., Goshorn, D., Goshorn, J., & Goshorn, L. (2008). Abnormal behavior-detection using sequential syntactical classification in a network of clustered cameras. *Proc. 2nd ACM/IEEE Int. Conf. Distrib. Smart Cameras*, 1–10. 10.1109/ICDSC.2008.4635732

Gouldson, A., Sudmant, A., Khreis, H., & Papargyropoulou, E. (2018). *The economic and social benefits of low-carbon cities: a systematic review of the evidence*. Coalition for Urban Transitions.

Granzer, W., Praus, F., & Kastner, W. (2010). Security in building automation systems. *IEEE Transactions on Industrial Electronics, 57*(11), 3622–3630. doi:10.1109/TIE.2009.2036033

Gregor, S., & Baskerville, R. (2012). The Fusion of Design Science and Social Science Research. *Information Systems Foundation Workshop*.

Gregor, S., & Jones, D. (2007). The Anatomy of a Design Theory. *Journal of the Association for Information Systems, 8*(5), 312–335. doi:10.17705/1jais.00129

Guan, Zhang, Xiao, & Yan. (2017). In-situ investigation of the effect of dust deposition on the performance of polycrystalline silicon photovoltaic modules. *Renew Energy, 101*, 1273–84.

Guinard, Trifa, & Wilde. (2010). A resource oriented architecture for the Web of Things. *Proc. Internet Things (IOT),* 1-8.

Guo, Tang, & Cai. (2013). Fusion strategy for single image DE hazing. *Int. J. Digit. Content Technol. its Appl, 7*, 19–28.

Guo, P., & Miao, Z. (2007). A home environment posture and behavior recognition system. *Proc. Int. Conf. Convergence Inf. Technol.*, 175–180. 10.1109/ICCIT.2007.35

Habenstein, A., D'Onofrio, S., Portmann, E., Stürmer, M., & Myrach, T. (2016). Open Smart City: Good Governance für smarte Städte. In A. Meier & E. Portmann (Eds.), *Smart City: Strategie, Governance und Projekte* (pp. 47–71). Wiesbaden: Springer Vieweg. doi:10.1007/978-3-658-15617-6_3

Hamaguchi, K., Ma, Y., Takada, M., Nishijima, T., & Shimura, T. (2012). Telecommunications systems in smart cities. *Hitachi Rev*, *61*(3), 152–158.

Hammad, Al-Abed, M., Al-Ghandoor, A., Al-Sardeah, A., & Al-Bashir, A. (2018). Modelling and analysis of dust and temperature effects on photovoltaic systems' performance and optimal cleaning frequency: Jordan case study. *Renewable & Sustainable Energy Reviews*, *82*, 2218–2234. doi:10.1016/j.rser.2017.08.070

Han, J. K., Park, B. S., Choi, Y. S., & Park, H. K. (2001). Genetic approach with a new representation for base station placement in mobile communications. *Proceedings of the 2001 IEEE Vehicular Technology Conference*, 4, 2703–2707.

Hegazy, A. A. (2001). Effect of dust accumulation on solar transmittance through glass covers of plate-type collectors. *Renewable Energy*, *22*(4), 525–540. doi:10.1016/S0960-1481(00)00093-8

He, K., Sun, J., & Tang, X. (2011). Single Image Haze Removal Using Dark Channel Prior. *IEEE Transactions on Pattern Analysis and Machine Intelligence*, *33*(12). PMID:20820075

Hillier, B. (2012). The City as a Socio-technical System: A Spatial Reformulation in the Light of the Levels Problem and the Parallel Problem. *Digital Urban Modeling and Simulation*, *242*, 24–48. doi:10.1007/978-3-642-29758-8_3

Hindustan Times. (2015, November 6). The stubble trouble: Gasping for air as smog reigns. *Hindustan Times*.

Hirsch Hadorn, G., Hoffmann-Riem, H., Biber-Klemm, S., Grossenbacher-Mansuy, W., Joye, D., Pohl, C., ... Zemp, E. (2008). *Handbook of Transdisciplinary Research*. Springer. doi:10.1007/978-1-4020-6699-3

Höller, J., Tsiatsis, V., Mulligan, C., Karnouskos, S., Avesand, S., & Boyle, D. (2014). *From Machine-to-Machine to the Internet of Things: Introduction to a New Age of Intelligence*. Amsterdam: Elsevier.

Hollingshead, A. B. (2001). Cognitive Interdependence and Convergent Expectations in Transactive Memory. *Journal of Personality and Social Psychology*, *81*(6), 1080–1089. doi:10.1037/0022-3514.81.6.1080 PMID:11761309

Howell, J. R., Siegel, R., & Menguc, M. P. (2010). *Thermal Radiation Heat Transfer* (5th ed.). New York: Taylor and Francis. doi:10.1201/9781439894552

Hsieh, C., & Hsu, S. (2007). A simple and fast surveillance system for human tracking and behavior analysis. *Proc. 3rd IEEE Int. Conf. SignalImage Technol. Internet-Based Syst.*, 812–818. 10.1109/SITIS.2007.128

Hughes, T. P. (1987). The Evolution of Large Technological Systems. In W. E. Bijker, T. P. Hughes, & T. J. Pinch (Eds.), *The Social Construction of Technological Systems. New Directions on the Sociology and History of Technology* (pp. 51–82). Cambridge, MA: MIT Press.

Hurni, T., Huber, T. L., & Dibbern, J. (2017). Swiss Software Industry Survey 2017. Current State, Emerging Trends & Long-term Developments in the Swiss Software Industry. *ICT Switzerland.* Retrieved from: https://ictswitzerland.ch/media/dateien/Studien/Swiss_Software_Industry_Survey_2017-Report.pdf

Hurwitz, J., Kaufman, M., & Bowles, A. (Eds.). (2015). *Cognitive Computing and Big Data Analytics.* Indianapolis, IN: John Wiley and Sons.

Hussain. (2017). An experimental study on effect of dust on power loss in solar photovoltaic module. Renewables: Wind, Water, and Solar, 4(9).

Hu, W., Tan, T., Wang, L., & Maybank, S. (2004). A survey on visual surveillance of object motion and behaviors. *IEEE Transactions on Systems, Man and Cybernetics. Part C, Applications and Reviews*, *34*(3), 334–352. doi:10.1109/TSMCC.2004.829274

Ibarra-Castanedo, C., Tarpani, J. R., & Maldague, X. P. (2013). Nondestructive testing with thermography. *European Journal of Physics*, *34*(6), S91–S109. doi:10.1088/0143-0807/34/6/S91

IEA. (2015). *Key world energy statistics.* Retrieved from https://www.iea.org/weo2017

India's solar capacity to cross 20GW in next 15 months: Piyush Goyal. (2017, April 6). *Economic Times.*

Information Society Technologies Advisory Group (ISTAG). (2011). *Orientations for EU ICT R&D & Innovation beyond 2013.* Available online http://cordis.europa.eu/fp7/ict/istag/documents/ stag_key_recommendations_beyond_2013_full.pdf

IPCC Fifth Assessment Report. (2013). *IPCC.* Retrieved from: http://ipcc-wg2.gov/oeAR5/images/uploads/WGIIAR5-Glossary_FGD.pdf

IRU. (2009). *Annual Report 2009.* Retrieved from http://www.iru.org/en_bookshop_item?id=214

Islam, S. M. R., Kwak, D., Kabir, H., Hossain, M., & Kwak, K. S. (2015). *Internet of Things for Health Care: A Comprehensive Survey.* doi:10.1109/ACCESS.2015.2437951

ISO/IEC. (2014). *Smart Cities: Preliminay Report.* Available online: http://www.iso.org/ iso/smart_cities_reportjtc1.pdf

Ito, M., Gutiérrez, K., Livingstone, S., Penual, B., Rhodes, J., Salen, K., ... Watkins, S. C. (Eds.). (2013). *Connected Learning: An Agenda for Research and Design.* Irvine, CA: Digital Media and Learning Research Hub.

ITU-T Focus Group on Smart Sustainable Cities (FG-SSC). (2014). *An Overview of Smart Sustainable Cities and the Role of Information and Communication Technologies.* Focus Group Technical Report.

Ivan, M. (n.d.). *Types of machine learning algorithms.* Retrieved from https://en.proft.me/2015/12/24/types-machine-learning-algorithms/

Jan, T., Piccardi, M., & Gunes, H. (2003). Suspicious behavior assessment for visual surveillance using neural network classifiers. *Proc. Int. Conf. Imaging Sci., Syst. Technol., 2*, 65–71.

Jiang, H., Lu, L., & Sun, K. (2011). Experimental investigation of the impact of airborne dust deposition on the performance of solar photovoltaic (PV) modules. *Atmospheric Environment, 45*(25), 4299–4304. doi:10.1016/j.atmosenv.2011.04.084

Jiang, J., Sun, H., & Chiu, W. Y. (2016). *Energy efficient massive MIMO system design for smart grid communications. In Proceedings of IEEE Communications Workshops* (pp. 337–341). ICC.

Jin, L., Xue, Y., Li, Q., & Feng, L. (2016). Integrating human mobility and social media for adolescent psychological stress detection. In *Proceedings of the International Conference on Database Systems for Advanced Applications*. Dallas, TX: Springer. 10.1007/978-3-319-32049-6_23

Jin-Yi, C., & Ying-Jia, H. (2011) Low-carbon urban planning of the resource-based cities. In *Multimedia Technology (ICMT), 2011 International Conference on* (pp. 1102-1105). IEEE.

Jobson, D. J., & Woodell, G. A. (1995). *Properties of a center/surround retinex, part 2: Surround design.* NASA Tech. Memo. 110188.

Jobson, Rahman, & Woodell. (n.d.). *A Multiscale Retinex for Bridging the Gap Between Color Images and the Human Observation of Scenes.* Academic Press.

Johansson-Stenman, O., & Martinsson, P. (2006). Honestly, why are you driving a BMW? *Journal of Economic Behavior & Organization*, *60*(2), 129–146. doi:10.1016/j.jebo.2004.08.006

Kaldellis, J., & Kapsali, M. (2011). Simulating the dust effect on the energy performance of photovoltaic generators based on experimental measurements. *Energy*, *36*(8), 5154–5161. doi:10.1016/j.energy.2011.06.018

Kalogirou, Agathokleous, R., & Panayiotou, G. (2013). On-site PV characterization and the effect of soiling on their performance. *Energy*, *51*, 439–446. doi:10.1016/j.energy.2012.12.018

Kaltenrieder, P., Altun, T., D'Onofrio, S., Portmann, E., & Myrach, E. (2016). Personal digital assistant 2.0 – A software prototype for cognitive cities. *IEEE International Conference on Fuzzy Systems (FUZZ-IEEE 2016)*. 10.1109/FUZZ-IEEE.2016.7737872

Kamel, M., Hamouda, W., & Youssef, A. (2016). Ultra-Dense Networks: A Survey. *IEEE Communications Surveys and Tutorials*, *18*(4), 2522–2545. doi:10.1109/COMST.2016.2571730

Karthika, P. B., & Sindhu, R. (2017). A simple haze removal algorithm to remove haze from videos. In *Computing Methodologies and Communication (ICCMC), 2017 International Conference on* (pp. 461-464). IEEE. 10.1109/ICCMC.2017.8282732

Kaźmierczak, J., Hilarowicz, A., & Wolny, M. (2018). *Introduction to the research project "smart city: a holistic approach."* Retrieved from: https://www.researchgate.net/publication/326147170_Smart_City_Holistic_Approach

Khan, A. A., Rehmani, M. H., & Rachedi, A. (2016). When Cognitive Radio Meets the Internet of Things? In *Proceedings of International Wireless Communications and Mobile Computing Conference (IWCMC)* (pp. 469-474). Academic Press. 10.1109/IWCMC.2016.7577103

Khan, A. A., Rehmani, M. H., & Rachedi, A. (2017). Cognitive-Radio-Based Internet of Things: Applications, Architectures, Spectrum Related Functionalities, and Future Research Directions. *IEEE Wireless Communications*, *24*(3), 17–25. doi:10.1109/MWC.2017.1600404

Khan, A. A., Rehmani, M. H., & Reisslein, M. (2016). Cognitive Radio for Smart Grids: Survey of Architectures, Spectrum Sensing Mechanisms, and Networking Protocols. *IEEE Communications Surveys and Tutorials*, *18*(1), 860–898. doi:10.1109/COMST.2015.2481722

Khedo, K. K., Perseedoss, R., & Mungur, A. (2010). *A wireless sensor network air pollution monitoring system.* arXiv preprint arXiv:1005.1737

Khurana, H., Hadley, M., Lu, N., & Frincke, D. A. (2010). Smart-grid security issues. *IEEE Security and Privacy*, *8*(1), 81–85. doi:10.1109/MSP.2010.49

Kitamura, R., Golob, T. F., Yamamoto, T., & Wu, G. (1999). *Accessibility and Auto Use in a Motorized Metropolis Center for Activity Systems Analysis*. Institute of Transportation Studies, UC Irvine.

Klugmann, & … . (2015). Degradation of electrical performance of a 326 crystalline photovoltaic module due to dust deposition in northern Poland. *Renewable Energy*, *78*, 418–426. doi:10.1016/j.renene.2015.01.018

Kortuem, Kawsar, Fitton, & Sundramoorthy. (2010). Smart objects as building blocks for the Internet of Things. IEEE Internet Comput., 14(1), 44-51.

Koshizuka, N., & Sakamura, K. (2010). Ubiquitous ID: Standards for Ubiquitous computing and the Internet of Things. *IEEE Pervasive Computing*, *9*(4), 98–101. doi:10.1109/MPRV.2010.87

Kosmatos, E. A., Tselikas, N. D., & Boucouvalas, A. C. (2011). Integrating RFIDs and smart objects into a Unified Internet of Things architecture. *Advances in Internet of Things*, *1*(01), 5–12. doi:10.4236/ait.2011.11002

Koutra, S., Ioakimidis, C.S., Gallas, M-A & Becue, V. (2018, March). Towards the Development of a Net-Zero Energy District Evaluation Approach: A Review of Sustainable Approaches and Assessment Tools. *Sustainable Cities and Society.*

Krasnova, H., Abramova, O., Notter, I., & Baumann, A. (2016). Why Phubbing is Toxic for Your Relationship: Understanding the Role of Smartphone Jealousy among "Generation Y" Users. *European Conference in Information Systems (ECIS 2016).*

Krugman, P. (2010) The finite world. *The New York Times,* 26.

Kummitha, R. K. R., & Crutzen, N. (2017). How Do We Understand Smart Cities? An Evolutionary Perspective. *Cities (London, England)*, *67*, 43–52. doi:10.1016/j.cities.2017.04.010

Lahiri, B., Bagavathiappan, S., Jayakumar, T., & Philip, J. (2012). Medical applications of infrared thermography: A review. *Infrared Physics & Technology*, *55*(4), 221–235. doi:10.1016/j.infrared.2012.03.007

Lao, W., Han, J., & de With, P. H. N. (2009). Automatic video-based human motion analyzer for consumer surveillance system. *IEEE Transactions on Consumer Electronics*, *55*(2), 591–598. doi:10.1109/TCE.2009.5174427

Lasobras, J., Alonso, R., Carretero, C., Carretero, E., & Imaz, E. (2014). Infrared Sensor-Based Temperature Control for Domestic Induction Cooktops. *Sensors (Basel)*, *14*(3), 5278–5295. doi:10.3390140305278 PMID:24638125

Leccese, F., Cagnetti, M., Calogero, A., Trinca, D., di Pasquale, S., Giarnetti, S., & Cozzella, L. (2014). A New Acquisition and Imaging System for Environmental Measurements: An Experience on the Italian Cultural Heritage. *Sensors (Basel)*, *14*(5), 9290–9312. doi:10.3390140509290 PMID:24859030

Leccese, F., Cagnetti, M., & Trinca, D. (2014). A Smart City Application: A Fully Controlled Street Lighting Isle Based on Raspberry-Pi Card, a ZigBee Sensor Network and WiMAX. *Sensors (Basel)*, *14*(12), 24408–24424. doi:10.3390141224408 PMID:25529206

Lee, C., Lim, K., & Woon, W. (2010). Statistical and entropy based multipurpose human motion analysis. *Proc. 2nd Int. Conf. Signal Process. Syst.*, 734–738.

Le, T. N., Chin, W. L., & Chen, H. H. (2017). Standardization and Security for Smart Grid Communications Based on Cognitive Radio Technologies—A Comprehensive Survey. *IEEE Communications Surveys and Tutorials*, *19*(1), 423–445. doi:10.1109/COMST.2016.2613892

Li, H., & Han, Z. (2010). Catching attacker(s) for collaborative spectrum sensing in cognitive radio systems: An abnormality detection approach. *Proc. IEEE Symp. New Frontiers Dyn. Spectr.*, 1–1. 10.1109/DYSPAN.2010.5457898

Li, J., Gong, S., & Xiang, T. (2009). On-the-fly global activity prediction and anomaly detection. *Proc. IEEE 12th Int. Conf. Comput Vision Workshops*, 1330–1337. 10.1109/ICCVW.2009.5457455

Liang, W., Ng, S. X., & Hanzo, L. (2017). Cooperative Overlay Spectrum Access in Cognitive Radio Networks. *IEEE Communications Surveys and Tutorials*, *19*(3), 1924–1944. doi:10.1109/COMST.2017.2690866

Li, Z., Kang, J., Yu, R., Ye, D., Deng, Q., & Zhang, Y. (2018). Consortium Blockchain for Secure Energy Trading in Industrial Internet of Things. *IEEE Transactions on Industrial Informatics*, *14*(8), 3690–3700.

Louis Berger Group Construction (GIDB). (2006). *A Report submitted to Gujarat Infrastructure Development Board*. Author.

Lugaric, L., & Krajcar, S. (2016). Transforming cities towards sustainable low-carbon energy systems using emergy synthesis for support in decision making. *Energy Policy*, *98*, 471–482. doi:10.1016/j.enpol.2016.09.028

Lu, H., Lu, L., & Wang, Y. (2016). Numerical investigation of dust pollution on a solar photovoltaic (PV) system mounted on an isolated building. *Applied Energy*, *180*, 27–36. doi:10.1016/j.apenergy.2016.07.030

Luong, M. P. (1992). Infrared thermography of fatigue in metals. *Proceedings of the SPIE*, 1682, 222–233. 10.1117/12.58539

LVM. (2017). Vehicle fleet renewal supports the reduction of emissions. *LVM*. Retrieved from: https://www.lvm.fi/en/-/vehicle-fleet-renewal-supports-the-reduction-of-emissions-955841

Madakam, S., Ramaswamy, R., & Tripathi, S. (2015). Internet of Things (IoT): A literature review. *Journal of Computer and Communications*, *3*(05), 164–173. doi:10.4236/jcc.2015.35021

Maghami, Hizam, H., Gomes, C., Radzi, M. A., Rezadad, M. I., & Hajighorbani, S. (2016). Power loss due to soiling on solar panel: A review. *Renewable & Sustainable Energy Reviews*, *59*, 1307–1316. doi:10.1016/j.rser.2016.01.044

Maghami, M., Hizam, H., Gomes, C., Hajighorbani, S., & Rezaei, N. (2015). Evaluation of the Southeast Asian Haze on Solar Generation Performance. *PLoS One*, *10*(8), e0135118. doi:10.1371/journal.pone.0135118 PMID:26275303

Maldague, X. (2001). *Theory and Practice of Infrared Technology for Nondestructive Testing*. New York: Wiley.

Malone, T. W. (Ed.). (2018). *Superminds: The Surprising Power of People and Computers Thinking Together*. Kindle Edition.

Malone, T. W., & Bernstein, M. S. (Eds.). (2015). *Handbook of Collective Intelligence*. Cambridge: MIT Press.

Mani. (2010). *Impact of dust on solar photovoltaic (PV) performance: Research status, challenges and recommendations*. Academic Press.

Mani, M., & Pillai, R. (2010). Impact of dust on solar photovoltaic (PV) performance: Research status, challenges and recommendations. *Renewable & Sustainable Energy Reviews*, *14*(9), 3124–3131. doi:10.1016/j.rser.2010.07.065

Mareis, C. (Ed.). (2014). *Theorien des Design*. Hamburg, Germany: Junius-Verlag.

Martiskainen, M., & Kivimaa, P. (2018). Creating innovative zero carbon homes in the United Kingdom—Intermediaries and champions in building projects. *Environmental Innovation and Societal Transitions*, *26*, 15–31. doi:10.1016/j.eist.2017.08.002

Marx, K. (Ed.). (1959). Das Kapital: Vol. I. Marx/Engels-Gesamtausgabe. Berlin: Dietz Verlag. (Originally Published 1867)

Mastekbayeva, G. A. (2000). Effect of dust on the transmittance of low density polyethylene glazing in a tropical climate. Sol. Energy, 68, 135-141.

Medaglia, C. M., & Serbanati, A. (2010). An Overview of Privacy and Security Issues in the Internet of Things. *The Internet of Things*, 389-395.

Mediate. (2011). *EU FP7 project - Methodology for describing the accessibility of transport in Europe*. Retrieved from http://www.mediate-project.eu/

Mehran, R., Oyama, A., & Shah, M. (2009). Abnormal crowd behavior detection using social force model. *Proc. IEEE Conf. Comput. Vision Pattern Recognit.*, 935–942. 10.1109/CVPR.2009.5206641

Menoufi, K. (2017). Dust Accumulation on the Surface of Photovoltaic Panels. *Introducing the Photovoltaic Soiling Index*, *9*, 963.

Meola, C. (2012). *Infrared Thermography: Recent Advances and Future Trends*. New York: Bentham Science. doi:10.2174/97816080514341120101

Meola, C., & Carlomagno, G. M. (2004). Recent advances in the use of infrared thermography. *Measurement Science & Technology*, *15*(9), R27–R58. doi:10.1088/0957-0233/15/9/R01

Meola, C., & Carlomagno, G. M. (2010). *Recent Advances in Non-Destructive Inspection*. New York: Nova Science Publisher Inc.

Metzger, S., Portmann, E., Finger, M., Habenstein, A., Riedle, A., & Witschi, R. (2018). Human Smart City – der Mensch im Zentrum. *Transforming Cities*, *1*, 62–67.

Meunier, H., Talbi, E., & Reininger, P. (2000). A multiobjective genetic algorithm for radio network optimization. *Proceedings of the 2000 Congress on Evolutionary Computation*, 1, 317–324. 10.1109/CEC.2000.870312

Miah, M. S., Schukat, M., & Barrett, E. (2018). An enhanced sum rate in the cluster based cognitive radio relay network using the sequential approach for the future Internet of Things. *Human-centric Computing and Information Sciences*, *8*(16), 1–27.

Minichiello, V., Aroni, R., & Hays, T. N. (Eds.). (2008). *In-Depth Interviewing: Principles, Techniques, Analysis*. Pearson Education Australia.

Miorandi, D., Sicari, S., De Pellegrini, F., & Chlamtac, I. (2012). Internet of things: Vision, applications and research challenges. *Ad Hoc Networks, 10*(7), 1497–1516. doi:10.1016/j.adhoc.2012.02.016

Mitola, J., III. (2000). *Cognitive Radio an Integrated Agent Architecture for Software Defined Radio* (Doctoral dissertation). KTH Royal Institute of Technology, Stockholm, Sweden.

Modest, M. F. (2013). *Radiative Heat Transfer*. Waltham, MA: Academic Press. doi:10.1016/B978-0-12-386944-9.50023-6

Mohamed, & … . (2012). Effect of dust accumulation on performance of photovoltaic solar modules in Sahara environment. *Journal of Basic and Applied Scientific Research, 2*(11), 11030–11036.

Mohammed, F., Idries, A., Mohamed, N., Al-Jaroodi, J., & Jawhar, I. (2014). UAVs for smart cities: Opportunities and challenges. *International Conference on Unmanned Aircraft Systems (ICUAS)*, 267-273. 10.1109/ICUAS.2014.6842265

Mohsin, L. (2016). Optimized cleaning and cooling for photovoltaic modules based on the output performance. *Thermal Science, 145.*

Mollmann, K., Karstadt, D., Pinno, F., & Vollmer, M. (2005). Selected critical applications for thermography: Convections in fluids, selective emitters and highly reflecting materials. *Proceedings of the Infrared Camera Calibration Conference*, 6, 161–173.

Moore, T., de haan, F.J., Horne, R., & Gleeson, B J. (Eds.). (2017). Urban sustainability transitions: Australian cases – International perspectives. Springer.

Morvaj, B., Lugaric, L., & Krajcar, S. (2011). Demonstrating smart buildings and smart grid features in a smart energy city. *Proceedings of the 3rd International Youth Conference on Energetics (IYCE)*, 1-8.

Moyser, R., & Uffer, S. (2016). From Smart to Cognitive: a Roadmap for the Adoption of Technology in Cities. In E. Portmann & M. Finger (Eds.), *Towards Cognitive Cities. Advances in Cognitive Computing and Its Application to the Governance of Large Urban Systems* (pp. 13–35). Heidelberg, Germany: Springer International Publishing. doi:10.1007/978-3-319-33798-2_2

Müller, S. M., D'Onofrio, S., & Portmann, E. (2018). Fuzzy Analogical Reasoning in Cognitive Cities – A Conceptual Framework for Urban Dialogue Systems. *20th International Conference on Enterprise Information Systems (ICEIS 2018).* 10.5220/0006816103530360

Murgante, B., & Borruso, G. (2013). Cities and smartness: a critical analysis of opportunities and risks. *Lect. Notes Comput. Sci., 7973*, 630–642. doi:10.1007/978-3-642-39646-5_46

Murgante, B., & Borruso, G. (2013). Smart cities: un'analisi critica delle opportunità e dei rischi. *GEO Media, 17*(3).

Nasirudin, M. A., Za'bah, U. N., & Sidek, O. (2011). Fresh water real-time monitoring system based on Wireless Sensor Network and GSM. *Proceedings of the IEEE Conference on Open Systems (ICOS2011)*, 354–357. 10.1109/ICOS.2011.6079290

National Urban Transport Policy. (2006). Ministry of Urban Development, Government of India.

Nawaz, T., Marcenaro, L., & Regazzoni, C. S. (2017). Defense against jamming attacks in wide-band radios using cyclic spectral analysis and compressed sensing. In *Proceedings of Ninth International Conference on Ubiquitous and Future Networks (ICUFN-09)* (pp. 874-879). Academic Press.

Nayar, S. K., & Narasimhan, S. G. (1999). Vision in bad weather. *Proc. IEEE Conf. Computer Vision*, 2, 820–827. 10.1109/ICCV.1999.790306

Newell, W. H. (2001). A Theory of Interdisciplinary Studies. *Issues in Integrative Studies*, *19*, 1–25.

Newton, P. (2017). Framing new retrofit models for regenerating Australia's fast-growing cities. In Retrofitting Cities for Tomorrow's World. Academic Press.

Newton, P. W. (2006). *Australian State of the Environment Report: Human Settlements Theme Commentary*. Department of Environment and Heritage, Canberra. Retrieved from www.environment.gov.au/soe

Newton, P. W., & Doherty, P. (2014). The challenges to urban sustainability and resilience. Resilient Sustainable Cities, 7(18), 7-18.

Newton, P.W. & Meyer, D. (2016). 11 Built Environment and Human Behaviour Challenges to Sustainable Consumption. *Ethics and morality in consumption: Interdisciplinary perspectives, 8*, 189.

Newton, P. W. (2007). Horizon 3 Planning: Meshing Liveability with Sustainability, Environment and Planning B. *Environmental Development, 34*, 571–575.

Newton, P. W. (2008). *Transitions: Pathways Towards Sustainable Urban Development in Australia*. Dordrecht: Springer.

Newton, P. W. (2012). Liveable and Sustainable? Socio-Technical Challenges for 21st Century Cities. *Journal of Urban Technology*, *19*(1), 81–102. doi:10.1080/10630732.2012.626703

Newton, P. W. (2017). Innovation for a Sustainable Low Carbon Built Environment. *Procedia Engineering*, *18*, 16–32. doi:10.1016/j.proeng.2017.04.161

Newton, P. W., & Bai, X. (2008). Transitioning to Sustainable Urban Development. In P. W. Newton (Ed.), *Transitions: Pathways to Sustainable Urban Development in Australia*. Dordrecht: Springer.

Newton, P. W., & Tucker, S. N. (2011). Pathways to Decarbonising the Housing Sector: A Scenario Approach. *Building Research and Information*, *39*(1), 34–50. doi:10.1080/09613218.2010.531085

Nie, L., Zhao, Y. L., Akbari, M., Shen, J., & Chua, T. S. (2015). Bridging the vocabulary gap between health seekers and healthcare knowledge. *IEEE Transactions on Knowledge and Data Engineering*, *27*(2), 396–409. doi:10.1109/TKDE.2014.2330813

Noethen, M., Wolter, K. J., & Meyendorf, N. (2010). Surface crack detection in ferritic and austenitic steel components using inductive heated thermography. *Proceedings of the 2010 33rd International Spring Seminar on IEEE Electronics Technology (ISSE)*, 249–254. 10.1109/ISSE.2010.5547297

Normile, D. (2016). China rethinks cities. *Science*, *352*(6288), 916–918. doi:10.1126cience.352.6288.916 PMID:27199411

Nunes, V., Gouveia, J.P., Rodrigues, A.M., & Simão, T. (2017). INSMART–towards the new distribution systems operators potential roles in low carbon future and integrated frameworks for smart cities. *CIRED-Open Access Proceedings Journal, 2017*(1), 2797-2799.

Oliveira, A., & Campolargo, M. (2015). From Smart Cities to Human Smart Cities. *48th International Conference on System Sciences*, 2336-2344. 10.1109/HICSS.2015.281

Orlikowski, W., & Iacono, C. (2001). Desperately Seeking the "IT" in IT Research – A Call to Theorizing the IT Artifact. *Information Systems Research, 12*(2), 121–134. doi:10.1287/isre.12.2.121.9700

Ortiz-Conde. (2014). *A review of diode and solar cell equivalent circuit model lumped parameter extraction procedures*. Academic Press.

Österle, H., Becker, J., Frank, U., Hess, T., Karagiannis, D., Krcmar, H., ... Sinz, E. J. (2011). Memorandum on design-oriented information systems research. *European Journal of Information Systems, 20*(1), 11–15. doi:10.1057/ejis.2010.55

Österle, H., & Otto, B. (2010). Consortium Research: A Method for Researcher-Practitioner Collaboration in Design-Oriented IS Research. *Business & Information Systems Engineering*, *2*(5), 283–293. doi:10.100712599-010-0119-3

Parthasarathy, S., & Sankaran, P. (2012). A RETINEX based haze removal method. In *Industrial and Information Systems (ICIIS), 2012 7th IEEE International Conference on* (pp. 1-6). IEEE.

Patorniti, N. P., Stevens, N. J., & Salmon, P. M. (2017). A Systems Approach to City Design: Exploring the Compatibility of Sociotechnical Systems. *Habitat International*, *66*, 42–48. doi:10.1016/j.habitatint.2017.05.008

Paudyal, & Shakya, S. R. (2016). Dust accumulation effects on efficiency of solar PV modules for off grid purpose: A case study of Kathmandu. *Solar Energy*, *135*, 103–110. doi:10.1016/j.solener.2016.05.046

Pavan, A. M. (2013). A comparison between BNN and regression polynomial methods for the evaluation of the effect of soiling in large scale photovoltaic plants. *Applied Energy*, *108*, 392–401. doi:10.1016/j.apenergy.2013.03.023

Peffers, K., Tuunanen, T., & Niehaves, B. (2018). Design Science Research Genres: Introduction to the Special Issue on Exemplars and Criteria for Applicable Design Science Research. *European Journal of Information Systems*, *27*(2), 129–139. doi:10.1080/0960085X.2018.1458066

Peffers, K., Tuunanen, T., Rothenberger, M. A., & Chatterjee, S. (2007). A Design Science Research Methodology for Information Systems Research. *Journal of Management Information Systems*, *24*(3), 45–78. doi:10.2753/MIS0742-1222240302

Peng, Y. & Bai, X. (2017). Experimenting towards a low-carbon city: Policy evolution and nested structure of innovation. *Journal of Cleaner Production, 174*, 201-212.

Peng, K. H. (2016). Ultra Dense Network: Challenges, Enabling Technologies and New Trends. *China Communications, 13*(2), 30–40.

Petersson, A. M., & Lundberg, J. (2016). Applying action design research (ADR) to develop concept generation and selection methods. *Procedia CIRP, 50*, 222–227. doi:10.1016/j.procir.2016.05.024

Petit, J., Stottelaar, B., Feiri, M., & Kargl, F. (2015). *Remote attacks on automated vehicles sensors: Experiments on camera and lidar*. Black Hat Europe.

Pfäffli, M., Habenstein, A., Metzger, S., & Portmann, E. (2018). Eine Architektur zur Transformation von Städten in Human Smart Cities. In M. Knoll (Ed.), *IT-Architektur*. HMD der Wirtschaftsinformatik. doi:10.136540702-018-00451-z

Physical Progress (Achievements). (2018). Ministry of New & Renewable Energy.

Picotti, G. (2017). Soiling of solar collectors– Modelling approaches for airborne dust and its interactions with surfaces. *Renewable & Sustainable Energy Reviews.*

Plekhov, O., Palin-Luc, T., Saintier, N., Uvarov, S., & Naimark, O. (2005). Fatigue crack initiation and growth in a 35CrMo4 steel investigated by infrared thermography. *Fatigue & Fracture of Engineering Materials & Structures, 28*(1-2), 169–178. doi:10.1111/j.1460-2695.2005.00856.x

Polzin, S., & Baltes, M. (2002). Bus Rapid Transit: A Viable Alternative? *Journal of Public Transportation, 5*(2), 47–69. doi:10.5038/2375-0901.5.2.3

Ponto, J. (2015). Understanding and Evaluating Survey Research. *Journal of the Advanced Practitioner in Oncology, 6*(2), 168–171. PMID:26649250

Portmann, E., & Metzger, S. (2017). PostGrid: Das smarte Netzwerk der Schweizerischen Post für eine intelligentere Stadt. *Informatik Spektrum, 40*(1), 88–99. doi:10.100700287-016-1012-3

Portmann, E., & Pletscher, C. (2016). *Transdisziplinarität als Antwort auf die digitale Transformation. In BeWL Departement Betriebswirtschaftslehre* (pp. 33–35). Universität Bern.

Pruteanu-Malinici, I., & Carin, L. (2008). Infinite hidden Markov models for unusual-event detection in video. *IEEE Transactions on Image Processing, 17*(5), 811–822. doi:10.1109/TIP.2008.919359 PMID:18390385

Pti. (2017, January 11). *Air pollution causes 1.2 million deaths in India annually; Delhi most polluted: Greenpeace report - Times of India.* Retrieved from https://timesofindia.indiatimes.com/india/air-pollution-causes-12-lakh-deaths-in-india-annually-delhi-most-polluted-greenpeace-report/articleshow/56478622.cms

Punch, K. F. (Ed.). (2013). *Introduction to Social Research: Quantitative and Qualitative Approaches*. SAGE.

Purohit, R., Turner, T. A., & Pascoe, D. D. (2008). Use of infrared imaging in veterinary medicine. In N. A. Diakides & J. D. Bronzino (Eds.), *Medical Infrared Imaging*. Boca Raton, FL: CRC Press.

Rahman, Jobson, & Woodell. (2004). Retinex processing for automatic image enhancement. *Journal of Electronic Imaging, 13*(1), 100-111.

Rao (2014). Influence of dust deposition on photovoltaic panel performance. *Energy Procedia, 54*, 690-700.

Rao, Bhardwaj, & Subba Rao. (n.d.). *Intelligent Transportation Systems (ITS): A Technology Solution For 21st Century Public Transport System Management.* Academic Press.

Rebato, M., Boccardi, F., Mezzavilla, M., Rangan, S., & Zorzi, M. (2017). Hybrid Spectrum Sharing in mmWave Cellular Networks. *IEEE Transactions on Cognitive Communications and Networking, 3*(2), 155–168. doi:10.1109/TCCN.2017.2707551

Reitmaier, T., & Sick, B. (2015). The responsibility weighted Mahalanobis kernel for semi-supervised training of support vector machines for classification. *Inform. Sci., 323*, 179–198. doi:10.1016/j.ins.2015.06.027

Rohde, C., Dütschke, E., Gigli, M. & Bles, M. (2012). *Behavioural Climate Change Mitigation Options Domain Report Housing.* Report Delft.

Romer, K., Ostermaier, B., Mattern, F., Fahrmair, M., & Kellerer, W. (2010). Real-time search for real-world entities: A survey. *Proc. IEEE*, 98*(*11), 1887-1902. 10.1109/JPROC.2010.2062470

Rophol, G. (1988). Allgemeine Technologie der Netzwerke. *Technikgeschichte, 56*, 153–162.

Rose, R. (2001). A smart technique for determining base-station locations in an urban environment. *IEEE Transactions on Vehicular Technology*, *50*(1), 43–47. doi:10.1109/25.917869

Rusek. (2013). Scaling Up MIMO: Opportunities and Challenges with Very Large Arrays. *IEEE Signal processing Magazine*, *30*, 40–60.

Ryan, C. (2016). *Visions and Pathways 2040*. Sydney: CRC for Low Carbon Living.

Safara, F., & Eftekhari-Moghadam, A. (2006). Knowledge discovery of traffic/people behaviors based on image mining approach. *Proc. Geom. Model. Imaging-New Trends*, 255–258. 10.1109/GMAI.2006.31

Said, S. A., & Walwil, H. M. (2014). Fundamental studies on dust fouling effects on PV module performance. *Solar Energy*, *107*, 328–337. doi:10.1016/j.solener.2014.05.048

Sakarapunthip. (2017). *Effects of dust accumulation and module cleaning on performance ratio of solar rooftop system and solar power plants*. Academic Press.

SamyKamkar. (2013). *SkyJack*. Retrieved from https://github.com/samyk/skyjack

Sarver, Al-Qaraghuli, A., & Kazmerski, L. L. (2013). A comprehensive review of the impact of dust on the use of solar energy: History, investigations, results, literature, and mitigation approaches. *Renewable & Sustainable Energy Reviews*, *22*, 698–733. doi:10.1016/j.rser.2012.12.065

Saykol, E., Bastan, M., Gudukbay, U., & Ulusoy, O. (2010). Keyframe labeling technique for surveillance event classification. *Optical Engineering (Redondo Beach, Calif.)*, *49*(11), 1–12. doi:10.1117/1.3509270

Sayyah, A., Horenstein, M. N., & Mazumder, M. K. (2014). Energy yield loss caused by dust deposition on photovoltaic panels. *Solar Energy*, *107*, 576–604. doi:10.1016/j.solener.2014.05.030

Schaller Consulting. (2002). *Bus Rapid Transit For New York City: Prepared for Transportation Alternatives and NYPIRG Straphangers Campaign*. Retrieved from http://www.straphangers.org/pokeyaward/ letter.html

Scheler, M. (Ed.). (1960). Die Wissensformen und die Gesellschaft. Bern: Francke. (Originally published 1926)

Schroten, A. (2012). *Behavioural Climate Change Mitigation Options: Domain Report Transport*. Report Delft.

Sein, M. K., Henfridsson, O., Purao, S., Rossi, M., & Lindgren, R. (2011). Action Design Research. *Management Information Systems Quarterly*, *35*(1), 37–56. doi:10.2307/23043488

Semaoui, S., Arab, A. H., Boudjelthia, E. K., Bacha, S., & Zeraia, H. (2015). Dust effect on optical transmittance of photovoltaic module glazing in a desert region. *Energy Procedia*, *74*, 1347–1357. doi:10.1016/j.egypro.2015.07.781

Seto, K., Sanchez-Rodriguez, R., & Fragkias, M. (2010). The new geography of contemporary urbanisation and the environment. *Annual Review of Environment and Resources*, *35*(1), 167–194. doi:10.1146/annurev-environ-100809-125336

Shah, M., Javed, O., & Shafique, K. (2007). Automated visual surveillance in realistic scenarios. *IEEE MultiMedia*, *14*(1), 30–39. doi:10.1109/MMUL.2007.3

Sham, F., Chen, N., & Long, L. (2008). Surface crack detection by flash thermography on concrete surface. Insight-Non-Destructive Test. *Cond. Monit.*, *50*, 240–243.

Sharif, M. H., Uyaver, S., & Djeraba, C. (2010). Crowd behavior surveillance using bhattacharyya distance metric. *Proc. CompIMAGE*, 311–323. 10.1007/978-3-642-12712-0_28

Shen, S., & Lok, T. (2014). Power control for underlay in-band spectrum sharing in device-to-device communications. In Proceedings of IEEE Wireless Communications and Networking Conference (WCNC-14) (pp. 1350-1355). IEEE.

Shi, D., Zurada, J., & Guan, J. (2015). A Neuro-fuzzy System with Semi-supervised Learning for Bad Debt Recovery in the Healthcare Industry. In *Proceedings of the 2015 48th Hawaii International Conference on System Sciences (HICSS)*. IEEE Press. 10.1109/HICSS.2015.376

Shukla, P. (n.d.). *Air Pollution in Select Indian Cities - teriin.org*. Retrieved from http://www.teriin.org/library/files/Pollution-level-in-India.pdf

Siemens, G. (2005). Connectivism: A learning theory for the digital age. *International Journal of Instructional Technology and Distance Learning*, 2(1), 3–10.

Simon, H. A. (Ed.). (1969). *The Sciences of the Artificial*. Cambridge, MA: The MIT Press.

Singh & Sood. (2017). A Review on Influence of Fog on Road Crash. *International Journal of Engineering Research & Technology, 6*(7). Retrieved from http://www.ijert.org

Skinner, Van Essen, Smokers, & Hill. (2010). *EU Transport GHG: routes to 2050.* Towards the decarbonisation of the EU's transport sector by 2050.

Smart Cities Mission. (n.d.). Retrieved from: http://smartcities.gov.in/content/

Smart Freight. (2011). *EU FP7 project.* Smart Freight Transport in Urban Areas. Retrieved from http://www.smartfreight.info/

Smith, R. C., Bossen, C., & Kanstrup, A. M. (2017). Participatory design in an era of participation. *CoDesign, 13*(2), 65–69. doi:10.1080/15710882.2017.1310466

Software-Defined GPS Signal Simulator. (n.d.). Retrieved from https://github.com/osqzss/gps-sdr-sim

Soucek, S., & Zucker, G. (2012). Current developments and challenges in building automation. *e & iElektrotechnik and Informations technik, 129*(4), 278-285.

Spradley, J. P. (Ed.). (2016). *Participant Observation.* Waveland Press.

Sridhara, K., Chandra, A., & Tripathi, P. S. M. (2008). Spectrum Challenges and Solutions by Cognitive Radio: An Overview. *Wireless Personal Communications, 45*(3), 281–291. doi:10.100711277-008-9465-6

Steffen, W., Broadgate, W., Deutch, L., Gaffney, O., & Ludwig, C. (2015). The trajectory of the Anthropocene: The great acceleration. *The Anthropocene Review, 2*(1), 81–98. doi:10.1177/2053019614564785

Stergiou, C., Psannis, K. E., Kim, B. G., & Gupta, B. (2018). Secure integration of IoT and cloud computing. *Future Generation Computer Systems, 78*, 964–975. doi:10.1016/j.future.2016.11.031

Sudevalayam, S., & Kulkarn, P. (2011). Energy Harvesting Sensor Nodes: Survey and Implications. *IEEE Communications Surveys and Tutorials, 13*(3), 443–461. doi:10.1109/SURV.2011.060710.00094

Sue, V. M., & Ritter, L. A. (Eds.). (2012). *Conducting Online Surveys.* SAGE. doi:10.4135/9781506335186

Sulaiman, S. A. (2014). *Influence of Dirt Accumulation on Performance of PV Panels.* Academic Press.

Sulami, M., Glatzer, I., Fattal, R., & Werman, M. (2014). Automatic recovery of the atmospheric light in hazy images. *Proc. IEEE Int. Conf. Comput. Photogr*, 1–11.

SUMPs Guidelines. (2011). *Developing and implementing a Sustainable Urban Mobility Plan*. Retrieved from http://www.mobilityplans.eu/docs/SUMP_guidelines_web0.pdf

Sun, C.-C., Lai, H.-C., Sheu, M.-H., & Huang, Y.-H. (2016). Single image fog removal algorithm based on an improved dark channel prior method. In *Intelligent Signal Processing and Communication Systems (ISPACS), 2016 International Symposium on* (pp. 1-4). IEEE.

Sun, C.-C., Heng-Chi, L. M.-H. S., & Huang, Y.-H. (2013). *Single Image Fog Removal Algorithm based on an Improved Dark Channel Prior Method*. Depart. of EE, NFU, Taiwan & Depart. of EE, NYUST, Taiwan.

Sun, Q. B., Liu, J., Li, S., Fan, C. X., & Sun, J. J. (2010). Internet of Things: Summarize on Concepts, Architecture and Key Technology Problem. *Journal of Beijing University of Posts and Telecommunications*, *3*(3), 1–9.

Suresh Kumar, E. (2013). Soiling and Dust Impact on the Efficiency and the Maximum Power Point in the Photovoltaic Modules (Vol. 2). Academic Press.

Swetha, G. D., & Murthy, G. R. (2017). Selective overlay mode operation for D2D communication in dense 5G cellular networks. In *Proceedings of IEEE Symposium on Computers and Communications (ISCC)* (pp. 704-709). IEEE.

Syafiq, A., Pandey, A. K., Adzman, N. N., & Rahim, N. A. (2018). Advances in approaches and methods for self-cleaning of solar photovoltaic panels. *Solar Energy*, *162*, 597–619. doi:10.1016/j.solener.2017.12.023

Tabacchi, M. E., Portmann, E., Seising, R., & Habenstein, A. (2018) Introduction. In E. Portmann, M.E. Tabacchi, R. Seising, & A. Habenstein (Eds.), Designing Cognitive Cities. Heidelberg, Germany: Springer International Publishing.

Talari, S., Khah, M. S., Siano, P., Loia, V., Tommasetti, A., & Catalão, J. P. S. (2017). A Review of Smart Cities Based on the Internet of Things Concept. *Energies*, *10*(4), 421. doi:10.3390/en10040421

Tan, L., & Wang, N. (2010). Future Internet: The Internet of Things. *Proc.3rd Int. Conf. Adv. Comput. Theory Eng. (ICACTE)*, 5, V5-376-V5-380.

Tan, R. T. (2008). Visibility in bad weather from a single image. *Proc. IEEE Conf. Comput. Vis. Pattern Recognition*, 1–8.

Tanesab, Parlevliet, D., Whale, J., & Urmee, T. (2017). Seasonal effect of dust on the degradation of PV modules performance deployed in different climate areas. *Renewable Energy*, *111*, 105–115. doi:10.1016/j.renene.2017.03.091

Tang, Y., Wang, X., & Lu, H. (2009). Intelligent video analysis technology for elevator cage abnormality detection in computer vision. *Proc. 4th Int. Conf. Comput. Sci. Convergence Inf. Technol.*, 1252–1258. 10.1109/ICCIT.2009.206

Tan, S., Yang, J., Yan, J., Lee, C., Hashim, H., & Chen, B. (2017). A holistic low carbon city indicator framework for sustainable development. *Applied Energy*, *185*(2), 1919–1930. doi:10.1016/j.apenergy.2016.03.041

Tarel, J.-P., & Hautière, N. (2009). Fast visibility restoration from a single color or gray level image. *Proc. IEEE Int. Conf.Comput. Vision*, 2201–2208. 10.1109/ICCV.2009.5459251

TEMS. (2012). *The EPOMM Modal Split Tool*. Retrieved from http://www.epomm.eu/tems/index.phtml

The University of Minnesota. (2019). *SMART Signal, Traffic Data Collection for Urban Traffic Flow*. Retrieved from http://license.umn.edu/technologies/z08214_smart-signal-traffic-data-collection-for-urban-traffic-flow

Theoleyre, F., Watteyne, T., Bianchi, G., & Tuna, V. C., Gungor, & Pang, A.C. (2015). Networking and communications for smart cities. *Computer Communications*, *58*(1-3).

Transport for London (TfL), (2003). *The case for investing in London's buses: Presenting the results of the London Buses Strategic Review*. Author.

Transport for London (TfL). (2006). *Bus priority at traffic signals keeps London's buses moving Selective Vehicle Detection (SVD)*. Author.

Transportation Research Board. (n.d.). *Bus Rapid Transit Volume 1: Case Studies in Bus Rapid Transit*. TCRP Report 90 Volume 1 BRT Case Studies. Retrieved from gulliver.trb.org/publications/tcrp/tcrp_rpt_90v1. pdf

Transports en Commun d'Agglo. (n.d.). Retrieved from http://www.tcar.fr/presentation/index.asp?rub_code=52&thm_id=317&gpl_id=)

Tschudi, F., & D'Onofrio, S. (2018). Wie wir durch unsere Denk- und Handlungsmuster beeinflusst werden. In A. Meier & R. Seising (Eds.), Vague Information Processing (pp. 567-571). Springer Fachmedien Wiesbaden.

Tsolakis, N., & Anthopoulos, L. (2015). Eco-cities: An integrated system dynamics framework and a concise research taxonomy. *Sustainable Cities and Society*, *17*, 1–14. doi:10.1016/j.scs.2015.03.002

U.S. Energy Information Administration. (2015). *Natural Gas*. Energy Information Agency. Retrieved August 26, 2015, from: http://www.eia.gov/dnav/ng/ng_pri_sum_a_EPG0_PRS_DMcf_a.htm

Uckelmann, D., Harrison, M., & Michahelles, F. (2011). An architectural approach towards the future internet of things. *Architecting the Internet Of things*, 1-24.

UN DESA. (2014). *2014 Revision of the world urbanization prospects*. New York: United Nations.

United Nations. (2011). World Urbanization Prospects: the 2011 Revision. New York: Author.

Usamentiaga, R., Venegas, P., Guerediaga, J., Vega, L., Molleda, J., & Bulnes, F. G. (2014). *Infrared Thermography for Temperature Measurement and Non-Destructive Testing Sensors*. Academic Press. doi:10.3390140712305 PMID:25014096

Usman, A., Gebremariam, A., Raza, U., & Granelli, F. (2015). A Software- Defined Device-to- device Communication Architecture for Public Safety Applications in 5G Networks. *IEEE Access: Practical Innovations, Open Solutions*, *3*, 1649–1654. doi:10.1109/ACCESS.2015.2479855

Vaishnavi, V., & Kuechler, W. (2015). *Design Science Research in Information Systems*. Retrieved from http://desrist.org/desrist/article.aspx

Valente, J., & Cardenas, A. A. (2017). Understanding security threats in consumer drones through the lens of the discovery quadcopter family. *Proceedings of the 2017 Workshop on Internet of Things Security and Privacy*, 31-36. 10.1145/3139937.3139943

Vallejo, D., Albusac, J., Jimenez, L., Gonzalez, C., & Moreno, J. (2009). A cognitive surveillance system for detecting incorrect traffic behaviors. *Expert Systems with Applications*, *36*(7), 10503–10511. doi:10.1016/j.eswa.2009.01.034

Varadarajan, J., & Odobez, J. (2009). Topic models for scene analysis and abnormality detection. *Proc. IEEE 12th Int. Conf. Comput. Vision Workshops*, 1338–1345. 10.1109/ICCVW.2009.5457456

Vasanth, K., & Sbert, J. (n.d.). Creating solutions for health through technology innovation. *Texas Instruments*. Available: http://www.ti.com/lit/wp/sszy006/sszy006.pdf

Vattapparamban, E., Güvenç, İ., Yurekli, A. İ., Akkaya, K., & Uluağaç, S. (2016). Drones for smart cities: Issues in cybersecurity, privacy, and public safety. *International Conference Wireless Communications and Mobile computing (IWCMC)*, 216-221. 10.1109/IWCMC.2016.7577060

Viale Pereira, G., Cunha, M. A., Lampoltshammer, T. J., Parycek, P., & Testa, M. G. (2017). Increasing collaboration and participation in smart city governance: A cross-case analysis of smart city initiatives. *Information Technology for Development*, *23*(3), 526–553. doi:10.1080/02681102.2017.1353946

Vivar, M., Herrero, R., Antón, I., Martínez-Moreno, F., Moretón, R., Sala, G., ... Smeltink, J. (2010). Effect of soiling in CPV systems. *Solar Energy*, *84*(7), 1327–1335. doi:10.1016/j.solener.2010.03.031

Vollmer, M., & Mollmann, K. P. (2011). *Infrared Thermal Imaging: Fundamentals, Research and Applications*. Weinheim, Germany: Wiley.

Walls, J. G., Widmeyer, G. R., & El Sawy, O. A. (1992). Building an Information Systems Design Theory for Vigilant EIS. *Information Systems Research*, *3*(1), 36–59. doi:10.1287/isre.3.1.36

Walls, J. G., Widmeyer, G. R., & El Sawy, O. A. (2004). Assessing Information System Design Theory in Perspective: How useful was our 1992 Initial Rendition? *Journal of Information Technology Theory and Application*, *6*(2), 43–58.

Wang, F., Jiang, M., Qian, C., Yang, S., Li, C., Zhang, H., . . . Tang, X. (2017). *Residual attention network for image classification*. arXiv preprint arXiv:1704.06904

Wang, S., & Chen, B. (2018). Three-Tier carbon accounting model for cities. *Applied Energy*, *229*, 163–175. doi:10.1016/j.apenergy.2018.07.109

Wang, W., Li, W., Guan, Q., & Qi, M. (2015, October). Multiscale single image dehazing based on adaptive wavelet fusion. *Mathematical Problems in Engineering*, *2015*, 1–14. doi:10.1155/2015/813507

Wang, X., Huang, C., & Zou, Z. (2016). The analysis of energy consumption and greenhouse gas emissions of a large-scale commercial building in Shanghai, China. *Advances in Mechanical Engineering*, *8*(2), 1–8. doi:10.1177/1687814016628395

Wang, Y., Chen, S., Xue, H., & Fu, Z. (2015). Semi-supervised classification learning by discrimination-aware manifold regularization. *Neurocomputing*, *147*, 299–306. doi:10.1016/j.neucom.2014.06.059

Wicaksono, H., Rogalski, S., & Kusnady, E. (2010). Knowledge-based intelligent energy management using building automation system. *IPEC, 2010 Conference Proceeding*, 1140-1145. 10.1109/IPECON.2010.5696994

Wikipedia. (n.d.). *Block Diagram.* Retrieved from https://en.wikipedia.org/wiki/Block_diagram

Willis, A. M. (2006). Ontological Designing. *Design Philosophy Papers*, *4*(2), 69–92. doi:10.2752/144871306X13966268131514

Winter, R. (2008). Design Science Research in Europe. *European Journal of Information Systems*, *17*(5), 470–475. doi:10.1057/ejis.2008.44

World Bank and Institute for Health Metrics and Evaluation. (2016). *The Cost of Air Pollution: Strengthening the economic case for action.* Available from: http://documents.worldbank.org/curated/en/781521473177013155/pdf/108141-REVISEDCost-of-PollutionWebCORRECTEDfile.pdf

World Health Organization. (2016). *WHO Global Urban Ambient Air Pollution Database (Update 2016).* WHO.

Wu, Q., Ding, G., Xu, Y., Feng, S., Du, Z., Wang, J., & Long, K. (2014). Cognitive Internet of Things: A New Paradigm Beyond Connection. *Internet of Things.*, *1*(2), 129–143. doi:10.1109/JIOT.2014.2311513

Wu, X., Gong, H., Chen, P., Zhong, Z., & Xu, Y. (2009). Surveillance robot utilizing video and audio information. *J. Intell. Robot. Syst.: Theory Appl.*, *55*(4–5), 403–421. doi:10.100710846-008-9297-3

Wu, Z., Tang, J., & Wang, D. (2016). Low carbon urban transitioning in Shenzhen: A multi-level environmental governance perspective. *Sustainability*, *8*(8), 720. doi:10.3390u8080720

WWF Living Planet Report 2014. (n.d.). Available: http://wwf.panda.org/about_our_earth/all_publications/living_planet_report/

Xiang, T., & Gong, S. (2005). Video behaviour profiling and abnormality detection without manual labelling. *Proc. 10th IEEE Int. Conf. Comput. Vision*, *2*, 1238–1245. 10.1109/ICCV.2005.248

Xie, B., Guo, F., & Cai, Z. (2010). Improved single image dehazing using dark channel prior and multi-scale retinex. In *Intelligent System Design and Engineering Application (ISDEA), 2010 International Conference on* (*vol. 1*, pp. 848-851). IEEE. 10.1109/ISDEA.2010.141

Xu, H., Hematian, Y. A., Griffith, D., & Golmie, N. (2018). *Performance Evaluation of Energy Efficiency with Sleep Mode in Ultra Dense Networks. In Proceedings of Computing, Networking and Communications* (pp. 747–751). Maui, HI: ICNC.

Xu, Z., Liu, X., & Chen, X. (2009). *Fog Removal from Video Sequences Using Contrast Limited Adaptive Histogram Equalization. In Information Science and Technology College Dalian Maritime University*. IEEE.

Yan, C., Xu, W., & Liu, J. (2016). *Can you trust autonomous vehicles: Contactless attacks against sensors of self-driving vehicle?* DEF CON.

Yang, X., Wang, X.-C., & Zhou, Z.-Y. (2018). Development path of Chinese low-carbon cities based on index evaluation. *Advances in Climate Change Research*, *9*(2), 144–153. doi:10.1016/j.accre.2018.05.004

Yan, Y., Chen, L., & Tjhi, W. C. (2013). Fuzzy semi-supervised co-clustering for text documents. *Fuzzy Sets and Systems*, *215*, 74–89. doi:10.1016/j.fss.2012.10.016

Yan, Y., Qian, Y., Sharif, H., & Tipper, D. (2012). A survey on cyber security for smart grid communications. *IEEE Communications Surveys and Tutorials*, *14*(4), 998–1010. doi:10.1109/SURV.2012.010912.00035

Yaqoob, I., Targio Hashem, I. A., Mehmood, Y., & Gani, A. (2017). Enabling Communication Technologies for Smart Cities. *IEEE Communications Magazine*, *55*(1), 112–120. doi:10.1109/MCOM.2017.1600232CM

Yun, M., & Yuxin, B. (2010). Research on the architecture and key technology of Internet of Things (IoT) applied on smart grid. *International Conference on in Advances in Energy Engineering (ICAEE)*, 69-72.

Zadeh, L. A. (1988). Fuzzy Logic. *IEEE Computer*, *21*(4), 83–93. doi:10.1109/2.53

Zeinab, K. A. M., & Elmustafa, S. A. A. (2017). Internet of Things applications, challenges and related future technologies. *World Scientific News*, *2*(67), 126–148.

Zhang, C., Chen, W., Chen, X., Yang, L., & Johnstone, J. (2010). A multiple instance learning and relevance feedback framework for retrieving abnormal incidents in surveillance videos. *Journal of Multimedia*, *5*(4), 310–321. doi:10.4304/jmm.5.4.310-321

Zhang, J., & Liu, Z. (2008). Detecting abnormal motion of pedestrian in video. *Proc. Int. Conf. Inf. Autom.*, 81–85.

Zhang, J., & Liu, Z. (2010). Abnormal behavior of pedestrian detection based on fuzzy theory. *Moshi Shibie Yu Rengong Zhineng/Pattern Recognit. Artificial Intelligence*, *23*(3), 421–427.

Zhang, K., Ni, J., Yang, K., Liang, X., Ren, J., & Shen, X. S. (2017). Security and privacy in smart city applications: Challenges and solutions. *IEEE Communications Magazine*, *55*(1), 122–129. doi:10.1109/MCOM.2017.1600267CM

Zhang, M., Zhao, H., Zheng, R., Wu, Q., & Wei, W. (2012). Cognitive internet of things: Concepts and application example. *International Journal of Computer Science Issues*, *9*(6), 151.

Zhang, X., Guan, N., Jia, Z., Qiu, X., & Luo, Z. (2015). Semi-supervised projective non-negative matrix factorization for cancer classification. *PLoS One*, *10*(9), e0138814. doi:10.1371/journal.pone.0138814 PMID:26394323

Zhao, Xiao, Yu, & Xu. (2015). Single image fog removal based on local extrema. *IEEE/CAA Journal of Automatica Sinica, 2*(2), 158-165.

Zhao, Q., & Sadler, B. M. (2007). A Survey of Dynamic Spectrum Access: Signal Processing, Networking and Regulatory Policy. *IEEE Signal Processing Magazine*, *55*(5), 2294–2309.

Zhicheng, W., & Jun, Z. (2008). Detecting pedestrian abnormal behavior based on fuzzy associative memory. *Proc. 4th Int. Conf. Nat. Comput.*, 143–147.

Zhou, J., Leppanen, T., Harjula, E., Ylianttila, M., Ojala, T., Yu, C., & Yang, L. T. (2013). Cloud things: A common architecture for integrating the internet of things with cloud computing. *IEEE 17th International Conference on Computer Supported Cooperative Work in Design (CSCWD),* 651-657.

Zhou, G., & Wu, Y. (2009). Anomalous event detection based on self-organizing map for supermarket monitoring. *Proc. Int. Conf. Inf. Eng. Comput. Sci.*, 1–4. 10.1109/ICIECS.2009.5364586

Zissis, G.J., & Wolfe, W.L. (1978). *The Infrared Handbook.* Technical report, DTIC document.

Zweng, A., & Kampel, M. (2010). Unexpected human behavior recognition in image sequences using multiple features. *Proc. 20th Int. Conf. Pattern Recognit.*, 368–371. 10.1109/ICPR.2010.98

Related References

To continue our tradition of advancing research on topics in the field of engineering, we have compiled a list of recommended IGI Global readings. These references will provide additional information and guidance to further enrich your knowledge and assist you with your own research and future publications.

Abawajy, J. H., Pathan, M., Rahman, M., Pathan, A., & Deris, M. M. (2013). *Network and traffic engineering in emerging distributed computing applications*. Hershey, PA: IGI Global. doi:10.4018/978-1-4666-1888-6

Abu-Faraj, Z. O. (2012). Bioengineering/biomedical engineering education. In Z. Abu-Faraj (Ed.), *Handbook of research on biomedical engineering education and advanced bioengineering learning: Interdisciplinary concepts* (pp. 1–59). Hershey, PA: IGI Global. doi:10.4018/978-1-4666-0122-2.ch001

Abu-Nimeh, S., & Mead, N. R. (2012). Combining security and privacy in requirements engineering. In T. Chou (Ed.), *Information assurance and security technologies for risk assessment and threat management: Advances* (pp. 273–290). Hershey, PA: IGI Global. doi:10.4018/978-1-61350-507-6.ch011

Abu-Taieh, E., El Sheikh, A., & Jafari, M. (2012). *Technology engineering and management in aviation: Advancements and discoveries*. Hershey, PA: IGI Global. doi:10.4018/978-1-60960-887-3

Achumba, I. E., Azzi, D., & Stocker, J. (2010). Low-cost virtual laboratory workbench for electronic engineering. *International Journal of Virtual and Personal Learning Environments, 1*(4), 1–17. doi:10.4018/jvple.2010100101

Achumba, I. E., Azzi, D., & Stocker, J. (2012). Low-cost virtual laboratory workbench for electronic engineering. In M. Thomas (Ed.), *Design, implementation, and evaluation of virtual learning environments* (pp. 201–217). Hershey, PA: IGI Global. doi:10.4018/978-1-4666-1770-4.ch014

Addo-Tenkorang, R., & Eyob, E. (2013). Engineer-to-order: A maturity concurrent engineering best practice in improving supply chains. In Industrial engineering: Concepts, methodologies, tools, and applications (pp. 1780-1796). Hershey, PA: IGI Global. doi:10.4018/978-1-4666-1945-6.ch095

Aguilera, A., & Davim, J. (2014). *Research developments in wood engineering and technology*. Hershey, PA: IGI Global. doi:10.4018/978-1-4666-4554-7

Aharoni, A., & Reinhartz-Berger, I. (2013). Semi-automatic composition of situational methods. In K. Siau (Ed.), *Innovations in database design, web applications, and information systems management* (pp. 335–364). Hershey, PA: IGI Global. doi:10.4018/978-1-4666-2044-5.ch013

Ahmad, M., Jung, L. T., & Zaman, N. (2014). A comparative analysis of software engineering approaches for sequence analysis. In *Software design and development: Concepts, methodologies, tools, and applications* (pp. 1093–1102). Hershey, PA: IGI Global. doi:10.4018/978-1-4666-4301-7.ch053

Ahrens, A., Bassus, O., & Zaščerinska, J. (2014). Enterprise 2.0 in engineering curriculum. In M. Cruz-Cunha, F. Moreira, & J. Varajão (Eds.), *Handbook of research on enterprise 2.0: Technological, social, and organizational dimensions* (pp. 599–617). Hershey, PA: IGI Global. doi:10.4018/978-1-4666-4373-4.ch031

Akbar, D. (2012). Community engagement in engineering education: Needs and learning outcomes. In M. Rasul (Ed.), *Developments in engineering education standards: Advanced curriculum innovations* (pp. 301–317). Hershey, PA: IGI Global. doi:10.4018/978-1-4666-0951-8.ch017

Alam, F., Subic, A., Plumb, G., Shortis, M., & Chandra, R. P. (2012). An innovative offshore delivery of an undergraduate mechanical engineering program. In M. Rasul (Ed.), *Developments in engineering education standards: Advanced curriculum innovations* (pp. 233–245). Hershey, PA: IGI Global. doi:10.4018/978-1-4666-0951-8.ch013

Ali, D. F., Patil, A., & Nordin, M. S. (2012). Visualization skills in engineering education: Issues, developments, and enhancement. In A. Patil, H. Eijkman, & E. Bhattacharyya (Eds.), *New media communication skills for engineers and IT professionals: Trans-national and trans-cultural demands* (pp. 175–203). Hershey, PA: IGI Global. doi:10.4018/978-1-4666-0243-4.ch011

Aljawarneh, S. (2013). Cloud security engineering: Avoiding security threats the right way. In S. Aljawarneh (Ed.), *Cloud computing advancements in design, implementation, and technologies* (pp. 147–153). Hershey, PA: IGI Global. doi:10.4018/978-1-4666-1879-4.ch010

Alkhatib, G. (2012). *Models for capitalizing on web engineering advancements: Trends and discoveries*. Hershey, PA: IGI Global. doi:10.4018/978-1-4666-0023-2

Allee, T., Handorf, A., & Li, W. (2010). Electrospinning: Development and biomedical applications. In A. Shukla & R. Tiwari (Eds.), *Intelligent medical technologies and biomedical engineering: Tools and applications* (pp. 48–78). Hershey, PA: IGI Global. doi:10.4018/978-1-61520-977-4.ch003

Alsmadi, I. (2014). Website performance measurement: Process and product metrics. In *Software design and development: Concepts, methodologies, tools, and applications* (pp. 1801–1827). Hershey, PA: IGI Global. doi:10.4018/978-1-4666-4301-7.ch086

Altarawneh, H., Alamaro, S., & El Sheikh, A. (2012). Web engineering and business intelligence: Agile web engineering development and practice. In A. Rahman El Sheikh & M. Alnoukari (Eds.), *Business intelligence and agile methodologies for knowledge-based organizations: Cross-disciplinary applications* (pp. 313–344). Hershey, PA: IGI Global. doi:10.4018/978-1-61350-050-7.ch015

Altarawneh, H., & El-Shiekh, A. (2010). Web engineering in small Jordanian web development firms: An XP based process model. In A. Tatnall (Ed.), *Web technologies: Concepts, methodologies, tools, and applications* (pp. 1696–1707). Hershey, PA: IGI Global. doi:10.4018/978-1-60566-982-3.ch091

Alzoabi, Z. (2014). Agile software: Body of knowledge. In *Software design and development: Concepts, methodologies, tools, and applications* (pp. 96–116). Hershey, PA: IGI Global. doi:10.4018/978-1-4666-4301-7.ch006

Andrade-Campos, A. (2013). Development of an optimization framework for parameter identification and shape optimization problems in engineering. In J. Davim (Ed.), *Dynamic methods and process advancements in mechanical, manufacturing, and materials engineering* (pp. 1–24). Hershey, PA: IGI Global. doi:10.4018/978-1-4666-1867-1.ch001

Andreatos, A. (2012). Educating the 21st century's engineers and IT professionals. In A. Patil, H. Eijkman, & E. Bhattacharyya (Eds.), *New media communication skills for engineers and IT professionals: Trans-national and trans-cultural demands* (pp. 132–159). Hershey, PA: IGI Global. doi:10.4018/978-1-4666-0243-4.ch009

Annamalai, C., & Ramayah, T. (2013). Reengineering for enterprise resource planning (ERP) systems implementation: An empirical analysis of assessing critical success factors (CSFs) of manufacturing organizations. In Industrial engineering: Concepts, methodologies, tools, and applications (pp. 791-806). Hershey, PA: IGI Global. doi:10.4018/978-1-4666-1945-6.ch044

Antchev, M. (2010). Other applications of converters and systems of converters. In *Technologies for electrical power conversion, efficiency, and distribution: Methods and processes* (pp. 270–299). Hershey, PA: IGI Global. doi:10.4018/978-1-61520-647-6.ch011

Anzelotti, G., & Valizadeh, M. (2010). Purpose-oriented small software: A case study for some engineering subjects. In R. Luppicini & A. Haghi (Eds.), *Cases on digital technologies in higher education: Issues and challenges* (pp. 164–178). Hershey, PA: IGI Global. doi:10.4018/978-1-61520-869-2.ch012

Anzelotti, G., & Valizadeh, M. (2010). Sights inside the virtual engineering education. In D. Russell & A. Haghi (Eds.), *Web-based engineering education: Critical design and effective tools* (pp. 160–174). Hershey, PA: IGI Global. doi:10.4018/978-1-61520-659-9.ch012

Asadi, M., Mohabbati, B., Gaševic, D., Bagheri, E., & Hatala, M. (2012). Developing semantically-enabled families of method-oriented architectures. *International Journal of Information System Modeling and Design*, *3*(4), 1–26. doi:10.4018/jismd.2012100101

Augusti, G., & Feyo de Azevedo, S. (2011). Qualification frameworks and field-specific approaches to quality assurance: Initiatives in engineering and technical education. *International Journal of Quality Assurance in Engineering and Technology Education*, *1*(1), 44–57. doi:10.4018/ijqaete.2011010104

Aung, Z., & Nyunt, K. K. (2014). Constructive knowledge management model and information retrieval methods for software engineering. In *Software design and development: Concepts, methodologies, tools, and applications* (pp. 253–269). Hershey, PA: IGI Global. doi:10.4018/978-1-4666-4301-7.ch014

Azad, A. K., Auer, M. E., & Harward, V. (2012). *Internet accessible remote laboratories: Scalable e-learning tools for engineering and science disciplines.* Hershey, PA: IGI Global. doi:10.4018/978-1-61350-186-3

Azar, A. T. (2013). Overview of biomedical engineering. In Bioinformatics: Concepts, methodologies, tools, and applications (pp. 1-28). Hershey, PA: IGI Global. doi:10.4018/978-1-4666-3604-0.ch001

Badr, K. B., Badr, A. B., & Ahmad, M. N. (2013). Phases in ontology building methodologies: A recent review. In M. Nazir Ahmad, R. Colomb, & M. Abdullah (Eds.), *Ontology-based applications for enterprise systems and knowledge management* (pp. 100–123). Hershey, PA: IGI Global. doi:10.4018/978-1-4666-1993-7.ch006

Baer, W., & Renfro, C. (2013). Information sources and collection planning for engineering. In S. Holder (Ed.), *Library collection development for professional programs: Trends and best practices* (pp. 128–144). Hershey, PA: IGI Global. doi:10.4018/978-1-4666-1897-8.ch008

Baghdadi, Y., & Kraiem, N. (2014). Business process modeling with services: reverse engineering databases. In R. Perez-Castillo & M. Piattini (Eds.), *Uncovering essential software artifacts through business process archeology* (pp. 177–200). Hershey, PA: IGI Global. doi:10.4018/978-1-4666-4667-4.ch007

Bakopoulou, A., Leyhausen, G., Geurtsen, W., & Koidis, P. (2013). Dental tissue engineering research and translational approaches towards clinical application. In A. Daskalaki (Ed.), *Medical advancements in aging and regenerative technologies: Clinical tools and applications* (pp. 279–312). Hershey, PA: IGI Global. doi:10.4018/978-1-4666-2506-8.ch013

Baporikar, N. (2012). Developing right graduate attributes through project-based teaching. In M. Rasul (Ed.), *Developments in engineering education standards: Advanced curriculum innovations* (pp. 64–79). Hershey, PA: IGI Global. doi:10.4018/978-1-4666-0951-8.ch004

Baraldi, E., & Nadin, G. (2012). "Network process re-engineering" in a home textile network: The importance of business relationships and actor bonds. In T. Choi (Ed.), *Fashion supply chain management: Industry and business analysis* (pp. 212–234). Hershey, PA: IGI Global. doi:10.4018/978-1-60960-756-2.ch012

Barbu, M. C., Hasener, J., & Bernardy, G. (2014). Modern testing of wood-based panels, process control, and modeling. In A. Aguilera & J. Davim (Eds.), *Research developments in wood engineering and technology* (pp. 90–130). Hershey, PA: IGI Global. doi:10.4018/978-1-4666-4554-7.ch003

Bas, T. G. (2013). Dual market(ing) in "bio-engineering high technology" new products: The risk of uncertainty and failure. *International Journal of Measurement Technologies and Instrumentation Engineering*, *3*(2), 63–74. doi:10.4018/ijmtie.2013040104

Bedi, P., Gandotra, V., & Singhal, A. (2014). Innovative strategies for secure software development. In *Software design and development: Concepts, methodologies, tools, and applications* (pp. 2099–2119). Hershey, PA: IGI Global. doi:10.4018/978-1-4666-4301-7.ch097

Bellatreche, L. (2010). *Data warehousing design and advanced engineering applications: Methods for complex construction*. Hershey, PA: IGI Global. doi:10.4018/978-1-60566-756-0

Bhattacharyya, S., & Dutta, P. (2013). *Handbook of research on computational intelligence for engineering, science, and business* (Vols. 1–2). Hershey, PA: IGI Global. doi:10.4018/978-1-4666-2518-1

Blicblau, A. S., & Richards, D. (2012). Development of "real world" project skills for engineering students. *International Journal of Quality Assurance in Engineering and Technology Education*, *2*(1), 1–13. doi:10.4018/ijqaete.2012010101

Boci, E. S., Sarkani, S., & Mazzuchi, T. A. (2013). Development of a complex geospatial/RF design model in support of service volume engineering design. In M. Bartolacci & S. Powell (Eds.), *Advancements and innovations in wireless communications and network technologies* (pp. 56–67). Hershey, PA: IGI Global. doi:10.4018/978-1-4666-2154-1.ch005

Boudreaux, A., & Primeaux, B. (2014). Modular game engine design. In *Software design and development: Concepts, methodologies, tools, and applications* (pp. 1179–1199). Hershey, PA: IGI Global. doi:10.4018/978-1-4666-4301-7.ch058

Boudriga, N., & Hamdi, M. (2014). *Security engineering techniques and solutions for information systems: Management and implementation*. Hershey, PA: IGI Global. doi:10.4018/978-1-61520-803-6

Brad, S. (2010). Competitive design of web-based courses in engineering education. In D. Russell & A. Haghi (Eds.), *Web-based engineering education: Critical design and effective tools* (pp. 119–148). Hershey, PA: IGI Global. doi:10.4018/978-1-61520-659-9.ch010

Brad, S. (2010). Designing effective web-based courses in engineering. In R. Luppicini & A. Haghi (Eds.), *Cases on digital technologies in higher education: Issues and challenges* (pp. 217–240). Hershey, PA: IGI Global. doi:10.4018/978-1-61520-869-2.ch016

Bradford, M., Gingras, R., & Hornby, J. (2010). Business process reengineering and ERP: Weapons for the global organization. In K. St.Amant (Ed.), *IT outsourcing: Concepts, methodologies, tools, and applications* (pp. 211–228). Hershey, PA: IGI Global. doi:10.4018/978-1-60566-770-6.ch012

Brennan, R. W., Hugo, R., & Rosehart, W. D. (2012). CDIO as an enabler for graduate attributes assessment: A Canadian case study. *International Journal of Quality Assurance in Engineering and Technology Education, 2*(2), 45–54. doi:10.4018/ijqaete.2012040105

Burnett, M. (2012). End-user software engineering and why it matters. In A. Dwivedi & S. Clarke (Eds.), *End-user computing, development, and software engineering: New challenges* (pp. 185–201). Hershey, PA: IGI Global. doi:10.4018/978-1-4666-0140-6.ch009

Burns, G. U., & Chisohlm, C. (2011). Engineering professional development related to sustainability of quality. *International Journal of Quality Assurance in Engineering and Technology Education, 1*(1), 15–29. doi:10.4018/ijqaete.2011010102

Byrne, D., Kelly, L., & Jones, G. J. (2014). Multiple multimodal mobile devices: Lessons learned from engineering lifelong solutions. In *Software design and development: Concepts, methodologies, tools, and applications* (pp. 2014–2032). Hershey, PA: IGI Global. doi:10.4018/978-1-4666-4301-7.ch093

Cervera, M., Albert, M., Torres, V., & Pelechano, V. (2012). A model-driven approach for the design and implementation of software development methods. *International Journal of Information System Modeling and Design, 3*(4), 86–103. doi:10.4018/jismd.2012100105

Chang, L., Levy, M., & Powell, P. (2011). Small firm process re-engineering success. In M. Tavana (Ed.), *Managing adaptability, intervention, and people in enterprise information systems* (pp. 138–155). Hershey, PA: IGI Global. doi:10.4018/978-1-60960-529-2.ch007

Chinyemba, F. (2011). Mobility of engineering and technology professionals and its impact on the quality of engineering and technology education: The case of Chinhoyi University of Technology, Zimbabwe. *International Journal of Quality Assurance in Engineering and Technology Education, 1*(2), 35–49. doi:10.4018/ijqaete.2011070104

Chiong, R. (2010). *Nature-inspired informatics for intelligent applications and knowledge discovery: Implications in business, science, and engineering*. Hershey, PA: IGI Global. doi:10.4018/978-1-60566-705-8

Chiprianov, V., Kermarrec, Y., & Rouvrais, S. (2014). Integrating DSLs into a software engineering process: Application to collaborative construction of telecom services. In *Software design and development: Concepts, methodologies, tools, and applications* (pp. 570–595). Hershey, PA: IGI Global. doi:10.4018/978-1-4666-4301-7.ch028

Chis, M. (2010). Introduction: A survey of the evolutionary computation techniques for software engineering. In M. Chis (Ed.), *Evolutionary computation and optimization algorithms in software engineering: Applications and techniques* (pp. 1–12). Hershey, PA: IGI Global. doi:10.4018/978-1-61520-809-8.ch001

Chiu, D. K. (2013). *Mobile and web innovations in systems and service-oriented engineering*. Hershey, PA: IGI Global. doi:10.4018/978-1-4666-2470-2

Chu, P. K., & Wu, S. (2012). Biomaterials. In Z. Abu-Faraj (Ed.), *Handbook of research on biomedical engineering education and advanced bioengineering learning: Interdisciplinary concepts* (pp. 238–283). Hershey, PA: IGI Global. doi:10.4018/978-1-4666-0122-2.ch006

Cimellaro, G. P. (2013). Optimal placement of controller for seismic structures. In N. Lagaros, V. Plevris, & C. Mitropoulou (Eds.), *Design optimization of active and passive structural control systems* (pp. 1–33). Hershey, PA: IGI Global. doi:10.4018/978-1-4666-2029-2.ch001

Clark, R., & Andrews, J. (2012). Engineering the future. In M. Rasul (Ed.), *Developments in engineering education standards: Advanced curriculum innovations* (pp. 143–155). Hershey, PA: IGI Global. doi:10.4018/978-1-4666-0951-8.ch008

Clark, T., & Willans, J. (2013). Software language engineering with XMF and XModeler. In M. Mernik (Ed.), *Formal and practical aspects of domain-specific languages: Recent developments* (pp. 311–340). Hershey, PA: IGI Global. doi:10.4018/978-1-4666-2092-6.ch011

Coll, R. K., & Zegwaard, K. E. (2012). Enculturation into engineering professional practice: Using legitimate peripheral participation to develop communication skills in engineering students. In A. Patil, H. Eijkman, & E. Bhattacharyya (Eds.), *New media communication skills for engineers and IT professionals: Trans-national and trans-cultural demands* (pp. 22–33). Hershey, PA: IGI Global. doi:10.4018/978-1-4666-0243-4.ch003

Coll, R. K., & Zegwaard, K. E. (2012). Enculturation into engineering professional practice: Using legitimate peripheral participation to develop communication skills in engineering students. In A. Patil, H. Eijkman, & E. Bhattacharyya (Eds.), *New media communication skills for engineers and IT professionals: Trans-national and trans-cultural demands* (pp. 22–33). Hershey, PA: IGI Global. doi:10.4018/978-1-4666-0243-4.ch003

Colomb, R. M. (2013). Representation of action is a primary requirement in ontologies for interoperating information systems. In M. Nazir Ahmad, R. Colomb, & M. Abdullah (Eds.), *Ontology-based applications for enterprise systems and knowledge management* (pp. 68–76). Hershey, PA: IGI Global. doi:10.4018/978-1-4666-1993-7.ch004

Cooklev, T. (2013). The role of standards in engineering education. In K. Jakobs (Ed.), *Innovations in organizational IT specification and standards development* (pp. 129–137). Hershey, PA: IGI Global. doi:10.4018/978-1-4666-2160-2.ch007

Costa, L., Loughran, N., & Grønmo, R. (2014). Model-driven engineering, services and interactive real-time applications. In *Software design and development: Concepts, methodologies, tools, and applications* (pp. 178–202). Hershey, PA: IGI Global. doi:10.4018/978-1-4666-4301-7.ch010

Dai, Y., Chakraborty, B., & Shi, M. (2011). *Kansei engineering and soft computing: Theory and practice*. Hershey, PA: IGI Global. doi:10.4018/978-1-61692-797-4

Daud, M. F., Taib, J. M., & Shariffudin, R. S. (2012). Assessing mechanical engineering undergraduates' conceptual knowledge in three dimensional computer aided design (3D CAD). In K. Yusof, N. Azli, A. Kosnin, S. Yusof, & Y. Yusof (Eds.), *Outcome-based science, technology, engineering, and mathematics education: Innovative practices* (pp. 350–363). Hershey, PA: IGI Global. doi:10.4018/978-1-4666-1809-1.ch017

Daugherty, A., Hires, W. E., & Braunstein, S. G. (2013). Collection development for the college of engineering at Louisiana State University libraries: Liaison responsibilities and duties. In S. Holder (Ed.), *Library collection development for professional programs: Trends and best practices* (pp. 291–305). Hershey, PA: IGI Global. doi:10.4018/978-1-4666-1897-8.ch017

Davim, J. (2013). *Dynamic methods and process advancements in mechanical, manufacturing, and materials engineering*. Hershey, PA: IGI Global. doi:10.4018/978-1-4666-1867-1 de Vere, I., & Melles, G. (2013). Integrating 'designerly' ways with engineering science: A catalyst for change within product design and development. In Industrial engineering: Concepts, methodologies, tools, and applications (pp. 56-78). Hershey, PA: IGI Global. doi:10.4018/978-1-4666-1945-6.ch005

del Sagrado Martinez, J., & del Aguila Cano, I. M. (2010). A Bayesian network for predicting the need for a requirements review. In F. Meziane & S. Vadera (Eds.), *Artificial intelligence applications for improved software engineering development: New prospects* (pp. 106–128). Hershey, PA: IGI Global. doi:10.4018/978-1-60566-758-4.ch006

Dhar-Bhattacharjee, S., & Takruri-Rizk, H. (2012). An Indo-British comparison. In C. Romm Livermore (Ed.), *Gender and social computing: Interactions, differences and relationships* (pp. 50–71). Hershey, PA: IGI Global. doi:10.4018/978-1-60960-759-3.ch004

Díaz, V. G., Lovelle, J. M., García-Bustelo, B. C., & Martinez, O. S. (2014). *Advances and applications in model-driven engineering*. Hershey, PA: IGI Global. doi:10.4018/978-1-4666-4494-6

Dogru, A., Senkul, P., & Kaya, O. (2012). Modern approaches to software engineering in the compositional era. In *Machine learning: Concepts, methodologies, tools and applications* (pp. 1903–1923). Hershey, PA: IGI Global. doi:10.4018/978-1-60960-818-7.ch803

Dogru, A. H., & Biçer, V. (2011). *Modern software engineering concepts and practices: Advanced approaches*. Hershey, PA: IGI Global. doi:10.4018/978-1-60960-215-4

Doll, W. J., & Deng, X. (2013). Antecedents of improvisation in IT-enabled engineering work. In A. Dwivedi & S. Clarke (Eds.), *Innovative strategies and approaches for end-user computing advancements* (pp. 242–264). Hershey, PA: IGI Global. doi:10.4018/978-1-4666-2059-9.ch013

Dominguez, U., & Magdaleno, J. (2011). Industrial training in engineering education in Spain. In P. Keleher, A. Patil, & R. Harreveld (Eds.), *Work-integrated learning in engineering, built environment and technology: Diversity of practice in practice* (pp. 72–84). Hershey, PA: IGI Global. doi:10.4018/978-1-60960-547-6.ch004

Dormido, S., Vargas, H., & Sánchez, J. (2013). AutomatL@bs consortium: A Spanish network of web-based labs for control engineering education. In *Industrial engineering: Concepts, methodologies, tools, and applications* (pp. 679–699). Hershey, PA: IGI Global. doi:10.4018/978-1-4666-1945-6.ch039

Dyro, J. F. (2012). Clinical engineering. In Z. Abu-Faraj (Ed.), *Handbook of research on biomedical engineering education and advanced bioengineering learning: Interdisciplinary concepts* (pp. 521–576). Hershey, PA: IGI Global. doi:10.4018/978-1-4666-0122-2.ch012

Easton, J. M., Davies, J. R., & Roberts, C. (2011). Ontology engineering the "what's", "why's", and "how's" of data exchange. *International Journal of Decision Support System Technology, 3*(1), 40–53. doi:10.4018/jdsst.2011010103

Eijkman, H., & Kayali, O. (2011). Addressing the politics of accreditation in engineering education: The benefits of soft systems thinking. *International Journal of Quality Assurance in Engineering and Technology Education, 1*(2), 1–10. doi:10.4018/ijqaete.2011070101

El-Khalili, N. H. (2013). Teaching agile software engineering using problem-based learning. *International Journal of Information and Communication Technology Education, 9*(3), 1–12. doi:10.4018/jicte.2013070101

Favre, L. M. (2010). MDA-based object-oriented reverse engineering. In L. Favre (Ed.), *Model driven architecture for reverse engineering technologies: Strategic directions and system evolution* (pp. 199–229). Hershey, PA: IGI Global. doi:10.4018/978-1-61520-649-0.ch010

Favre, L. M. (2010). Reverse engineering and MDA: An introduction. In L. Favre (Ed.), *Model driven architecture for reverse engineering technologies: Strategic directions and system evolution* (pp. 1–14). Hershey, PA: IGI Global. doi:10.4018/978-1-61520-649-0.ch001

Fenton, N., Hearty, P., Neil, M., & Radlinski, L. (2010). Software project and quality modelling using Bayesian networks. In F. Meziane & S. Vadera (Eds.), *Artificial intelligence applications for improved software engineering development: New prospects* (pp. 1–25). Hershey, PA: IGI Global. doi:10.4018/978-1-60566-758-4.ch001

Ferreira, R., Brisolara, L., Mattos, J. C., Spech, E., & Cota, E. (2010). Engineering embedded software: From application modeling to software synthesis. In L. Gomes & J. Fernandes (Eds.), *Behavioral modeling for embedded systems and technologies: Applications for design and implementation* (pp. 245–270). Hershey, PA: IGI Global. doi:10.4018/978-1-60566-750-8.ch010

Ferreira da Silva Oliveira, M., Wehrmeister, M. A., Assis do Nascimento, F., & Pereira, C. E. (2010). High-level design space exploration of embedded systems using the model-driven engineering and aspect-oriented design approaches. In L. Gomes & J. Fernandes (Eds.), *Behavioral modeling for embedded systems and technologies: Applications for design and implementation* (pp. 114–146). Hershey, PA: IGI Global. doi:10.4018/978-1-60566-750-8.ch005

Ferris, T. L. (2013). Engineering design as research. In *Industrial engineering: Concepts, methodologies, tools, and applications* (pp. 1766–1779). Hershey, PA: IGI Global. doi:10.4018/978-1-4666-1945-6.ch094

Frank, M. (2010). Capacity for engineering systems thinking (CEST): Literature review, principles for assessing and the reliability and validity of an assessing tool. In M. Hunter (Ed.), *Strategic information systems: Concepts, methodologies, tools, and applications* (pp. 1171–1183). Hershey, PA: IGI Global. doi:10.4018/978-1-60566-677-8.ch076

Fraser, D. (2012). A case study of curriculum development in engineering: Insights gained over two decades. In K. Yusof, N. Azli, A. Kosnin, S. Yusof, & Y. Yusof (Eds.), *Outcome-based science, technology, engineering, and mathematics education: Innovative practices* (pp. 27–49). Hershey, PA: IGI Global. doi:10.4018/978-1-4666-1809-1.ch002

Fraser, D. (2012). Curriculum initiatives to help engineering students learn and develop. In K. Yusof, N. Azli, A. Kosnin, S. Yusof, & Y. Yusof (Eds.), *Outcome-based science, technology, engineering, and mathematics education: Innovative practices* (pp. 85–106). Hershey, PA: IGI Global. doi:10.4018/978-1-4666-1809-1.ch005

Fries, T. P. (2014). Reengineering structured legacy system documentation to UML object-oriented artifacts. In *Software design and development: Concepts, methodologies, tools, and applications* (pp. 749–771). Hershey, PA: IGI Global. doi:10.4018/978-1-4666-4301-7.ch036

Funabashi, M. (2014). Transdisciplinary science and technology and service systems. In M. Kosaka & K. Shirahada (Eds.), *Progressive trends in knowledge and system-based science for service innovation* (pp. 101–127). Hershey, PA: IGI Global. doi:10.4018/978-1-4666-4663-6.ch006

Gaetano, L., Puppato, D., & Balestra, G. (2012). Modeling clinical engineering activities to support healthcare technology management. In A. Kolker & P. Story (Eds.), *Management engineering for effective healthcare delivery: Principles and applications* (pp. 113–131). Hershey, PA: IGI Global. doi:10.4018/978-1-60960-872-9.ch005

Galvão, T. A., Neto, F., & Bonates, M. F. (2013). eRiskGame: A persistent browser-based game for supporting project-based learning in the risk management context. In Industrial engineering: Concepts, methodologies, tools, and applications (pp. 1243-1259). Hershey, PA: IGI Global. doi:10.4018/978-1-4666-1945-6.ch067

Garousi, V., Shahnewaz, S., & Krishnamurthy, D. (2013). UML-driven software performance engineering: A systematic mapping and trend analysis. In V. Díaz, J. Lovelle, B. García-Bustelo, & O. Martínez (Eds.), *Progressions and innovations in model-driven software engineering* (pp. 18–64). Hershey, PA: IGI Global. doi:10.4018/978-1-4666-4217-1.ch002

Ge, W., Yang, N., Wang, W., & Li, J. (2011). Interfacial interactions: Drag. In S. Pannala, M. Syamlal, & T. O'Brien (Eds.), *Computational gas-solids flows and reacting systems: Theory, methods and practice* (pp. 128–177). Hershey, PA: IGI Global. doi:10.4018/978-1-61520-651-3.ch004

Génova, G., Llorens, J., & Morato, J. (2014). Software engineering research: The need to strengthen and broaden the classical scientific method. In *Software design and development: Concepts, methodologies, tools, and applications* (pp. 1639–1658). Hershey, PA: IGI Global. doi:10.4018/978-1-4666-4301-7.ch079

Genvigir, E. C., & Vijaykumar, N. L. (2010). Requirements traceability. In M. Ramachandran & R. de Carvalho (Eds.), *Handbook of research on software engineering and productivity technologies: Implications of globalization* (pp. 102–120). Hershey, PA: IGI Global. doi:10.4018/978-1-60566-731-7.ch008

Ghosh, S. (2010). Online automated essay grading system as a web based learning (WBL) tool in engineering education. In D. Russell & A. Haghi (Eds.), *Web-based engineering education: Critical design and effective tools* (pp. 53–62). Hershey, PA: IGI Global. doi:10.4018/978-1-61520-659-9.ch005

Gill, A. Q., & Bunker, D. (2013). SaaS requirements engineering for agile development. In X. Wang, N. Ali, I. Ramos, & R. Vidgen (Eds.), *Agile and lean service-oriented development: Foundations, theory, and practice* (pp. 64–93). Hershey, PA: IGI Global. doi:10.4018/978-1-4666-2503-7.ch004

Giraldo, F. D., Villegas, M. L., & Collazos, C. A. (2014). The use of HCI approaches into distributed CSCL activities applied to software engineering courses. In *Software design and development: Concepts, methodologies, tools, and applications* (pp. 2033–2050). Hershey, PA: IGI Global. doi:10.4018/978-1-4666-4301-7.ch094

Girardi, R., & Leite, A. (2011). Knowledge engineering support for agent-oriented software reuse. In M. Ramachandran (Ed.), *Knowledge engineering for software development life cycles: Support technologies and applications* (pp. 177–195). Hershey, PA: IGI Global. doi:10.4018/978-1-60960-509-4.ch010

Goodhew, P. (2013). Why get your engineering programme accredited? In *Industrial engineering: Concepts, methodologies, tools, and applications* (pp. 18–20). Hershey, PA: IGI Global. doi:10.4018/978-1-4666-1945-6.ch002

Goossenaerts, J., Possel-Dölken, F., & Popplewell, K. (2011). Vision, trends, gaps and a broad roadmap for future engineering. In I. Management Association (Ed.), Global business: Concepts, methodologies, tools and applications (pp. 2229-2243). Hershey, PA: IGI Global. doi:10.4018/978-1-60960-587-2.ch802

Goyal, S. B., & Prakash, N. (2013). Functional method engineering. *International Journal of Information System Modeling and Design*, *4*(1), 79–103. doi:10.4018/jismd.2013010104

Gunasekaran, A., & Shea, T. (2010). Requirements management for ERP projects. In *Organizational advancements through enterprise information systems: Emerging applications and developments* (pp. 29–45). Hershey, PA: IGI Global. doi:10.4018/978-1-60566-968-7.ch003

Gustavsson, I., Claesson, L., Nilsson, K., Zackrisson, J., Zubia, J. G., & Jayo, U. H. … Claesson, I. (2012). The VISIR open lab platform. In A. Azad, M. Auer, & V. Harward (Eds.) Internet accessible remote laboratories: Scalable e-learning tools for engineering and science disciplines (pp. 294-317). Hershey, PA: IGI Global. doi:10.4018/978-1-61350-186-3.ch015

Habib, M. K., & Davim, J. (2013). *Engineering creative design in robotics and mechatronics*. Hershey, PA: IGI Global. doi:10.4018/978-1-4666-4225-6

Haghi, A. K. (2013). *Methodologies and applications for chemoinformatics and chemical engineering*. Hershey, PA: IGI Global. doi:10.4018/978-1-4666-4010-8

Haghi, A. K., & Noroozi, B. (2010). Adapting engineering education to the new century. In D. Russell & A. Haghi (Eds.), *Web-based engineering education: Critical design and effective tools* (pp. 30–41). Hershey, PA: IGI Global. doi:10.4018/978-1-61520-659-9.ch003

Haghpanahi, M., Nikkhoo, M., & Peirovi, H. A. (2010). Computer aided tissue engineering from modeling to manufacturing. In A. Lazakidou (Ed.), *Biocomputation and biomedical informatics: Case studies and applications* (pp. 75–88). Hershey, PA: IGI Global. doi:10.4018/978-1-60566-768-3.ch004

Hepsø, I. L., Rindal, A., & Waldal, K. (2013). The introduction of a hand-held platform in an engineering and fabrication company. In T. Rosendahl & V. Hepsø (Eds.), *Integrated operations in the oil and gas industry: Sustainability and capability development* (pp. 246–260). Hershey, PA: IGI Global. doi:10.4018/978-1-4666-2002-5.ch015

Hernández-López, A., Colomo-Palacios, R., García-Crespo, Á., & Cabezas-Isla, F. (2013). Software engineering productivity: Concepts, issues and challenges. In J. Wang (Ed.), *Perspectives and techniques for improving information technology project management* (pp. 69–79). Hershey, PA: IGI Global. doi:10.4018/978-1-4666-2800-7.ch006

Herrmann, A., & Morali, A. (2012). Interplay of security requirements engineering and reverse engineering in the maintenance of undocumented software. In J. Rech & C. Bunse (Eds.), *Emerging technologies for the evolution and maintenance of software models* (pp. 57–91). Hershey, PA: IGI Global. doi:10.4018/978-1-61350-438-3.ch003

Hochrainer, M., & Ziegler, F. (2013). Tuned liquid column gas damper in structural control: The salient features of a general purpose damping device and its application in buildings, bridges, and dams. In N. Lagaros, V. Plevris, & C. Mitropoulou (Eds.), *Design optimization of active and passive structural control systems* (pp. 150–179). Hershey, PA: IGI Global. doi:10.4018/978-1-4666-2029-2.ch007

Hochstein, L., Schott, B., & Graybill, R. B. (2013). Computational engineering in the cloud: Benefits and challenges. In A. Dwivedi & S. Clarke (Eds.), *Innovative strategies and approaches for end-user computing advancements* (pp. 314–332). Hershey, PA: IGI Global. doi:10.4018/978-1-4666-2059-9.ch017

Höhn, S., Lowis, L., Jürjens, J., & Accorsi, R. (2010). Identification of vulnerabilities in web services using model-based security. In C. Gutiérrez, E. Fernández-Medina, & M. Piattini (Eds.), *Web services security development and architecture: Theoretical and practical issues* (pp. 1–32). Hershey, PA: IGI Global. doi:10.4018/978-1-60566-950-2.ch001

Holton, D. L., & Verma, A. (2010). Designing animated simulations and web-based assessments to improve electrical engineering education. In D. Russell & A. Haghi (Eds.), *Web-based engineering education: Critical design and effective tools* (pp. 77–95). Hershey, PA: IGI Global. doi:10.4018/978-1-61520-659-9.ch007

Honnutagi, A. R., Sonar, R., & Babu, S. (2012). Quality accreditation system for Indian engineering education using knowledge management and system dynamics. *International Journal of Quality Assurance in Engineering and Technology Education*, *2*(3), 47–61. doi:10.4018/ijqaete.2012070105

Hua, G. B. (2013). Business process re-engineering. In *Implementing IT business strategy in the construction industry* (pp. 118–140). Hershey, PA: IGI Global. doi:10.4018/978-1-4666-4185-3.ch006

Hussein, B., Hage-Diab, A., Hammoud, M., Kawtharani, A., El-Hage, H., & Haj-Ali, A. (2014). Management response to improve the educational performance of engineering students: The case of the lebanese international university. In G. Khoury & M. Khoury (Eds.), *Cases on management and organizational behavior in an Arab context* (pp. 91–110). Hershey, PA: IGI Global. doi:10.4018/978-1-4666-5067-1.ch006

Hussey, M., Wu, B., & Xu, X. (2011). Co-operation models for industries and software education institutions. In *Software industry-oriented education practices and curriculum development: Experiences and lessons* (pp. 39–56). Hershey, PA: IGI Global. doi:10.4018/978-1-60960-797-5.ch003

Hussey, M., Wu, B., & Xu, X. (2011). Curriculum issues in industry oriented software engineering education. In *Software industry-oriented education practices and curriculum development: Experiences and lessons* (pp. 153–165). Hershey, PA: IGI Global. doi:10.4018/978-1-60960-797-5.ch010

Hussey, M., Wu, B., & Xu, X. (2011). Industry oriented curriculum and syllabus creation for software engineering series courses in the school of software. In *Software industry-oriented education practices and curriculum development: Experiences and lessons* (pp. 98–109). Hershey, PA: IGI Global. doi:10.4018/978-1-60960-797-5.ch006

Ilyas, Q. M. (2013). Ontology augmented software engineering. In K. Buragga & N. Zaman (Eds.), *Software development techniques for constructive information systems design* (pp. 406–413). Hershey, PA: IGI Global. doi:10.4018/978-1-4666-3679-8.ch023

Islam, S., Mouratidis, H., Kalloniatis, C., Hudic, A., & Zechner, L. (2013). Model based process to support security and privacy requirements engineering. *International Journal of Secure Software Engineering*, *3*(3), 1–22. doi:10.4018/jsse.2012070101

Jahan, K., Everett, J. W., Tang, G., Farrell, S., Zhang, H., Wenger, A., & Noori, M. (2010). Use of living systems to teach basic engineering concepts. In D. Russell & A. Haghi (Eds.), *Web-based engineering education: Critical design and effective tools* (pp. 96–107). Hershey, PA: IGI Global. doi:10.4018/978-1-61520-659-9.ch008

Jakovljevic, M. (2013). A conceptual model of creativity, invention, and innovation (MCII) for entrepreneurial engineers. In S. Buckley & M. Jakovljevic (Eds.), *Knowledge management innovations for interdisciplinary education: Organizational applications* (pp. 66–87). Hershey, PA: IGI Global. doi:10.4018/978-1-4666-1969-2.ch004

Jaroucheh, Z., Liu, X., & Smith, S. (2014). A software engineering framework for context-aware service-based processes in pervasive environments. In *Software design and development: Concepts, methodologies, tools, and applications* (pp. 71–95). Hershey, PA: IGI Global. doi:10.4018/978-1-4666-4301-7.ch005

Joiner, R., Iacovides, I., Darling, J., Diament, A., Drew, B., Duddley, J., ... Gavin, C. (2013). Racing academy: A case study of a digital game for supporting students learning of physics and engineering. In Y. Baek & N. Whitton (Eds.), *Cases on digital game-based learning: Methods, models, and strategies* (pp. 509–523). Hershey, PA: IGI Global. doi:10.4018/978-1-4666-2848-9.ch026

Jordan, M. E. (2014). Interweaving the digital and physical worlds in collaborative project-based learning experiences. In D. Loveless, B. Griffith, M. Bérci, E. Ortlieb, & P. Sullivan (Eds.), *Academic knowledge construction and multimodal curriculum development* (pp. 265–282). Hershey, PA: IGI Global. doi:10.4018/978-1-4666-4797-8.ch017

Juang, Y. (2010). WIRE: A highly interactive blended learning for engineering education. In D. Russell & A. Haghi (Eds.), *Web-based engineering education: Critical design and effective tools* (pp. 149–159). Hershey, PA: IGI Global. doi:10.4018/978-1-61520-659-9.ch011

Kadry, S. (2013). *Diagnostics and prognostics of engineering systems: Methods and techniques*. Hershey, PA: IGI Global. doi:10.4018/978-1-4666-2095-7

Kamboj, A., Kumar, S., & Singh, H. (2012). Design and development of hybrid stir casting process. *International Journal of Applied Industrial Engineering, 1*(2), 1–6. doi:10.4018/ijaie.2012070101

Kamthan, P. (2010). A methodology for integrating the social web environment in software engineering education. In S. Dasgupta (Ed.), *Social computing: Concepts, methodologies, tools, and applications* (pp. 457–471). Hershey, PA: IGI Global. doi:10.4018/978-1-60566-984-7.ch031

Kamthan, P. (2010). A social web perspective of software engineering education. In S. Murugesan (Ed.), *Handbook of research on web 2.0, 3.0, and X.0: Technologies, business, and social applications* (pp. 472–495). Hershey, PA: IGI Global. doi:10.4018/978-1-60566-384-5.ch026

Kamthan, P. (2010). On the prospects and concerns of pattern-oriented web engineering. In G. Alkhatib & D. Rine (Eds.), *Web engineering advancements and trends: Building new dimensions of information technology* (pp. 97–128). Hershey, PA: IGI Global. doi:10.4018/978-1-60566-719-5.ch006

Kamthan, P. (2011). Using the social web environment for software engineering education. In L. Tomei (Ed.), *Online courses and ICT in education: Emerging practices and applications* (pp. 23–45). Hershey, PA: IGI Global. doi:10.4018/978-1-60960-150-8.ch003

Kamthan, P. (2013). An exploration of the social web environment for collaborative software engineering education. In N. Karacapilidis, M. Raisinghani, & E. Ng (Eds.), *Web-based and blended educational tools and innovations* (pp. 1–23). Hershey, PA: IGI Global. doi:10.4018/978-1-4666-2023-0.ch001

Kapos, G., Dalakas, V., Nikolaidou, M., & Anagnostopoulos, D. (2014). An integrated framework to simulate SysML models using DEVS simulators. In P. Fonseca i Casas (Ed.), *Formal languages for computer simulation: Transdisciplinary models and applications* (pp. 305–332). Hershey, PA: IGI Global. doi:10.4018/978-1-4666-4369-7.ch010

Karlsson, F., & Ågerfalk, P. J. (2011). Towards structured flexibility in information systems development: Devising a method for method configuration. In K. Siau (Ed.), *Theoretical and practical advances in information systems development: Emerging trends and approaches* (pp. 214–238). Hershey, PA: IGI Global. doi:10.4018/978-1-60960-521-6.ch010

Karpati, P., Sindre, G., & Matulevicius, R. (2012). Comparing misuse case and mal-activity diagrams for modelling social engineering attacks. *International Journal of Secure Software Engineering, 3*(2), 54–73. doi:10.4018/jsse.2012040103

Keleher, P., & Patil, A. (2012). Conducting an effective residential school for an undergraduate materials science and engineering course. *International Journal of Quality Assurance in Engineering and Technology Education, 2*(3), 41–46. doi:10.4018/ijqaete.2012070104

Ker, H. W. (2012). Engineering education and attitudes toward mathematics: A comparative study. *International Journal of Quality Assurance in Engineering and Technology Education, 2*(1), 63–76. doi:10.4018/ijqaete.2012010105

Kinsner, W. (2012). Challenges in the design of adoptive, intelligent and cognitive systems. In Y. Wang (Ed.), *Software and intelligent sciences: New transdisciplinary findings* (pp. 47–67). Hershey, PA: IGI Global. doi:10.4018/978-1-4666-0261-8.ch004

Kirikova, M. (2011). Domain modeling approaches in IS engineering. In J. Osis & E. Asnina (Eds.), *Model-driven domain analysis and software development: Architectures and functions* (pp. 388–406). Hershey, PA: IGI Global. doi:10.4018/978-1-61692-874-2.ch018

Kljajic, M., & Farr, J. V. (2010). Importance of systems engineering in the development of information systems. In D. Paradice (Ed.), *Emerging systems approaches in information technologies: Concepts, theories, and applications* (pp. 51–66). Hershey, PA: IGI Global. doi:10.4018/978-1-60566-976-2.ch004

Kljajic, M., & Farr, J. V. (2010). The role of systems engineering in the development of information systems. In M. Hunter (Ed.), *Strategic information systems: Concepts, methodologies, tools, and applications* (pp. 369–381). Hershey, PA: IGI Global. doi:10.4018/978-1-60566-677-8.ch026

Knoell, H. D. (2010). User participation in the quality assurance of requirements specifications: An evaluation of traditional models and animated systems engineering techniques. In M. Hunter (Ed.), *Strategic information systems: Concepts, methodologies, tools, and applications* (pp. 1623–1638). Hershey, PA: IGI Global. doi:10.4018/978-1-60566-677-8.ch106

Kof, L. (2010). From textual scenarios to message sequence charts. In F. Meziane & S. Vadera (Eds.), *Artificial intelligence applications for improved software engineering development: New prospects* (pp. 83–105). Hershey, PA: IGI Global. doi:10.4018/978-1-60566-758-4.ch005

Köhler, B., Gluchow, M., & Brügge, B. (2014). Teaching basic software engineering to senior high school students. In *K-12 education: Concepts, methodologies, tools, and applications* (pp. 1634–1649). Hershey, PA: IGI Global. doi:10.4018/978-1-4666-4502-8.ch094

Kolker, A. (2012). Efficient managerial decision-making in healthcare settings: Examples and fundamental principles. In A. Kolker & P. Story (Eds.), *Management engineering for effective healthcare delivery: Principles and applications* (pp. 1–45). Hershey, PA: IGI Global. doi:10.4018/978-1-60960-872-9.ch001

Kolker, A., & Story, P. (2012). *Management engineering for effective healthcare delivery: Principles and applications*. Hershey, PA: IGI Global. doi:10.4018/978-1-60960-872-9

Kornyshova, E., Deneckère, R., & Claudepierre, B. (2013). Towards method component contextualization. In J. Krogstie (Ed.), *Frameworks for developing efficient information systems: Models, theory, and practice* (pp. 337–368). Hershey, PA: IGI Global. doi:10.4018/978-1-4666-4161-7.ch015

Krishnan, S. (2012). Problem-based learning curricula in engineering. In M. Rasul (Ed.), *Developments in engineering education standards: Advanced curriculum innovations* (pp. 23–40). Hershey, PA: IGI Global. doi:10.4018/978-1-4666-0951-8.ch002

Ku, H., & Thonglek, S. (2011). Running a successful practice school: Challenges and lessons learned. In P. Keleher, A. Patil, & R. Harreveld (Eds.), *Work-integrated learning in engineering, built environment and technology: Diversity of practice in practice* (pp. 131–163). Hershey, PA: IGI Global. doi:10.4018/978-1-60960-547-6.ch007

Lacuesta, R., Fernández-Sanz, L., & Romay, M. D. (2014). Requirements specification as basis for mobile software quality assurance. In *Software design and development: Concepts, methodologies, tools, and applications* (pp. 719–732). Hershey, PA: IGI Global. doi:10.4018/978-1-4666-4301-7.ch034

Lai, A., Zhang, C., & Busovaca, S. (2013). 2-SQUARE: A web-based enhancement of SQUARE privacy and security requirements engineering. *International Journal of Software Innovation, 1*(1), 41–53. doi:10.4018/ijsi.2013010104

Lane, J. A., Petkov, D., & Mora, M. (2010). Software engineering and the systems approach: A conversation with Barry Boehm. In M. Hunter (Ed.), *Strategic information systems: Concepts, methodologies, tools, and applications* (pp. 333–337). Hershey, PA: IGI Global. doi:10.4018/978-1-60566-677-8.ch024

Lansiquot, R. D. (2013). *Cases on interdisciplinary research trends in science, technology, engineering, and mathematics: Studies on urban classrooms*. Hershey, PA: IGI Global. doi:10.4018/978-1-4666-2214-2

Laribi, S., Le Bris, A., Huang, L. M., Olsson, P., & Guillemoles, J. F. (2013). Phononic engineering for hot carrier solar cells. In L. Fara & M. Yamaguchi (Eds.), *Advanced solar cell materials, technology, modeling, and simulation* (pp. 214–242). Hershey, PA: IGI Global. doi:10.4018/978-1-4666-1927-2.ch012

Lee, C. K., & Sidhu, M. S. (2013). Computer-aided engineering education: New learning approaches and technologies. In V. Wang (Ed.), *Handbook of research on teaching and learning in K-20 education* (pp. 317–340). Hershey, PA: IGI Global. doi:10.4018/978-1-4666-4249-2.ch019

Lee, J., Ma, S., Lee, S., Wu, C., & Lee, C. L. (2014). Towards a high-availability-driven service composition framework. In *Software design and development: Concepts, methodologies, tools, and applications* (pp. 1498–1520). Hershey, PA: IGI Global. doi:10.4018/978-1-4666-4301-7.ch073

Leng, J., Rhyne, T., & Sharrock, W. (2012). Visualization: Future technology and practices for computational science and engineering. In J. Leng & W. Sharrock (Eds.), *Handbook of research on computational science and engineering: Theory and practice* (pp. 381–413). Hershey, PA: IGI Global. doi:10.4018/978-1-61350-116-0.ch016

Leng, J., & Sharrock, W. (2012). *Handbook of research on computational science and engineering: Theory and practice* (Vols. 1–2). Hershey, PA: IGI Global. doi:10.4018/978-1-61350-116-0

Li, Z. J. (2013). Prototyping of robotic systems in surgical procedures and automated manufacturing processes. In *Industrial engineering: Concepts, methodologies, tools, and applications* (pp. 1969–1987). Hershey, PA: IGI Global. doi:10.4018/978-1-4666-1945-6.ch106

Lladó, C. M., Bonet, P., & Smith, C. U. (2014). Towards a multi-formalism multi-solution framework for model-driven performance engineering. In M. Gribaudo & M. Iacono (Eds.), *Theory and application of multi-formalism modeling* (pp. 34–55). Hershey, PA: IGI Global. doi:10.4018/978-1-4666-4659-9.ch003

Loo, A. (2013). *Distributed computing innovations for business, engineering, and science*. Hershey, PA: IGI Global. doi:10.4018/978-1-4666-2533-4

Ma, Z. M. (2011). Engineering design knowledge management. In D. Schwartz & D. Te'eni (Eds.), *Encyclopedia of knowledge management* (2nd ed.; pp. 263–269). Hershey, PA: IGI Global. doi:10.4018/978-1-59904-931-1.ch025

Maiti, C. K., & Maiti, A. (2013). Teaching technology computer aided design (TCAD) online. In Industrial engineering: Concepts, methodologies, tools, and applications (pp. 1043-1063). Hershey, PA: IGI Global. doi:10.4018/978-1-4666-1945-6.ch057

Male, S. A. (2012). Generic engineering competencies required by engineers graduating in Australia: The competencies of engineering graduates (CEG) project. In M. Rasul (Ed.), *Developments in engineering education standards: Advanced curriculum innovations* (pp. 41–63). Hershey, PA: IGI Global. doi:10.4018/978-1-4666-0951-8.ch003

Malmqvist, J. (2012). A comparison of the CDIO and EUR-ACE quality assurance systems. *International Journal of Quality Assurance in Engineering and Technology Education, 2*(2), 9–22. doi:10.4018/ijqaete.2012040102

Maree, D. J., & Maree, M. (2010). Factors contributing to the success of women working in science, engineering and technology (SET) careers. In A. Cater-Steel & E. Cater (Eds.), *Women in engineering, science and technology: Education and career challenges* (pp. 183–210). Hershey, PA: IGI Global. doi:10.4018/978-1-61520-657-5.ch009

Marichal, G. N., & González, E. J. (2012). Intelligent MAS in system engineering and robotics. In *Machine learning: Concepts, methodologies, tools and applications* (pp. 175–182). Hershey, PA: IGI Global. doi:10.4018/978-1-60960-818-7.ch204

Martinez, L., Favre, L., & Pereira, C. (2013). Architecture-driven modernization for software reverse engineering technologies. In V. Díaz, J. Lovelle, B. García-Bustelo, & O. Martínez (Eds.), *Progressions and innovations in model-driven software engineering* (pp. 288–307). Hershey, PA: IGI Global. doi:10.4018/978-1-4666-4217-1.ch012

Matanovic, D. (2014). The macondo 252 disaster: Causes and consequences. In D. Matanovic, N. Gaurina-Medjimurec, & K. Simon (Eds.), *Risk analysis for prevention of hazardous situations in petroleum and natural gas engineering* (pp. 115–131). Hershey, PA: IGI Global. doi:10.4018/978-1-4666-4777-0.ch006

Matanovic, D., Gaurina-Medjimurec, N., & Simon, K. (2014). *Risk analysis for prevention of hazardous situations in petroleum and natural gas engineering*. Hershey, PA: IGI Global. doi:10.4018/978-1-4666-4777-0

Matsuo, T., & Fujimoto, T. (2012). Analogical thinking based instruction method in IT professional education. In R. Colomo-Palacios (Ed.), *Professional advancements and management trends in the IT sector* (pp. 95–108). Hershey, PA: IGI Global. doi:10.4018/978-1-4666-0924-2.ch007

Mazo, R., Salinesi, C., Diaz, D., Djebbi, O., & Lora-Michiels, A. (2012). Constraints: The heart of domain and application engineering in the product lines engineering strategy. *International Journal of Information System Modeling and Design, 3*(2), 33–68. doi:10.4018/jismd.2012040102

Mead, N. R. (2010). Benefits and challenges in the use of case studies for security requirements engineering methods. *International Journal of Secure Software Engineering, 1*(1), 74–91. doi:10.4018/jsse.2010102005

Méausoone, P., & Aguilera, A. (2014). Inventory of experimental works on cutting tools' life for the wood industry. In A. Aguilera & J. Davim (Eds.), *Research developments in wood engineering and technology* (pp. 320–342). Hershey, PA: IGI Global. doi:10.4018/978-1-4666-4554-7.ch009

Medlin, B. D., & Cazier, J. (2011). Obtaining patient's information from hospital employees through social engineering techniques: An investigative study. In H. Nemati (Ed.), *Pervasive information security and privacy developments: Trends and advancements* (pp. 77–89). Hershey, PA: IGI Global. doi:10.4018/978-1-61692-000-5.ch006

Meziane, F., & Vadera, S. (2010). *Artificial intelligence applications for improved software engineering development: new prospects*. Hershey, PA: IGI Global. doi:10.4018/978-1-60566-758-4

Meziane, F., & Vadera, S. (2012). Artificial intelligence in software engineering: Current developments and future prospects. In *Machine learning: Concepts, methodologies, tools and applications* (pp. 1215–1236). Hershey, PA: IGI Global. doi:10.4018/978-1-60960-818-7.ch504

Milanovic, N. (2011). *Engineering reliable service oriented architecture: Managing complexity and service level agreements*. Hershey, PA: IGI Global. doi:10.4018/978-1-60960-493-6

Miller, W. S., & Summers, J. D. (2013). Tool and information centric design process modeling: Three case studies. In *Industrial engineering: Concepts, methodologies, tools, and applications* (pp. 1613–1637). Hershey, PA: IGI Global. doi:10.4018/978-1-4666-1945-6.ch086

Mishra, B., & Shukla, K. K. (2014). Data mining techniques for software quality prediction. In *Software design and development: Concepts, methodologies, tools, and applications* (pp. 401–428). Hershey, PA: IGI Global. doi:10.4018/978-1-4666-4301-7.ch021

Mishra, S. (2011). Social and ethical concerns of biomedical engineering research and practice. In A. Shukla & R. Tiwari (Eds.), *Biomedical engineering and information systems: Technologies, tools and applications* (pp. 54–80). Hershey, PA: IGI Global. doi:10.4018/978-1-61692-004-3.ch003

Moeller, D. P., & Sitzmann, D. (2012). Online computer engineering: combining blended e-learning in engineering with lifelong learning. In M. Rasul (Ed.), *Developments in engineering education standards: Advanced curriculum innovations* (pp. 194–215). Hershey, PA: IGI Global. doi:10.4018/978-1-4666-0951-8.ch011

Moeller, D. P., & Vakilzadian, H. (2012). Technology-enhanced learning standard through integration of modeling and simulation into engineering study programs. In M. Rasul (Ed.), *Developments in engineering education standards: Advanced curriculum innovations* (pp. 157–177). Hershey, PA: IGI Global. doi:10.4018/978-1-4666-0951-8.ch009

Mohan Baral, L., Kifor, C. V., Bondrea, I., & Oprean, C. (2012). Introducing problem based learning (PBL) in textile engineering education and assessing its influence on six sigma project implementation. *International Journal of Quality Assurance in Engineering and Technology Education, 2*(4), 38–48. doi:10.4018/ijqaete.2012100104

Moore, S., May, D., & Wold, K. (2012). Developing cultural competency in engineering through transnational distance learning. In R. Hogan (Ed.), *Transnational distance learning and building new markets for universities* (pp. 210–228). Hershey, PA: IGI Global. doi:10.4018/978-1-4666-0206-9.ch013

Mora, M., Gelman, O., Frank, M., Paradice, D. B., & Cervantes, F. (2010). Toward an interdisciplinary engineering and management of complex IT-intensive organizational systems: A systems view. In D. Paradice (Ed.), *Emerging systems approaches in information technologies: Concepts, theories, and applications* (pp. 1–24). Hershey, PA: IGI Global. doi:10.4018/978-1-60566-976-2.ch001

Mora, M., O'Connor, R., Raisinghani, M. S., Macías-Luévano, J., & Gelman, O. (2013). An IT service engineering and management framework (ITS-EMF). In P. Ordóñez de Pablos & R. Tennyson (Eds.), *Best practices and new perspectives in service science and management* (pp. 76–91). Hershey, PA: IGI Global. doi:10.4018/978-1-4666-3894-5.ch005

Moser, T., Biffl, S., Sunindyo, W. D., & Winkler, D. (2013). Integrating production automation expert knowledge across engineering domains. In N. Bessis (Ed.), *Development of distributed systems from design to application and maintenance* (pp. 152–167). Hershey, PA: IGI Global. doi:10.4018/978-1-4666-2647-8.ch009

Mouratidis, H. (2011). *Software engineering for secure systems: Industrial and research perspectives*. Hershey, PA: IGI Global. doi:10.4018/978-1-61520-837-1

Mourtos, N. J. (2012). Defining, teaching, and assessing engineering design skills. *International Journal of Quality Assurance in Engineering and Technology Education, 2*(1), 14–30. doi:10.4018/ijqaete.2012010102

Moustafa, A. (2012). Damage assessment of inelastic structures under simulated critical earthquakes. In V. Plevris, C. Mitropoulou, & N. Lagaros (Eds.), *Structural seismic design optimization and earthquake engineering: Formulations and applications* (pp. 128–151). Hershey, PA: IGI Global. doi:10.4018/978-1-4666-1640-0.ch006

Nasir, M. H., Alias, N. A., Fauzi, S. S., & Massatu, M. H. (2012). Implementation of the personal software process in academic settings and current support tools. In S. Fauzi, M. Nasir, N. Ramli, & S. Sahibuddin (Eds.), *Software process improvement and management: Approaches and tools for practical development* (pp. 117–148). Hershey, PA: IGI Global. doi:10.4018/978-1-61350-141-2.ch007

Nhlabatsi, A., Nuseibeh, B., & Yu, Y. (2012). Security requirements engineering for evolving software systems: A survey. In K. Khan (Ed.), *Security-aware systems applications and software development methods* (pp. 108–128). Hershey, PA: IGI Global. doi:10.4018/978-1-4666-1580-9.ch007

Noroozi, B., Valizadeh, M., & Sorial, G. A. (2010). Designing of e-learning for engineering education in developing countries: Key issues and success factors. In D. Russell & A. Haghi (Eds.), *Web-based engineering education: Critical design and effective tools* (pp. 1–19). Hershey, PA: IGI Global. doi:10.4018/978-1-61520-659-9.ch001

Oh Navarro, E. (2011). On the role of learning theories in furthering software engineering education. In *Instructional design: Concepts, methodologies, tools and applications* (pp. 1645–1666). Hershey, PA: IGI Global. doi:10.4018/978-1-60960-503-2.ch709

Onwubiko, C. (2014). Modelling situation awareness information and system requirements for the mission using goal-oriented task analysis approach. In *Software design and development: Concepts, methodologies, tools, and applications* (pp. 460–478). Hershey, PA: IGI Global. doi:10.4018/978-1-4666-4301-7.ch023

Osis, J., & Asnina, E. (2011). Is modeling a treatment for the weakness of software engineering? In J. Osis & E. Asnina (Eds.), *Model-driven domain analysis and software development: Architectures and functions* (pp. 1–14). Hershey, PA: IGI Global. doi:10.4018/978-1-61692-874-2.ch001

Othman, R., & Awang, Z. (2012). Using multiple methods in assessing oral communication skills in the final year project design course of an undergraduate engineering program. In K. Yusof, N. Azli, A. Kosnin, S. Yusof, & Y. Yusof (Eds.), *Outcome-based science, technology, engineering, and mathematics education: Innovative practices* (pp. 263–287). Hershey, PA: IGI Global. doi:10.4018/978-1-4666-1809-1.ch013

Ozcelik, Y. (2010). IT-enabled reengineering: Productivity impacts. In K. St.Amant (Ed.), *IT outsourcing: Concepts, methodologies, tools, and applications* (pp. 371–376). Hershey, PA: IGI Global. doi:10.4018/978-1-60566-770-6.ch022

Ožvoldová, M., & Schauer, F. (2012). Remote experiments in freshman engineering education by integrated e-learning. In A. Azad, M. Auer, & V. Harward (Eds.), *Internet accessible remote laboratories: Scalable e-learning tools for engineering and science disciplines* (pp. 60–83). Hershey, PA: IGI Global. doi:10.4018/978-1-61350-186-3.ch004

Paay, J., Pedell, S., Sterling, L., Vetere, F., & Howard, S. (2011). The benefit of ambiguity in understanding goals in requirements modelling. *International Journal of People-Oriented Programming*, *1*(2), 24–49. doi:10.4018/ijpop.2011070102

Palmer, S., & Holt, D. (2010). Online discussion in engineering education: Student responses and learning outcomes. In L. Shedletsky & J. Aitken (Eds.), *Cases on online discussion and interaction: Experiences and outcomes* (pp. 105–122). Hershey, PA: IGI Global. doi:10.4018/978-1-61520-863-0.ch005

Pandian, A., Ismail, S. A., & Abdullah, A. S. (2012). Communication framework to empower 21st century engineers and IT professionals. In A. Patil, H. Eijkman, & E. Bhattacharyya (Eds.), *New media communication skills for engineers and IT professionals: Trans-national and trans-cultural demands* (pp. 34–54). Hershey, PA: IGI Global. doi:10.4018/978-1-4666-0243-4.ch004

Pasupathy, K. S. (2010). Systems engineering and health informatics: Context, content, and implementation. In S. Kabene (Ed.), *Healthcare and the effect of technology: Developments, challenges and advancements* (pp. 123–144). Hershey, PA: IGI Global. doi:10.4018/978-1-61520-733-6.ch009

Pasupathy, K. S. (2010). Transforming healthcare: Leveraging the complementarities of health informatics and systems engineering. *International Journal of Healthcare Delivery Reform Initiatives*, *2*(2), 35–55. doi:10.4018/jhdri.2010040103

Pasupathy, K. S. (2011). Systems engineering and health informatics. In *Clinical technologies: Concepts, methodologies, tools and applications* (pp. 1684–1705). Hershey, PA: IGI Global. doi:10.4018/978-1-60960-561-2.ch606

Patel, C., & Ramachandran, M. (2010). Story card process improvement framework for agile requirements. In M. Ramachandran & R. de Carvalho (Eds.), *Handbook of research on software engineering and productivity technologies: Implications of globalization* (pp. 61–54). Hershey, PA: IGI Global. doi:10.4018/978-1-60566-731-7.ch006

Patro, C. S. (2012). A study on adaptability of total quality management in engineering education sector. *International Journal of Quality Assurance in Engineering and Technology Education*, *2*(4), 25–37. doi:10.4018/ijqaete.2012100103

Peña de Carrillo, C. I., Choquet, C., Després, C., Iksal, S., Jacoboni, P., Lekira, A., … Thi-Ngoc, D. P. (2014). Engineering and reengineering of technology enhanced learning scenarios using context awareness processes. In Software design and development: Concepts, methodologies, tools, and applications (pp. 1289-1313). Hershey, PA: IGI Global. doi:10.4018/978-1-4666-4301-7.ch063

Peng, F., & Li, H. (2011). A steganalysis method for 2D engineering graphics based on the statistic of geometric features. *International Journal of Digital Crime and Forensics*, *3*(2), 35–40. doi:10.4018/jdcf.2011040103

Pérez, J. L., Rabuñal, J., & Abella, F. M. (2012). Soft computing techniques in civil engineering: Time series prediction. In *Computer engineering: Concepts, methodologies, tools and applications* (pp. 1982–1997). Hershey, PA: IGI Global. doi:10.4018/978-1-61350-456-7.ch811

Pérez-Castillo, R., Rodríguez de Guzmán, I. G., & Piattini, M. (2011). Architecture-driven modernization. In A. Dogru & V. Biçer (Eds.), *Modern software engineering concepts and practices: Advanced approaches* (pp. 75–103). Hershey, PA: IGI Global. doi:10.4018/978-1-60960-215-4.ch004

Pérez-Castillo, R., Rodríguez de Guzmán, I. G., & Piattini, M. (2012). Model-driven reengineering. In J. Rech & C. Bunse (Eds.), *Emerging technologies for the evolution and maintenance of software models* (pp. 200–229). Hershey, PA: IGI Global. doi:10.4018/978-1-61350-438-3.ch008

Petkov, D., Edgar-Nevill, D., Madachy, R., & O'Connor, R. (2010). Information systems, software engineering, and systems thinking: Challenges and opportunities. In M. Hunter (Ed.), *Strategic information systems: Concepts, methodologies, tools, and applications* (pp. 315–332). Hershey, PA: IGI Global. doi:10.4018/978-1-60566-677-8.ch023

Petkov, D., Edgar-Nevill, D., Madachy, R., & O'Connor, R. (2012). Towards a wider application of the systems approach in information systems and software engineering. In *Computer engineering: Concepts, methodologies, tools and applications* (pp. 1627–1645). Hershey, PA: IGI Global. doi:10.4018/978-1-61350-456-7.ch701

Petre, L., Sere, K., & Troubitsyna, E. (2012). *Dependability and computer engineering: Concepts for software-intensive systems*. Hershey, PA: IGI Global. doi:10.4018/978-1-60960-747-0

Poels, G., Decreus, K., Roelens, B., & Snoeck, M. (2013). Investigating goal-oriented requirements engineering for business processes. *Journal of Database Management, 24*(2), 35–71. doi:10.4018/jdm.2013040103

Polgar, J., & Adamson, G. (2011). *New generation of portal software and engineering: Emerging technologies*. Hershey, PA: IGI Global. doi:10.4018/978-1-60960-571-1

Power, C., Freire, A. P., & Petrie, H. (2011). Integrating accessibility evaluation into web engineering processes. In G. Alkhatib (Ed.), *Web engineered applications for evolving organizations: Emerging knowledge* (pp. 315–339). Hershey, PA: IGI Global. doi:10.4018/978-1-60960-523-0.ch018

Praeg, C. (2011). Framework for IT service value engineering: Managing value and IT service quality. In C. Praeg & D. Spath (Eds.), *Quality management for IT services: Perspectives on business and process performance* (pp. 274–297). Hershey, PA: IGI Global. doi:10.4018/978-1-61692-889-6.ch016

Prescott, J., & Bogg, J. (2013). *Gendered occupational differences in science, engineering, and technology careers*. Hershey, PA: IGI Global. doi:10.4018/978-1-4666-2107-7

Prpic, J. K., & Moore, G. (2012). E-portfolios as a quantitative and qualitative means of demonstrating learning outcomes and competencies in engineering. In K. Yusof, N. Azli, A. Kosnin, S. Yusof, & Y. Yusof (Eds.), *Outcome-based science, technology, engineering, and mathematics education: Innovative practices* (pp. 124–154). Hershey, PA: IGI Global. doi:10.4018/978-1-4666-1809-1.ch007

Puteh, M. M., & Ismail, K. (2012). Quality assurance through innovation policy: The pedagogical implications on engineering education. In *Human resources management: Concepts, methodologies, tools, and applications* (pp. 40–49). Hershey, PA: IGI Global. doi:10.4018/978-1-4666-1601-1.ch004

Puthanpurayil, A. M., Dhakal, R. P., & Carr, A. J. (2013). Optimal passive damper positioning techniques: State-of-the-art. In N. Lagaros, V. Plevris, & C. Mitropoulou (Eds.), *Design optimization of active and passive structural control systems* (pp. 85–111). Hershey, PA: IGI Global. doi:10.4018/978-1-4666-2029-2.ch004

Ramachandran, M. (2010). Knowledge engineering support for software requirements, architectures and components. In F. Meziane & S. Vadera (Eds.), *Artificial intelligence applications for improved software engineering development: New prospects* (pp. 129–145). Hershey, PA: IGI Global. doi:10.4018/978-1-60566-758-4.ch007

Ramachandran, M. (2011). *Knowledge engineering for software development life cycles: Support technologies and applications*. Hershey, PA: IGI Global. doi:10.4018/978-1-60960-509-4

Ramachandran, M., & de Carvalho, R. (2010). *Handbook of research on software engineering and productivity technologies: Implications of globalization*. Hershey, PA: IGI Global. doi:10.4018/978-1-60566-731-7

Rasul, M., Nouwens, F., Swift, R., Martin, F., & Greensill, C. V. (2012). Assessment of final year engineering projects: A pilot investigation on issues and best practice. In M. Rasul (Ed.), *Developments in engineering education standards: Advanced curriculum innovations* (pp. 80–104). Hershey, PA: IGI Global. doi:10.4018/978-1-4666-0951-8.ch005

Razali, Z. B., & Trevelyan, J. (2012). An evaluation of students' practical intelligence and ability to diagnose equipment faults. In K. Yusof, N. Azli, A. Kosnin, S. Yusof, & Y. Yusof (Eds.), *Outcome-based science, technology, engineering, and mathematics education: Innovative practices* (pp. 328–349). Hershey, PA: IGI Global. doi:10.4018/978-1-4666-1809-1.ch016

Reimann, D. (2011). Shaping interactive media with the sewing machine: Smart textile as an artistic context to engage girls in technology and engineering education. *International Journal of Art, Culture and Design Technologies, 1*(1), 12–21. doi:10.4018/ijacdt.2011010102

Rikkilä, J. (2014). Agile, lean, and service-oriented development, continuum, or chasm. In *Software design and development: concepts, methodologies, tools, and applications* (pp. 132–163). Hershey, PA: IGI Global. doi:10.4018/978-1-4666-4301-7.ch008

Rocci, L. (2010). Engineering and environmental technoethics. In R. Luppicini (Ed.), *Technoethics and the evolving knowledge society: Ethical issues in technological design, research, development, and innovation* (pp. 146–162). Hershey, PA: IGI Global. doi:10.4018/978-1-60566-952-6.ch008

Rodrigues, D., Estrella, J. C., Monaco, F. J., Branco, K. R., Antunes, N., & Vieira, M. (2012). Engineering secure web services. In V. Cardellini, E. Casalicchio, K. Castelo Branco, J. Estrella, & F. Monaco (Eds.), *Performance and dependability in service computing: Concepts, techniques and research directions* (pp. 360–380). Hershey, PA: IGI Global. doi:10.4018/978-1-60960-794-4.ch016

Rodríguez, J., Fernández-Medina, E., Piattini, M., & Mellado, D. (2011). A security requirements engineering tool for domain engineering in software product lines. In N. Milanovic (Ed.), *Non-functional properties in service oriented architecture: Requirements, models and methods* (pp. 73–92). Hershey, PA: IGI Global. doi:10.4018/978-1-60566-794-2.ch004

Rojko, A., Zürcher, T., Hercog, D., & Stebler, R. (2012). Implementation of remote laboratories for industrial education. In A. Azad, M. Auer, & V. Harward (Eds.), *Internet accessible remote laboratories: Scalable e-learning tools for engineering and science disciplines* (pp. 84–107). Hershey, PA: IGI Global. doi:10.4018/978-1-61350-186-3.ch005

Rosado, D. G., Mellado, D., Fernandez-Medina, E., & Piattini, M. G. (2013). *Security engineering for cloud computing: Approaches and tools*. Hershey, PA: IGI Global. doi:10.4018/978-1-4666-2125-1

Rose, S., Lauder, M., Schlereth, M., & Schürr, A. (2011). A multidimensional approach for concurrent model-driven automation engineering. In J. Osis & E. Asnina (Eds.), *Model-driven domain analysis and software development: Architectures and functions* (pp. 90–113). Hershey, PA: IGI Global. doi:10.4018/978-1-61692-874-2.ch005

Russell, C. (2012). Conceptual mapping, visualisation, and systems thinking in engineering. In A. Patil, H. Eijkman, & E. Bhattacharyya (Eds.), *New media communication skills for engineers and IT professionals: Trans-national and trans-cultural demands* (pp. 72–93). Hershey, PA: IGI Global. doi:10.4018/978-1-4666-0243-4.ch006

Russell, D., & Haghi, A. (2010). *Web-based engineering education: Critical design and effective tools*. Hershey, PA: IGI Global. doi:10.4018/978-1-61520-659-9

Russo, B., Scotto, M., Sillitti, A., & Succi, G. (2010). Improving agile methods. In B. Russo, M. Scotto, A. Sillitti, & G. Succi (Eds.), *Agile technologies in open source development* (pp. 189–231). Hershey, PA: IGI Global. doi:10.4018/978-1-59904-681-5.ch012

Russo, B., Scotto, M., Sillitti, A., & Succi, G. (2010). Requirements management. In B. Russo, M. Scotto, A. Sillitti, & G. Succi (Eds.), *Agile technologies in open source development* (pp. 268–286). Hershey, PA: IGI Global. doi:10.4018/978-1-59904-681-5.ch015

Russo, M. T. (2012). Humanities in engineering education. In M. Rasul (Ed.), *Developments in engineering education standards: Advanced curriculum innovations* (pp. 285–300). Hershey, PA: IGI Global. doi:10.4018/978-1-4666-0951-8.ch016

Saidane, A., & Guelfi, N. (2014). Towards test-driven and architecture model-based security and resilience engineering. In *Software design and development: Concepts, methodologies, tools, and applications* (pp. 2072–2098). Hershey, PA: IGI Global. doi:10.4018/978-1-4666-4301-7.ch096

Sala, N. (2012). Virtual reality in architecture, in engineering and beyond. In E. Abu-Taieh, A. El Sheikh, & M. Jafari (Eds.), *Technology engineering and management in aviation: Advancements and discoveries* (pp. 336–345). Hershey, PA: IGI Global. doi:10.4018/978-1-60960-887-3.ch020

Salzmann, C., Gillet, D., Esquembre, F., Vargas, H., Sánchez, J., & Dormido, S. (2012). Web 2.0 open remote and virtual laboratories in engineering education. In A. Okada, T. Connolly, & P. Scott (Eds.), *Collaborative learning 2.0: Open educational resources* (pp. 369–390). Hershey, PA: IGI Global. doi:10.4018/978-1-4666-0300-4.ch020

Samaras, G. (2012). Human-centered systems engineering: Managing stakeholder dissonance in healthcare delivery. In A. Kolker & P. Story (Eds.), *Management engineering for effective healthcare delivery: Principles and applications* (pp. 148–171). Hershey, PA: IGI Global. doi:10.4018/978-1-60960-872-9.ch007

Sampaio, A. Z., Henriques, P. G., Cruz, C. O., & Martins, O. P. (2011). Interactive models based on virtual reality technology used in civil engineering education. In G. Vincenti & J. Braman (Eds.), *Teaching through multi-user virtual environments: Applying dynamic elements to the modern classroom* (pp. 387–413). Hershey, PA: IGI Global. doi:10.4018/978-1-61692-822-3.ch021

Sanabria, S. J., Furrer, R., Neuenschwander, J., Niemz, P., & Sennhauser, U. (2014). Bonding defect imaging in glulam with novel air-coupled ultrasound testing. In A. Aguilera & J. Davim (Eds.), *Research developments in wood engineering and technology* (pp. 221–246). Hershey, PA: IGI Global. doi:10.4018/978-1-4666-4554-7.ch006

Sánchez de la Rosa, J. L., Miranda, S. A., & González, C. S. (2013). Evaluation of transversal competences of the engineering students and their relation to the enterprise requirements. In K. Patel & S. Vij (Eds.), *Enterprise resource planning models for the education sector: Applications and methodologies* (pp. 1–17). Hershey, PA: IGI Global. doi:10.4018/978-1-4666-2193-0.ch001

Sarirete, A., & Chikh, A. (2012). A knowledge management process in communities of practice of engineering based on the SECI model for knowledge. In E. Ng, N. Karacapilidis, & M. Raisinghani (Eds.), *Evaluating the impact of technology on learning, teaching, and designing curriculum: Emerging trends* (pp. 134–149). Hershey, PA: IGI Global. doi:10.4018/978-1-4666-0032-4.ch009

Sass, L. (2013). Direct building manufacturing of homes with digital fabrication. In *Industrial engineering: Concepts, methodologies, tools, and applications* (pp. 1231–1242). Hershey, PA: IGI Global. doi:10.4018/978-1-4666-1945-6.ch066

Sauvet, B., Boukhicha, M., Balan, A., Hwang, G., Taverna, D., Shukla, A., & Régnier, S. (2012). Selective pick-and-place of thin film by robotic micromanipulation. *International Journal of Intelligent Mechatronics and Robotics, 2*(3), 24–37. doi:10.4018/ijimr.2012070103

Schnabel, M. A. (2013). Learning parametric designing. In *Industrial engineering: Concepts, methodologies, tools, and applications* (pp. 197–210). Hershey, PA: IGI Global. doi:10.4018/978-1-4666-1945-6.ch013

Schrödl, H., & Wind, S. (2013). Requirements engineering for cloud application development. In A. Bento & A. Aggarwal (Eds.), *Cloud computing service and deployment models: Layers and management* (pp. 137–150). Hershey, PA: IGI Global. doi:10.4018/978-1-4666-2187-9.ch007

Schulz, T., Radlinski, L., Gorges, T., & Rosenstiel, W. (2011). Software process model using dynamic bayesian networks. In M. Ramachandran (Ed.), *Knowledge engineering for software development life cycles: Support technologies and applications* (pp. 289–310). Hershey, PA: IGI Global. doi:10.4018/978-1-60960-509-4.ch016

Sevillian, D. B. (2013). Aircraft development and design: Enhancing product safety through effective human factors engineering design solutions. In *Industrial engineering: Concepts, methodologies, tools, and applications* (pp. 858–886). Hershey, PA: IGI Global. doi:10.4018/978-1-4666-1945-6.ch048

Shan, T. C., & Hua, W. W. (2010). Strategic technology engineering planning. In M. Hunter (Ed.), *Strategic information systems: Concepts, methodologies, tools, and applications* (pp. 414–434). Hershey, PA: IGI Global. doi:10.4018/978-1-60566-677-8.ch029

Sharma, A., & Maurer, F. (2014). A roadmap for software engineering for the cloud: Results of a systematic review. In *Software design and development: Concepts, methodologies, tools, and applications* (pp. 1–16). Hershey, PA: IGI Global. doi:10.4018/978-1-4666-4301-7.ch001

Sharma, N., Singh, K., & Goyal, D. (2014). Software engineering, process improvement, and experience management: Is the nexus productive? Clues from the Indian giants. In *Software design and development: Concepts, methodologies, tools, and applications* (pp. 1401–1414). Hershey, PA: IGI Global. doi:10.4018/978-1-4666-4301-7.ch068

Shirk, S., Arreola, V., Wobig, C., & Russell, K. (2012). Girls' e-mentoring in science, engineering, and technology based at the University of Illinois at Chicago women in science and engineering (WISE) program. In Computer engineering: Concepts, Methodologies, tools and applications (pp. 1144-1163). Hershey, PA: IGI Global. doi:10.4018/978-1-61350-456-7.ch505

Shukla, A., & Tiwari, R. (2011). *Biomedical engineering and information systems: Technologies, tools and applications*. Hershey, PA: IGI Global. doi:10.4018/978-1-61692-004-3

Sidhu, M. S. (2010). Challenges and trends of TAPS packages in enhancing engineering education. In M. Sidhu (Ed.), *Technology-assisted problem solving for engineering education: Interactive multimedia applications* (pp. 158–166). Hershey, PA: IGI Global. doi:10.4018/978-1-60566-764-5.ch011

Sidhu, M. S. (2010). Evaluation of TAPS packages. In M. Sidhu (Ed.), *Technology-assisted problem solving for engineering education: Interactive multimedia applications* (pp. 128–147). Hershey, PA: IGI Global. doi:10.4018/978-1-60566-764-5.ch009

Sidhu, M. S. (2010). Technology assisted problem solving packages: A new approach to learning, visualizing, and problem solving in engineering. In M. Sidhu (Ed.), *Technology-assisted problem solving for engineering education: Interactive multimedia applications* (pp. 69–90). Hershey, PA: IGI Global. doi:10.4018/978-1-60566-764-5.ch006

Sidhu, M. S. (2010). *Technology-assisted problem solving for engineering education: Interactive multimedia applications*. Hershey, PA: IGI Global. doi:10.4018/978-1-60566-764-5

Sidhu, M. S., & Kang, L. C. (2012). Emerging trends and technologies for enhancing engineering education: An overview. In L. Tomei (Ed.), *Advancing education with information communication technologies: Facilitating new trends* (pp. 320–330). Hershey, PA: IGI Global. doi:10.4018/978-1-61350-468-0.ch026

Sidhu, M. S., & Kang, L. C. (2012). New trends and futuristic information communication technologies for engineering education. In S. Chhabra (Ed.), *ICTs for advancing rural communities and human development: Addressing the digital divide* (pp. 251–262). Hershey, PA: IGI Global. doi:10.4018/978-1-4666-0047-8.ch017

Simonette, M. J., & Spina, E. (2014). Enabling IT innovation through soft systems engineering. In M. Pańkowska (Ed.), *Frameworks of IT prosumption for business development* (pp. 64–72). Hershey, PA: IGI Global. doi:10.4018/978-1-4666-4313-0.ch005

Singh, T., Verma, M. K., & Singh, R. (2014). Role of emotional intelligence in academic achievement: An empirical study on engineering students. In P. Ordóñez de Pablos & R. Tennyson (Eds.), *Strategic approaches for human capital management and development in a turbulent economy* (pp. 255–263). Hershey, PA: IGI Global. doi:10.4018/978-1-4666-4530-1.ch016

Smolnik, S., Teuteberg, F., & Thomas, O. (2012). *Semantic technologies for business and information systems engineering: Concepts and applications*. Hershey, PA: IGI Global. doi:10.4018/978-1-60960-126-3

Solemon, B., Sahibuddin, S., & Ghani, A. A. (2014). Requirements engineering process improvement and related models. In *Software design and development: Concepts, methodologies, tools, and applications* (pp. 203–218). Hershey, PA: IGI Global. doi:10.4018/978-1-4666-4301-7.ch011

Sorial, G. A., & Noroozi, B. (2012). Improvement of engineering students education by e-learning. In *Virtual learning environments: Concepts, methodologies, tools and applications* (pp. 870–883). Hershey, PA: IGI Global. doi:10.4018/978-1-4666-0011-9.ch414

Srinivasan, A., López-Ribot, J. L., & Ramasubramanian, A. K. (2011). Microfluidic applications in vascular bioengineering. In A. Shukla & R. Tiwari (Eds.), *Biomedical engineering and information systems: Technologies, tools and applications* (pp. 1–30). Hershey, PA: IGI Global. doi:10.4018/978-1-61692-004-3.ch001

Steinbuch, R. (2013). Evolutionary optimization of passive compensators to improve earthquake resistance. In N. Lagaros, V. Plevris, & C. Mitropoulou (Eds.), *Design optimization of active and passive structural control systems* (pp. 250–273). Hershey, PA: IGI Global. doi:10.4018/978-1-4666-2029-2.ch011

Stephenson, S. V., & Sage, A. P. (2010). Information and knowledge perspectives in systems engineering and management for innovation and productivity through enterprise resource planning. In M. Hunter (Ed.), *Strategic information systems: Concepts, methodologies, tools, and applications* (pp. 338–368). Hershey, PA: IGI Global. doi:10.4018/978-1-60566-677-8.ch025

Strasser, T., Zoitl, A., & Rooker, M. (2013). Zero-downtime reconfiguration of distributed control logic in industrial automation and control. In *Industrial engineering: Concepts, methodologies, tools, and applications* (pp. 2024–2051). Hershey, PA: IGI Global. doi:10.4018/978-1-4666-1945-6.ch109

Sudeikat, J., & Renz, W. (2010). Building complex adaptive systems: On engineering self-organizing multi-agent systems. In M. Hunter (Ed.), *Strategic information systems: Concepts, methodologies, tools, and applications* (pp. 767–787). Hershey, PA: IGI Global. doi:10.4018/978-1-60566-677-8.ch050

Sugiyama, S., & Burgess, L. (2012). Principle for engineering service based system by swirl computing. In X. Liu & Y. Li (Eds.), *Advanced design approaches to emerging software systems: Principles, methodologies and tools* (pp. 48–60). Hershey, PA: IGI Global. doi:10.4018/978-1-60960-735-7.ch003

Sun, Z., Han, J., Dong, D., & Zhao, S. (2010). Engineering of experience based trust for e-commerce. In M. Wang & Z. Sun (Eds.), *Handbook of research on complex dynamic process management: Techniques for adaptability in turbulent environments* (pp. 342–367). Hershey, PA: IGI Global. doi:10.4018/978-1-60566-669-3.ch014

Sunindyo, W. D., Moser, T., Winkler, D., Mordinyi, R., & Biffl, S. (2013). Workflow validation framework in collaborative engineering environments. In A. Loo (Ed.), *Distributed computing innovations for business, engineering, and science* (pp. 285–299). Hershey, PA: IGI Global. doi:10.4018/978-1-4666-2533-4.ch015

Szeto, A. Y. (2014). Assistive technology and rehabilitation engineering. In *Assistive technologies: Concepts, methodologies, tools, and applications* (pp. 277–331). Hershey, PA: IGI Global. doi:10.4018/978-1-4666-4422-9.ch015

Takruri-Rizk, H., Sappleton, N., & Dhar-Bhattacharjee, S. (2010). Progression of UK women engineers: Aids and hurdles. In A. Cater-Steel & E. Cater (Eds.), *Women in engineering, science and technology: Education and career challenges* (pp. 280–300). Hershey, PA: IGI Global. doi:10.4018/978-1-61520-657-5.ch013

Tekinerdogan, B., & Aksit, M. (2011). A comparative analysis of software engineering with mature engineering disciplines using a problem-solving perspective. In A. Dogru & V. Biçer (Eds.), *Modern software engineering concepts and practices: Advanced approaches* (pp. 1–18). Hershey, PA: IGI Global. doi:10.4018/978-1-60960-215-4.ch001

Tellioglu, H. (2012). About representational artifacts and their role in engineering. In G. Viscusi, G. Campagnolo, & Y. Curzi (Eds.), *Phenomenology, organizational politics, and IT design: The social study of information systems* (pp. 111–130). Hershey, PA: IGI Global. doi:10.4018/978-1-4666-0303-5.ch007

Terawaki, Y., Takahashi, Y., Kodama, Y., & Yana, K. (2013). The development of educational environment suited to the Japan-specific educational service using requirements engineering techniques: Case study of running Sakai with PostgreSQL. In V. Kumar & F. Lin (Eds.), *System and technology advancements in distance learning* (pp. 261–270). Hershey, PA: IGI Global. doi:10.4018/978-1-4666-2032-2.ch019

Thomas, K. D., & Muga, H. E. (2012). Sustainability: The new 21st century general education requirement for engineers. In M. Rasul (Ed.), *Developments in engineering education standards: Advanced curriculum innovations* (pp. 263–284). Hershey, PA: IGI Global. doi:10.4018/978-1-4666-0951-8.ch015

Trigo, B., Olguin, G., & Matai, P. (2010). The use of applets in an engineering chemistry course: Advantages and new ideas. In D. Russell & A. Haghi (Eds.), *Web-based engineering education: Critical design and effective tools* (pp. 108–118). Hershey, PA: IGI Global. doi:10.4018/978-1-61520-659-9.ch009

Tsadimas, A., Nikolaidou, M., & Anagnostopoulos, D. (2014). Model-based system design using SysML: The role of the evaluation diagram. In P. Fonseca i Casas (Ed.), *Formal languages for computer simulation: Transdisciplinary models and applications* (pp. 236–266). Hershey, PA: IGI Global. doi:10.4018/978-1-4666-4369-7.ch008

Unhelkar, B., Ghanbary, A., & Younessi, H. (2010). *Collaborative business process engineering and global organizations: Frameworks for service integration*. Hershey, PA: IGI Global. doi:10.4018/978-1-60566-689-1

Uziak, J., Oladiran, M. T., & Kommula, V. P. (2012). Integrating general education courses into engineering curriculum: Students' perspective. In M. Rasul (Ed.), *Developments in engineering education standards: Advanced curriculum innovations* (pp. 247–262). Hershey, PA: IGI Global. doi:10.4018/978-1-4666-0951-8.ch014

Valizadeh, M., Anzelotti, G., & Salehi, A. S. (2010). Web-based approaches in engineering education. In R. Luppicini & A. Haghi (Eds.), *Cases on digital technologies in higher education: Issues and challenges* (pp. 241–256). Hershey, PA: IGI Global. doi:10.4018/978-1-61520-869-2.ch017

Valizadeh, M., Anzelotti, G., & Salehi, S. (2010). Web-based training: An applicable tool for engineering education. In D. Russell & A. Haghi (Eds.), *Web-based engineering education: Critical design and effective tools* (pp. 186–198). Hershey, PA: IGI Global. doi:10.4018/978-1-61520-659-9.ch014

Valverde, R., Toleman, M., & Cater-Steel, A. (2010). Design science: A case study in information systems re-engineering. In M. Hunter (Ed.), *Strategic information systems: Concepts, methodologies, tools, and applications* (pp. 490–503). Hershey, PA: IGI Global. doi:10.4018/978-1-60566-677-8.ch033

Vaquero, L. M., Rodero-Merino, L., Cáceres, J., Chapman, C., Lindner, M., & Galán, F. (2012). Principles, methodology and tools for engineering cloud computing systems. In X. Liu & Y. Li (Eds.), *Advanced design approaches to emerging software systems: Principles, methodologies and tools* (pp. 250–273). Hershey, PA: IGI Global. doi:10.4018/978-1-60960-735-7.ch012

Vargas, E. P. (2014). Quality, improvement and measurements in high risk software. In *Software design and development: Concepts, methodologies, tools, and applications* (pp. 733–748). Hershey, PA: IGI Global. doi:10.4018/978-1-4666-4301-7.ch035

Vasant, P. (2013). *Meta-heuristics optimization algorithms in engineering, business, economics, and finance*. Hershey, PA: IGI Global. doi:10.4018/978-1-4666-2086-5

Venkatraman, S. (2013). Software engineering research gaps in the cloud. *Journal of Information Technology Research*, *6*(1), 1–19. doi:10.4018/jitr.2013010101

Villazón-Terrazas, B. C., Suárez-Figueroa, M., & Gómez-Pérez, A. (2012). A pattern-based method for re-engineering non-ontological resources into ontologies. In A. Sheth (Ed.), *Semantic-enabled advancements on the web: Applications across industries* (pp. 17–54). Hershey, PA: IGI Global. doi:10.4018/978-1-4666-0185-7.ch002

Vinod, J. S. (2010). Dem simulations in geotechnical earthquake engineering education. *International Journal of Geotechnical Earthquake Engineering*, *1*(1), 61–69. doi:10.4018/jgee.2010090804

Virdi, J. (2014). Business risk analysis: Obsolescence management in requirements engineering. In *Software design and development: Concepts, methodologies, tools, and applications* (pp. 1736–1763). Hershey, PA: IGI Global. doi:10.4018/978-1-4666-4301-7.ch083

Walk, S., Pöschko, J., Strohmaier, M., Andrews, K., Tudorache, T., Noy, N. F., ... Musen, M. A. (2013). PragmatiX: An interactive tool for visualizing the creation process behind collaboratively engineered ontologies. *International Journal on Semantic Web and Information Systems, 9*(1), 45–78. doi:10.4018/jswis.2013010103 PMID:24465189

Wang, J. (2013). Architects and engineers. In *Challenging ICT applications in architecture, engineering, and industrial design education* (pp. 48–64). Hershey, PA: IGI Global. doi:10.4018/978-1-4666-1999-9.ch003

Wang, J. (2013). *Challenging ICT applications in architecture, engineering, and industrial design education*. Hershey, PA: IGI Global. doi:10.4018/978-1-4666-1999-9

Wang, J. (2013). Pedagogy and curriculum in architecture and engineering. In *Challenging ICT applications in architecture, engineering, and industrial design education* (pp. 65–92). Hershey, PA: IGI Global. doi:10.4018/978-1-4666-1999-9.ch004

Wang, J. (2013). Professionalism in architecture and engineering. In *Challenging ICT applications in architecture, engineering, and industrial design education* (pp. 137–156). Hershey, PA: IGI Global. doi:10.4018/978-1-4666-1999-9.ch007

Wang, J. (2013). Reviewing engineers and introducing industrial designers. In *Challenging ICT applications in architecture, engineering, and industrial design education* (pp. 111–136). Hershey, PA: IGI Global. doi:10.4018/978-1-4666-1999-9.ch006

Wang, Y. (2010). On cognitive properties of human factors and error models in engineering and socialization. In Y. Wang (Ed.), *Discoveries and breakthroughs in cognitive informatics and natural intelligence* (pp. 93–109). Hershey, PA: IGI Global. doi:10.4018/978-1-60566-902-1.ch006

Wang, Y. (2012). Convergence of software science and computational intelligence: A new transdisciplinary research field. In Y. Wang (Ed.), *Software and intelligent sciences: New transdisciplinary findings* (pp. 1–13). Hershey, PA: IGI Global. doi:10.4018/978-1-4666-0261-8.ch001

Wang, Y., & Patel, S. (2012). Exploring the cognitive foundations of software engineering. In Y. Wang (Ed.), *Software and intelligent sciences: New transdisciplinary findings* (pp. 232–251). Hershey, PA: IGI Global. doi:10.4018/978-1-4666-0261-8.ch014

Wei, J., Chen, J., & Zhu, Q. (2010). Service science, management and engineering education: A unified model for university. *International Journal of Service Science, Management, Engineering, and Technology, 1*(2), 51–69. doi:10.4018/jssmet.2010040104

Weigang, L., Barros, A. D., & Romani de Oliveira, I. (2010). *Computational models, software engineering, and advanced technologies in air transportation: Next generation applications*. Hershey, PA: IGI Global. doi:10.4018/978-1-60566-800-0

Westh Nicolajsen, H. (2010). Limitations and perspectives on use of e-services in engineering consulting. In *Electronic services: Concepts, methodologies, tools and applications* (pp. 1280–1295). Hershey, PA: IGI Global. doi:10.4018/978-1-61520-967-5.ch078

Wilson, R., & Younis, H. (2013). *Business strategies for electrical infrastructure engineering: Capital project implementation*. Hershey, PA: IGI Global. doi:10.4018/978-1-4666-2839-7

Winter, A. (2010). The smart women – smart state strategy: A policy on women's participation in science, engineering and technology in Queensland, Australia. In A. Cater-Steel & E. Cater (Eds.), *Women in engineering, science and technology: Education and career challenges* (pp. 1–20). Hershey, PA: IGI Global. doi:10.4018/978-1-61520-657-5.ch001

Winter, R., & van Beijnum, I. (2014). Inter-domain traffic engineering using the origin preference attribute. In M. Boucadair & D. Binet (Eds.), *Solutions for sustaining scalability in internet growth* (pp. 18–38). Hershey, PA: IGI Global. doi:10.4018/978-1-4666-4305-5.ch002

Worth, D., Greenough, C., & Chin, S. (2014). Pragmatic software engineering for computational science. In *Software design and development: Concepts, methodologies, tools, and applications* (pp. 663–694). Hershey, PA: IGI Global. doi:10.4018/978-1-4666-4301-7.ch032

Wu, J. (2013). *Biomedical engineering and cognitive neuroscience for healthcare: Interdisciplinary applications*. Hershey, PA: IGI Global. doi:10.4018/978-1-4666-2113-8

Wu, Y., & Koszalka, T. A. (2013). Instructional design of an advanced interactive discovery environment: Exploring team communication and technology use in virtual collaborative engineering problem solving. In *Industrial engineering: Concepts, methodologies, tools, and applications* (pp. 117–136). Hershey, PA: IGI Global. doi:10.4018/978-1-4666-1945-6.ch009

Yadav, V. (2011). Research review: Globally distributed requirements engineering and agility. *International Journal of Innovation in the Digital Economy*, *2*(1), 1–11. doi:10.4018/jide.2011010101

Yang, J. J., Liu, J. F., Kurokawa, T., Kitamura, N., Yasuda, K., & Gong, J. P. (2013). Tough double-network hydrogels as scaffolds for tissue engineering: Cell behavior in vitro and in vivo test. In J. Wu (Ed.), *Technological advancements in biomedicine for healthcare applications* (pp. 213–222). Hershey, PA: IGI Global. doi:10.4018/978-1-4666-2196-1.ch023

Yeoh, W., Gao, J., & Koronios, A. (2010). Empirical investigation of critical success factors for implementing business intelligence systems in multiple engineering asset management organisations. In M. Hunter (Ed.), *Strategic information systems: Concepts, methodologies, tools, and applications* (pp. 2039–2063). Hershey, PA: IGI Global. doi:10.4018/978-1-60566-677-8.ch129

Yerrick, R., Lund, C., & Lee, Y. (2013). Online simulator use in the preparing chemical engineers. *International Journal of Online Pedagogy and Course Design*, *3*(2), 1–24. doi:10.4018/ijopcd.2013040101

Yilmaz, A. E., & Yilmaz, I. B. (2012). Natural language processing techniques in requirements engineering. In *Computer engineering: Concepts, methodologies, tools and applications* (pp. 533–545). Hershey, PA: IGI Global. doi:10.4018/978-1-61350-456-7.ch303

Yong, E. (2012). Literature review skills for undergraduate engineering students in large classes. In K. Yusof, N. Azli, A. Kosnin, S. Yusof, & Y. Yusof (Eds.), *Outcome-based science, technology, engineering, and mathematics education: Innovative practices* (pp. 240–261). Hershey, PA: IGI Global. doi:10.4018/978-1-4666-1809-1.ch012

Yun, H., Xu, J., Xiong, J., & Wei, M. (2013). A knowledge engineering approach to develop domain ontology. In V. Kumar & F. Lin (Eds.), *System and technology advancements in distance learning* (pp. 55–70). Hershey, PA: IGI Global. doi:10.4018/978-1-4666-2032-2.ch004

Zardari, S., Faniyi, F., & Bahsoon, R. (2013). Using obstacles for systematically modeling, analysing, and mitigating risks in cloud adoption. In I. Mistrik, A. Tang, R. Bahsoon, & J. Stafford (Eds.), *Aligning enterprise, system, and software architectures* (pp. 275–296). Hershey, PA: IGI Global. doi:10.4018/978-1-4666-2199-2.ch014

Zašcerinska, J., Ahrens, A., & Bassus, O. (2012). Enterprise 2.0 and 3.0 in education: Engineering and business students' view. In M. Cruz-Cunha, P. Gonçalves, N. Lopes, E. Miranda, & G. Putnik (Eds.), *Handbook of research on business social networking: Organizational, managerial, and technological dimensions* (pp. 472–494). Hershey, PA: IGI Global. doi:10.4018/978-1-61350-168-9.ch025

Zašcerinska, J., Ahrens, A., & Bassus, O. (2012). Enterprise 2.0 and 3.0 in education: Engineering and business students' view. In M. Cruz-Cunha, P. Gonçalves, N. Lopes, E. Miranda, & G. Putnik (Eds.), *Handbook of research on business social networking: Organizational, managerial, and technological dimensions* (pp. 472–494). Hershey, PA: IGI Global. doi:10.4018/978-1-61350-168-9.ch025

Zhang, D. (2012). Machine learning and value-based software engineering. In Y. Wang (Ed.), *Software and intelligent sciences: New transdisciplinary findings* (pp. 287–301). Hershey, PA: IGI Global. doi:10.4018/978-1-4666-0261-8.ch017

Zhang, H., Xu, C., Su, W., & Luo, H. (2014). Routing optimization for inter-domain traffic engineering under identifier network. In M. Boucadair & D. Binet (Eds.), *Solutions for sustaining scalability in internet growth* (pp. 127–147). Hershey, PA: IGI Global. doi:10.4018/978-1-4666-4305-5.ch007

Zhong, Y. (2013). Processing of 3D unstructured measurement data for reverse engineering. In S. Sirouspour (Ed.), *Advanced engineering and computational methodologies for intelligent mechatronics and robotics* (pp. 118–127). Hershey, PA: IGI Global. doi:10.4018/978-1-4666-3634-7.ch008

Zhou, Z., Wang, H., & Lou, P. (2010). *Manufacturing intelligence for industrial engineering: Methods for system self-organization, learning, and adaptation.* Hershey, PA: IGI Global. doi:10.4018/978-1-60566-864-2

Zhu, L., Jayaram, U., & Kim, O. (2011). Online semantic knowledge management for product design based on product engineering ontologies. *International Journal on Semantic Web and Information Systems*, 7(4), 36–61. doi:10.4018/jswis.2011100102

About the Contributors

Kiran Ahuja has done outstanding research in the fields of optical and wireless communication. She received her B.Tech & M.Tech degree in Electronics and Communication Engineering from Punjab Technical University, Jalandhar, India and Ph. D in Electronics & Communication Engineering in wireless communication and networks field from Thapar University, Patiala, Punjab, India. She completed her post-doc from NIT, Jalandhar under N-PDF scheme of SERB, India. She is working as faculty at DAVIET, Jalandhar since 2006.

Tarannum Awasthi completed her Bachelor of Technology in Electronics and Communication Engineering from Dr. B. R. Amdedkar National Institute of Technology Jalandhar, India. Her interest areas include image processing and computer vision.

Indu Bala is working as an associate professor in department of Electronics & Communication Engineering at Lovely Professional University, Jalandhar, India. She has more than 14 years of teaching experience. Her research interests include cognitive radio networks, dynamic spectrum access, interference mitigation and power optimization for next generation communication networks.

Parul Bhoriwal has done Btech from Surendera groups of institution, Sriganganagar and pursuing MTech in NIT Jalandhar, Punjab. Her research interest includes the wireless communication.

Rashmi Chawla is an Electronic Engineer with over 15 years of rich experience in diversified research including Solar, Internet of Things, Wireless Sensor Networks and Signal Processing. Currently, working as Assistant Professor in Electronics Engineering department at J.C.Bose University of Science and Technology, YMCA, Faridabad, India.

Subhankar Das, is a Ph.D. in Marketing-consumer behavior, UGC NET (Qualifying exam for Assistant Professor) qualified faculty having 8 SCOPUS indexed publications in management, 2 SCOPUS indexed acceptance (Inderscience & Espacios publications) and 3 research papers in review (All are in SCOPUS with Sage, Inderscience & Espacios publications) till date. He has qualified IELTS (International English Language Testing System) with overall band of 6.5 in 2017. He has an authored Book in review for Digital marketing with CRC press Routledge, India. He has won 4 International Awards in field of study & research from reputed organisations. He has also got certified by IIM Lucknow, University of British Columbia & IIM Bangalore, Manipal & Google for Case based pedagogy and digital branding. He is also a Member of Center of Excellence for Tata Institute of Social Science TISS-SVE for 'Management & Entrepreneurship and Professional Skills' Vertical, Mumbai. Apart from above research work He has also 1 International case, 3 journals, 9 conference proceeding publications in national repute. He is having 10 years of progressive teaching experience.

Sara D'Onofrio got her bilingual Bachelor's degree in Business Administration from the University of Fribourg, and her Master's degree in Business Administration with specialization in Information Systems from the University of Bern. Currently, she is doing her PhD in Computer Science at the Human-IST Institute of the University of Fribourg. Besides her PhD, she is working as a IT-Trainee at the Swiss Post. Her research interests lie in the area of cognitive computing and smart cities. She has already participated in various conferences in Europe, South America and Canada.

Astrid Habenstein leads the Transdisciplinary Research Centre Smart Swiss Capital Region (TRCSSCR) at the University of Bern, Switzerland. She graduated at the University of Bielefeld, Germany, in History and Philosophy (2005) and received her PhD in Ancient History at the University of Bern (2012). Astrid Habenstein worked as Research Assistant at the University of Bern and as Lecturer at the Universities of Bern and Basel.

Tadesse Hailu Ayane received his B.Sc in Electrical Engineering degree from Jimma University in 2008 and Masters in Electronics and Computer Engineering degree from Addis Ababa University in 2010. He is a member of Ethiopian Society of Electrical Engineer since 2009. He worked in Midroc Gold Mine from 2008-2009 as Electrical Engineer. He has an experience of teaching in different Governmental universities with lecturer position since 2010. Currently, he is a senior lecturer and Department head of Electronics& communication Engineering, Adama science and technology University, Ethiopia. His research interest includes IC and nanotechnology, signal processing, electronic automation, satellite technology.

Rachna Jain, currently working as Assistant Professor(Computer Science Department) in Bharati Vidyapeeth's College of Engineering (GGSIPU) since 2007. She did her PHD from Banasthali Vidyapith in Computer Science in 2017.She received ME degree in year 2011 from Delhi college of engineering (Delhi University) with specialisation in Computer Technology and Applications. She did her B.tech (Computer Science) in 2006 from N.C College of Engineering, Kurukshetra University. Her current research interests are Cloud Computing, Image Processing, Network and information security, Swarm Intelligence, Big Data and IoT. A total of 12+ Years of Academic / Research Experience with more than 20+ Publications in various National, International Conferences cum International Journals(Scopus/ISI/SCI) of High Repute.

Rajiv Kapoor, Ph.D in Electronics and Communication Engineering from Panjab University (Punjab Engineering College), Chandigarh has worked in industries of repute and also worked in Engineering institutions of repute. He is presently a professor in Electronics and Communication Engineering Department at Delhi Technological University (Formerly Delhi College of Engineering). He has published more than 100 papers in the various Journals/Conferences of repute and has filed 6 patents also. He has undertaken 21 R & D projects and is not only a strong believer but a practitioner for "MAKE in INDIA". He has an endeavor to design new products for Industries and work for society.

Pavneet Kaur has done her B.Tech from DAV Institute of Engineering and Technology in ECE branch. She has done her M. Tech from NIT Jalandhar. Her research interest is in field of eye tracking technology and gamification.

Tanveer Kaur has done her Btech from Dr. B. R. Ambedkar National Institute of Technology (Jalandhar). Her interests include image processing and video processing using MATLAB. She also has keen interest in microcontrollers.

Arun Khosla was born in Punjab, India. He received the BE degree from Thapar University, India, M. Tech from NIT Kurukshetra, and Ph.D. degree from Kurukshetra University, India. His research areas include Artificial Intelligence, Bio Medical instrumentation. He is working as a Professor in the department of Electronics and Communication Engineering, NIT Jalandhar.

Bandana Mahapatra has done her PhD in Computer Science & Engineering, in Security in AdHoc Network. Her specialization ranges from Adhoc Networks, Artificial intelligence, Swarm Optimization, etc.

Anand Nayyar (Academician, Researcher, Author, Writer, Inventor, Innovator, Scientist, Consultant, and Orator) received MCA from Punjabi University, Patiala in 2008 with Gold Medal and Distinction. He received his M.Phil in Computer Science in 2009 in 1st Division from V.M.U, Tamil Nadu. He received his M.Tech in Information Technology in 2011 with Distinction cum University Topper. He received his MBA in Information Systems from Sikkim Manipal University with 1st Division. He has done his Ph.D. in Computer Science from Desh Bhagat University, Mandi Gobindgarh in 2017 in Wireless Sensor Networks, Swarm Intelligence, and Network Simulation.

Edy Portmann is Swiss Post Professor of Computer Science at the Human-IST Institute of the University of Fribourg. His trans-disciplinary research focuses on cognitive computing and its application to cities. After an apprenticeship as an electrician, he studied information systems, business as well as economics and, later, got a doctorate in computer science. Among others, he worked for Swisscom, PwC and EY. In addition, Edy Portmann was also a researcher at the universities of Singapore, Berkeley and Bern.

Davinder Rathee was born in Rohtak, India. Professional, he is NET Qualified and having rich teaching experience. He earned his BSC from MDU Rohtak, MSC Electronics Science from KUK, MTech from MDU and PhD from Guru Jambeshwar University of Science & Technology, Hisar. He is having total experience of 14 years in teaching & Research in India and abroad. He published more than 35 papers in international reputed Journals and Conferences. He was selected as Associate Professor in Maharaja Agrasen University, Baddi H.P. presently he is working as A.P. in school of ECE, ADMA University of Science & Technology, ADMA, Ethiopia. His area of interest is VLSI, Microelectronics, Robotics and IoT, Electrical Vehicles.

Indu Saini received her B. Tech degree in Electronics and Communication Engineering from Guru Nanak Dev University, India, in 1994 and then obtained her M. Tech (by Research) and PhD degree in Electronics & Communication Engineering from Dr B. R. Ambedkar National Institute of Technology Jalandhar, where she is also serving as Assistant Professor in Electronics & Communication Engineering Department since 2002. Her professional research interests are Very Large-Scale Integration (VLSI) design, Biomedical Signal/Image Processing, and Machine Learning Algorithms.

Milin Saini completed BTech in Electronics and Communication Engineering from Dr. B. R. Ambedkar National Institute of Technology Jalandhar in 2018. She has interest in computer vision, artificial intelligence and wireless communication.

Anubhav Singh is pursuing B.Tech from BhartiVidyapeeth's College of engineering in Computer Science where he writes chapter contents. He has interest field in Machine Learning, Cloud Computing and Java with Data Structure algorithms. He recently completed a chapter in Security Concern in Cloud Computing.

Gajendra Singh received M.Tech Degree in Electronics & Communication from Dr. B. R Ambedkar National Institute of technology, Jalandhar, India, in 2012, and currently pursuing PhD in anomaly detection in computer vision from Dr. B. R Ambedkar National Institute of technology, Jalandhar, India. His research interests includes machine learning, deep learning and pattern recognition.

Ghanshyam Singh received PhD degree in Electronics Engineering from the Indian Institute of Technology, Banaras Hindu University, Varanasi, India, in 2000. He was associated with Central Electronics Engineering Research Institute, Pilani, and Institute for Plasma Research, Gandhinagar, India, respectively, where he was Research Scientist. He had also worked as an Assistant Professor at Electronics and Communication Engineering Department, Nirma University of Science and Technology, Ahmedabad, India. He was a Visiting Researcher at the Seoul National University, Seoul, South Korea. He also worked as assistant Professor, Associate Professor and full Professor in the Department of Electronics and Communication Engineering at Jaypee University of Information Technology, Solan, India. At present, he is full Professor with the Department of Electrical and Electronics Engineering, APK Campus, University of Johannesburg, S Africa. He is an author/co-author of more than 250 scientific papers of the refereed Journal and International Conferences. His research and teaching interests include RF/Microwave Engineering, Millimeter/THz Wave Antennas and its Applications in Communication and Imaging, Next Generation Communication Systems (OFDM and Cognitive Radio), and Nanophotonics. He has more than 19 years of teaching and research experience in the area of Electromagnetic/Microwave Engineering, Wireless Communication and Nanophotonics. He has supervised 14 Ph. D. and 45 M. Tech. Theses. He has worked as a reviewer for several reputed Journals and Conferences. He is author of four books published by Springer.

Parulpreet Singh received his Bachelor of Technology from IK Gujral Punjab Technical University, Jalandhar, Punjab and Master of Technology from Lovely Professional University, Jalandhar, Punjab with specialization in Electronics and Communication Engineering. He is presently pursuing his Ph.D. in the field of Wireless Sensor Networks from National Institute of Technology, Jalandhar, Punjab and working as an Assistant Professor in EEE department at Lovely Professional University. His research area of interest is MANETs and Wireless Sensor Networks.

Neetu Sood works as assistant professor in the Department of Electronics and Communication Engineering at Dr B R Ambedkar National Institute of Technology, Jalandhar in India since 2007. Her research interest includes wireless communication system design, OFDM based systems, channel modelling & its simulations, Biomedical signal processing and plant signalling.

Index

N

O

P

Q

R

S

T

U

V

www.ingramcontent.com/pod-product-compliance
Lightning Source LLC
LaVergne TN
LVHW081344300926
849161LV00008B/2057